# Treatise on Adhesion and Adhesives

## Volume 6

*edited by*

**Robert L. Patrick**

MARCEL DEKKER, INC. New York and Basel

ISBN: 0-8247-7587-2



MARCEL DEKKER, INC.
270 Madison Avenue, New York, New York 10016

Current printing (last digit):
10 9 8 7 6 5 4 3 2 1

PRINTED IN THE UNITED STATES OF AMERICA

# *Memorial*

## ROBERT L. PATRICK, Ph.D.
1920-1985

After a relatively short illness, Robert L. Patrick, a major contributor to both the science and technology of adhesion, passed away in Wells, Maine, on February 12, 1985.

Bob, as he was affectionately known to his wide circle of friends both in and out of science, had been blessed with an exceptionally large dose of enthusiasm for life and an indomitable spirit which permitted him to sail through his occasional rough waters of frustration without losing direction or perspective. So it was in his final time. A pragmatist almost to a fault, he lived a realist's life in a real world, bending with the tides of today yet retaining a perspective for the goals of tomorrow. As one of his friends expressed it, "Bob would have poorly suffered a lengthy illness at an enforced part-throttle."

After completing his undergraduate work at Bowdoin College, for which he retained a lifelong dedicated connection through its alumni association, he saw wartime service with the 15th Air Force in Italy. This experience itself was to prove the basis of many shared reminiscences later at the University of Pittsburgh. He completed his graduate work at the University of Indiana, where he held a DuPont Fellowship (1949-1950), with a doctorate in chemistry. He worked at several industrial companies including Goodrich and Continental Can Company, and then started his own company—Alpha Research and Development, Inc.—which was later merged into the Certain-Teed Corporation. After three years at the University of Pittsburgh, where he was Director of the Center for Adhesion Science and Technology, and during which time he became an avid Pitt Panther and Steeler fan to the delight of the lucky attendees of his memorable pre-game parties and "expert commentaries" (which were, of course, never in doubt but frequently in error), he joined the Harris Research Laboratories of Gillette as Director of Research.

Early in 1981 he and his wife, Carol, moved to Maine to follow one of their favorite avocations in the world of art, artifacts and antiques. Their shops in Wells, Maine, not only provided a repository for their own personal collection of paintings and limited edition prints that they had collected over the years, but also were a vehicle to implement their broader artistic leanings. Nevertheless, Bob retained his technological interests through continued consultancies in industry and government.

My own knowledge of Bob first came through one of the early papers bridging adhesion technology and continuum mechanics—the classic contribution of Ripling, Mostovoy, and Patrick given at the 66th Annual Meeting of ASTM in Atlantic City, New Jersey, in 1963. Similar to the experiences of several of his friends, I was hardly prepared for the human dynamo I later met face-to-face one very hot day in Salt Lake City. In response to my weather complaints, I was immediately subjected to a serious and erudite extemporaneous lecture upon the virtues of the Utah climate. I can hear it yet: "But, Max, it's a dry heat." From that time on I had the personal pleasure of observing and frequently sharing his ebullient enthusiasm and joie de vivre that so characterized his approach to life. The breadth and depth of his general knowledge was truly extraordinary and he used it to superb advantage, particularly in his contributions to the Gordon Research Conferences on Adhesion where, I understand, he had the longest unbroken attendance on record. Bob was a primary catalyst in changing the emphasis at the Conferences from mainly technology to one containing more science. He personally opened the door to invitations to several of us from the field of continuum mechanics in order to add additional perspectives to those of his other colleagues in physical, polymer and organic chemistry, in rheology, polymer physics and adhesives technology. In view of his long-time contribution to both the objectives and esprit of the Gordon Research Conferences on Adhesion, he was elected Permanant Honorary Chairman at the 1984 meeting.

Bob was one of the twelve founding members of The Adhesion Society and edited its newsletter for several years. His talks and published papers numbered in the scores. In addition, he edited the series of books entitled *Treatise on Adhesives and Adhesion*, published by Marcel Dekker, Inc., beginning in 1967. He was working on the sixth volume of that series when he passed away.

Outside of his technical interests and those in the arts which I have previously mentioned, he had diverse talents and outside activities which were reflected in backpacking, golfing, skiing—and, as he became legendary at the Gordon Conferences, a fierce and sometimes controversial antagonist in the annual sailing regattas! Many of his friends and some of his acquaintances well recall and will long remember with awe his occasional unbridled excursions into community affairs, or his righteously indignant representations as a consumer advocate which followed his opening line, "Well, in that case, let me speak to your president." The resulting Patrick stories are legion and remain part of his precious heritage to many of us.

Patrick was a great motivator, impetuous despot, enthusiastic participator, talented engineering scientist, dedicated constructive extrovert and, above all, a warm human being. It is a privilege to have this opportunity to salute his memory on behalf of his many colleagues, associates, and friends.

M. L. Williams
University of Pittsburgh
Pittsburgh, Pennsylvania

# *Preface*

We are most pleased to be the guest editors of this special volume of the Treatise on Adhesion and Adhesives. Robert Patrick is listed as editor, not only because he initiated the planning for this volume—in fact, he accumulated the first three manuscripts and selected approximately half the authors—but, more importantly, because of his dominant role in the adhesive community. He was a regular attendee at the Gordon Research Conferences on the Science of Adhesion held annually in New Hampshire; he had chaired this conference and numerous conference sessions. Those of us who had the pleasure of attending with him and his lovely and gracious wife Carol, will never forget his pointed and thoughtful contributions to the technical sessions. Even more memorable are the social activities he organized and directed. We were always impressed with how he could organize such elaborate activities as the regatta and the Thursday Night Awards Ceremony while making absolutely certain they in no way detracted from the professional program. Bob was one of the primary founders of the Adhesion Society and served as editor of its newsletter until poor health forced him to pass on these responsibilities. He also attended each of the Adhesion Society's annual meetings.

We feel this volume is particularly appropriate for honoring Bob, as his primary interests in adhesives were in the testing, inspection, and characterization areas, which are the book's central theme. The chapters in this treatise might be subdivided into three areas: (1) testing and mechanical evaluation, (2) surface treatments and analysis, and (3) adhesive characterization.

The topic of adhesive testing is introduced in the first chapter, by Bob's close friend and frequent collaborator, J. R. Huntsberger, on interpreting peel tests. This is followed by a related treatise by G. R. Hamed on "peel of rubbers." Chapter 3, by G. P. Anderson and K. L. DeVries, describes systematic analysis and testing of adhesives as a function of various geometric parameters on relatively standard testing geometries. Chapter 4, by W. Bascom and D. Hunston, explores fracture toughness for epoxies and elastomer modified epoxies. The next chapter, by L. T. Drzal, describes some of the effects interfacial adhesion has on the properties of composites. In our opinion, it is appropriate that these chapters build so strongly on the concepts of fracture mechanics. Bob was among the first to recognize the potential for fracture mechanics analysis, and commented on it in his publications in the 1960s.

The last two chapters bring the power of modern analytical instruments to bear on adhesive problems. H. R. Anderson and K. B. Sachdev explore the use of X-ray photo-electrospectroscopy for understanding adhesion phenomena. F. J. Boerio explores the effect of silane coupling agents on adhesive bond quality. He uses surface analysis techniques to explore the chemical nature of the interphase and to investigate its effects on physical bonding by relating his observations with wet strength and bond integrity.

G. P. Anderson and K. L. DeVries

# *Contributors*

Garron P. Anderson, Morton Thiokol, Inc., Brigham City, Utah

Herbert R. Anderson, Jr., IBM, Hopewell Junction, New York

W. D. Bascom, Hercules Inc. Bacchus Works, Magna, Utah*

F. J. Boerio, University of Cincinnati, Cincinnati, Ohio

K. L. DeVries, University of Utah, Salt Lake City, Utah

Lawrence T. Drzal, Michigan State University, East Lansing, Michigan

Gary R. Hamed, University of Akron, Akron, Ohio

James R. Huntsberger, E. I. du Pont de Nemours & Company, Wilmington, Delaware

D. L. Hunston, Hercules Inc. Bacchus Works, Magna, Utah

Krishna G. Sachdev, IBM, Hopewell Junction, New York

*Current affiliation: Department of Materials Science and Engineering, University of Utah.

# *Contents*

# *Contents of Other Volumes*

*editor*
*Robert L. Patrick*

# One

## Interpreting Peel Tests

JAMES R. HUNTSBERGER
*Consultant, Greenville, Delaware*

### I. INTRODUCTION

Current understanding of adhesion has evolved largely over the past 45 years. Much of the progress resulted directly from intelligible quantification and interpretation of data. As Lord Kelvin stated in the mid-nineteenth century, "Until we can quantify and express something in numbers our knowledge of it is meager."

The aim of this chapter is to emphasize the importance of proper interpretation of data. The intent is to provide to the newcomer to the science of adhesion a background for enlightened evaluation of adhesive performance and to others a useful review of the many factors, properties, and conditions that interact to give data which, without proper interpretation, may be not only useless but also misleading.

Adhesion or adherence between two materials results from molecular attractions acting across the interface or interphase. This has been measured directly by only a few workers for very special systems, and its measurement has required both sophisticated experimentation and interpretation of data. What are commonly called tests for adhesion are in reality tests of joint strength. Values obtained are measures of the force or work required to fracture joints or separate adherends in a given manner, under given conditions. Test values tell nothing directly about the adhesion between phases or components of joints. Even when samples are compared it is not necessarily valid that those giving higher values exhibit greater adhesion.

This chapter treats the many variables that influence test data. The nature of the test itself has an overwhelming impact. The geometry of the sample, how and at what rate the sample is strained, the environmental conditions, and often how the sample was prepared and its subsequent history determine performance. These various factors influence tests of all kinds, but their relative importance is often different for different tests. We will be concerned with peeling.

## II. PEEL TESTS

Several features make peel tests especially interesting and useful: (1) Mechanical characteristics make peeling amenable to approximate analyses of both distribution of stress and work of peeling. (2) When sample dimensions are large compared to the dimensions of the stressed regions, the macroscopic area of the bond does not influence the data. (3) In steady-state peeling, a single test may comprise hundreds or thousands of data points, depending on how values are obtained and recorded. (4) The dependence of peel strength on peel angle allows investigation of the influence of stress geometry. Note, however, that failure stresses cannot be obtained from peel tests.

### A. Analysis of Peel Adhesion

*Cases Involving Small Strains and Elastic Deformation*

In 1951 DeBruyne [1] stated that a rational analysis of peeling had yet to be made. Bikerman [2] in 1954 treated right-angle peeling of a flexible ribbon R adhered by a Hookean adhesive to a rigid substrate. He proposed the relation

$$P = 0.3799b\sigma_0\left(\frac{E_r}{E_a}\right)^{1/4} t_a^{1/4} t_r^{3/4} \tag{1}$$

where P is the peel force, b the width of the ribbon, $\sigma_0$ the tensile strength of the adhesive, $E_r$ and $E_a$ the moduli of the ribbon and the adhesive, $t_a$ the unstrained thickness of the adhesive, and $t_r$ the thickness of the ribbon.

Within the next 6 years several criticisms, refinements, and extensions of Bikerman's treatment appeared, along with more comprehensive and sophisticated developments [3-11].

Kaelble's work and that of Jouwersma have given gratifying results when used as the bases for analysis of practical problems, which will be discussed later. Bikerman's equation (Equation 1) neglected the moment arm m shown in Figure 1.1. Jouwersma included m to treat the special case of right-angle peeling, obtaining an equation differing from (1) by a factor $1/(1 + \beta x_0)$, where $\beta = (3E_r/E_a t_r^3 t_a)^{1/4}$:

$$P = \frac{0.3799b\sigma_0(E_r/E_a)^{1/4} t_a^{1/4} t_r^{3/4}}{1 + \beta x_0} \tag{2}$$

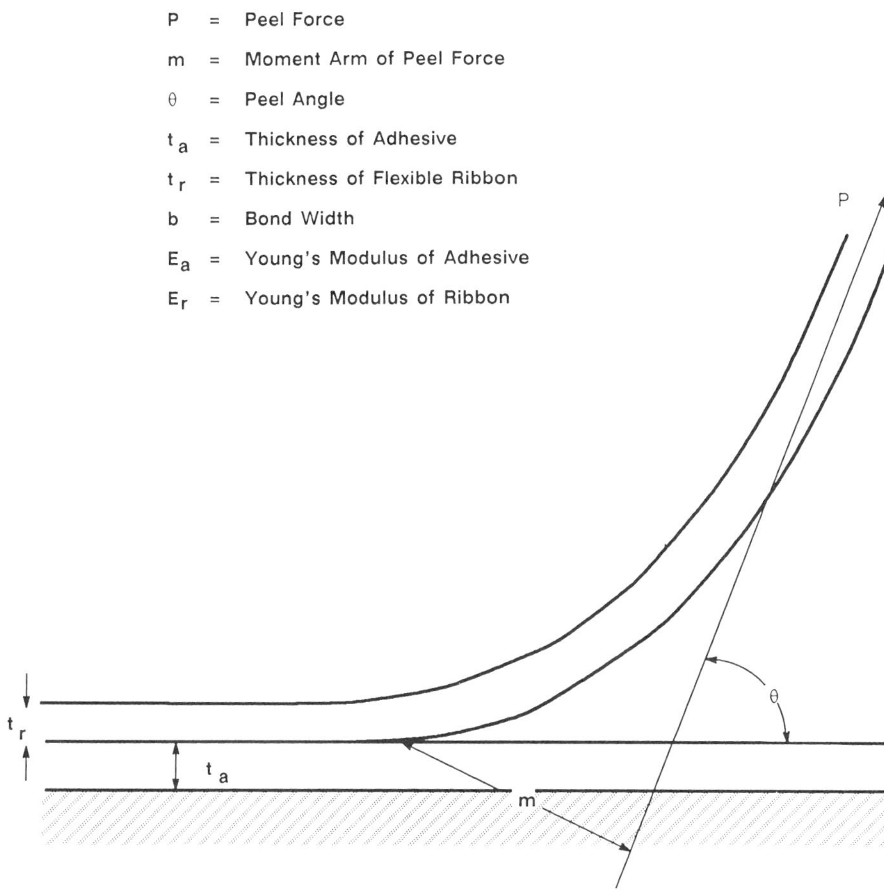


**Figure 1.1** Schematic model for analysis of peel.

The Jouwersma equation, like Bikerman's, assumes linear elastic materials and failure at the critical strain of the adhesive layer. In Equation (2) the moment arm m is equal to $x_0$ and is negative with respect to a coordinate system using $x = 0$ at the point of adhesive failure or detachment.

Kaelble's more comprehensive treatment based on Figure 1.1 includes the dependence of peel force on the angle of peeling. He analyzed both cleavage and shear stress distributions and used the combined stresses during steady-state peel.

$$P = (P_c^2 + P_s^2)^{1/2} \tag{3}$$

His final result for peel force is

$$P = bt_a \left[ \frac{K\sigma_0}{(2E_a)^{1/2}} + 3^{1/2} \cos\theta \frac{\tau_0}{(2G_a)^{1/2}} \right]^2 (1 - \cos\theta)^{-1} \tag{4}$$

where $\tau_0$ is the boundary shear stress and $G_a$ the shear modulus of the adhesive. The cleavage stress factor K is a function of $\theta$ and is given by

$$K = \beta m_c (\beta m_c + \sin\theta)^{-1} \tag{5}$$

One of the advantages of peel testing is that results can be analyzed in terms of work of peeling as well as distribution of strains or stresses. Steady-state peeling is analogous to crack extension in a constant-compliance mode. The angular dependence can be expressed in terms of continuum mechanics as a result of mixed mode I (cleavage) and mode II (shear) loadings [12-15].

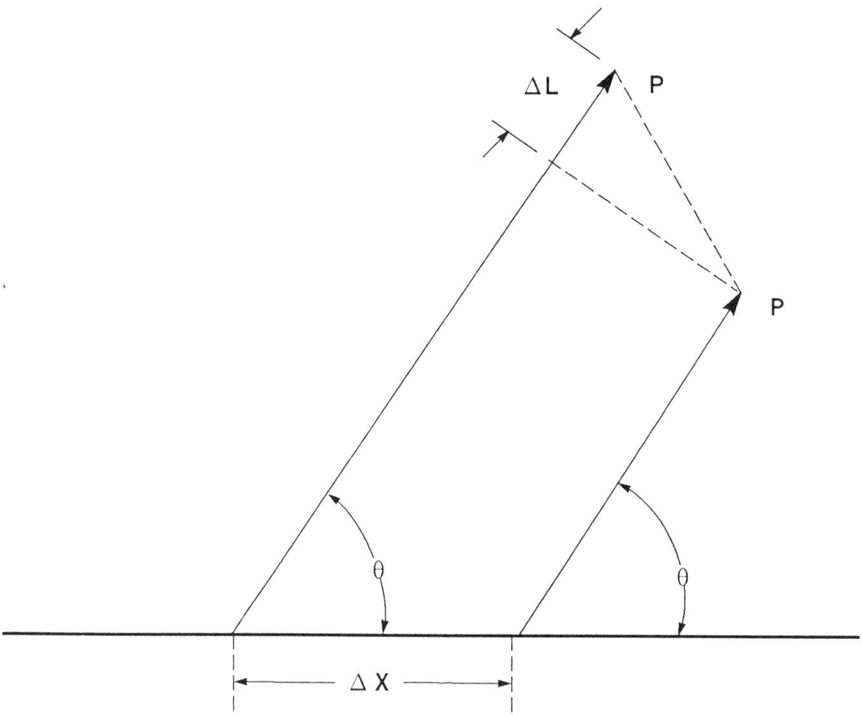


Figure 1.2 Schematic model for work of peel.

From mechanics, the nonrecoverable work of peeling is simply the product of the peel force and the displacement $\Delta L$ in the direction of the force resulting from detachment of a length $\Delta x$. From Figure 1.2 it is seen that $\Delta L$ is equal to $(1 - \cos\theta)\Delta x$ and

$$W_{NR} = P(1 - \cos\theta)\Delta x \quad (6)$$

Substituting W from Equation (6) into the Kaelble equation (Equation 4) gives the nonrecoverable work $W_{NR}$ as the sum of the contributions

$$W_{NR} = bt_a\left[\frac{K\sigma_0}{(2E_a)^{1/2}} + 3^{1/2}\cos\theta\,\frac{\tau_0}{(2G)^{1/2}}\right]\Delta x \quad (7)$$

As peeling also includes recoverable work of elastic deformation of the flexible ribbon, this can be included in an expression for the total work of peeling:

$$W = W_{NR} + W_R \quad (8)$$

Kaelble gives the following expression for $W_R$:

$$W_R = \frac{b}{4}\left[\frac{\sigma_0^2(1 - K)^2(t_r t_a)^{1/2}}{3(E_a E_r)^{1/2}} + \frac{2\tau_0^2 t_a}{G}\right]\Delta x \quad (9)$$

It must be remembered that Kaelble's equations are based on linear elastic responses of both adherend or ribbon and adhesive.

*Cases Involving Small Plastic Deformations*

More general equations based on energy criteria allow for yielding and viscous energy dissipation in both adherend and adhesive.

When plastic flow is confined to small regions in the vicinity of the fracture or debond terminus and bulk materials respond elastically, equations based on linear elastic behavior are still useful.

When a peel specimen is sufficiently long that portions of both the unbonded sample and the debonded strip exhibit zero curvature, the curved portion during steady-state peel or at constant applied force will have the same stored energy independent of debond length.

The energy stored per unit length in the zero curvature portion is not affected by the debonded length. The change in stored energy in the zero curvature portion after debonding a length $\Delta x$ is recoverable (for linear elastic response) and is given by

$$W_R = P^2\Delta x/2bt_r E_r \quad (10)$$

The total work of peeling is the product of the applied force and the distance of movement in the direction of the force. The load will move $\Delta x - \Delta x \cos\theta$ as the debond propagates a distance $\Delta x$ plus a distance due to the strain e of the debonded segment produced by the force P. The total work of peeling is

$$W = P(\Delta x - \Delta x \cos\theta) + Pe\Delta x = P\Delta x(\lambda - \cos\theta) \tag{11}$$

where $\lambda$ is the extension ratio $(1 + e)$.

The energy $(\Delta\Gamma)$ required to debond or fracture a distance $\Delta x$ is given by

$$\Delta\Gamma = \Delta x b W_{SEP} \tag{12}$$

where $W_{SEP}$ is the work per unit area required for creating debonded or fractured surfaces. The symbol $\gamma$ has often been used for this energy and has been identified as surface energy. Since $\gamma$ is also used to symbolize thermodynamic surface energies, the distinction often becomes clouded. For truly brittle fracture with purely elastic responses $W_{SEP}$ would, in fact, be identical to $2\gamma$ (the energy required to fracture a unit area of the bond, leading to two surfaces of unit area) or $\gamma_a + \gamma_r - \gamma_{ar}$ (the sum of the two new interfacial energies formed by separating a unit area of bonded surface minus the interfacial energy of the bond).

The work of separation $(W_{SEP})$ per unit area is equal to the total work of peeling minus the recoverable work. This can be written as

$$W_{SEP} = \frac{P}{b}\left(\frac{1 + \lambda}{2} - \cos\theta\right) \tag{13}$$

For a $90^\circ$ peel and small extensions, the work of separation approaches

$$W_{SEP} \sim \frac{P}{b} \tag{14}$$

or

$$P \sim bW_{SEP} \tag{14a}$$

For zero angle peel tests the work of separation approaches

$$W_{SEP} \sim \frac{P^2}{2b^2 t_r E_r} \tag{15}$$

and

$$P \sim 1.414 b t_r^{1/2} E_r^{1/2} W_{SEP}^{1/2} \tag{15a}$$

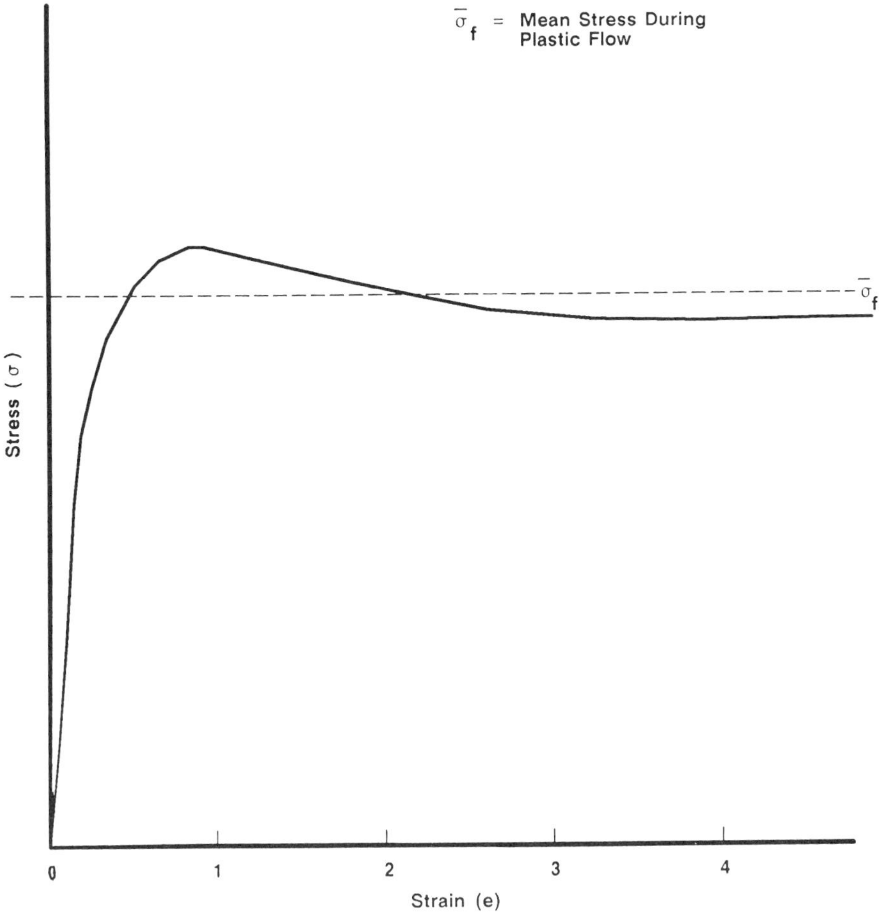


Figure 1.3 Stress-strain behavior for polymers exhibiting large plastic flow.

Equations (14) and (15) are applicable when no adherend or ribbon is present and the adhesive alone is peeled from the rigid substrate. (Then the values for $t_a$ and $E_a$ are used.) Anderson et al. [15] and Gent and Hamed [16] treated such cases.

*Cases Involving Adhesives Exhibiting Large Plastic Deformations*

Although the equations based on energy criteria permit some viscous dissipation, it seems desirable to find an expression for cases involving large plastic deformations. Many practical adhesives owe their success to their ability to undergo such deformation, which

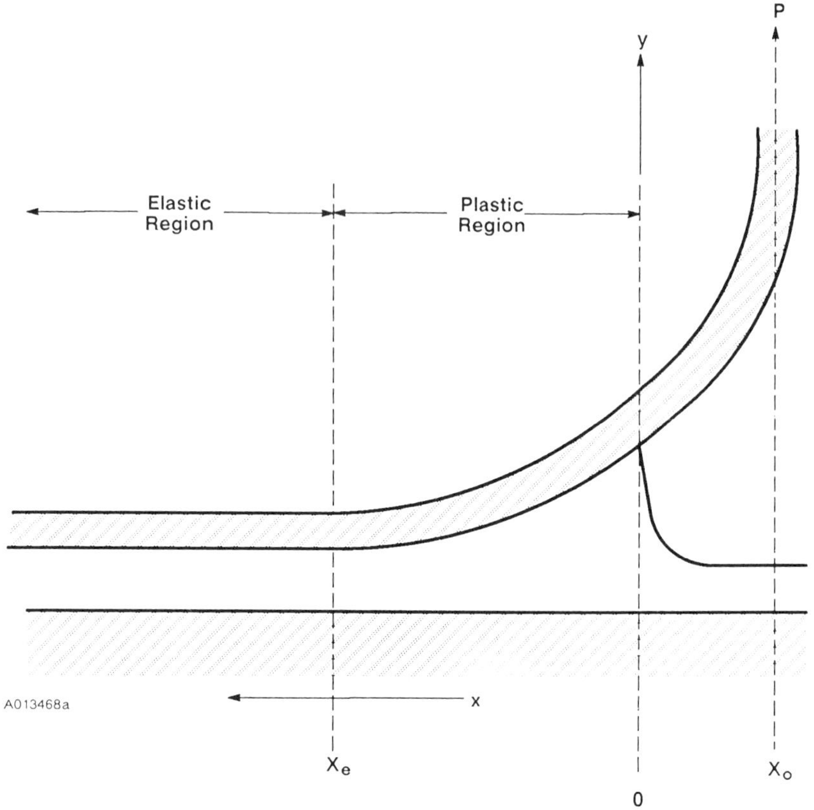


**Figure 1.4** Schematic model for peel of adhesives exhibiting large plastic flow.

not only diminishes stress concentration but also consumes large amounts of energy. The approximate treatment that follows is based on viscoelastic responses which approach those of Figure 1.3. It is assumed that virtually all of the fracture energy per unit area is given by the product of the mean stress during plastic flow ($\bar{\sigma}_f$) times the elongation at break ($e_b t_a$).

Using the model of Figure 1.4 for 90° peel, the deflection of the ribbon is given in terms of the bending moment M:

$$E_r I \frac{d^2 y}{dx^2} = M \tag{16}$$

The moment at any point in the bonded region is

$$M = -P(-x_0 + x) + \bar{\sigma}_f bx^2/2 \tag{17}$$

Substituting Equation (17) into (16), integrating, and evaluating the integration constants by assuming that $dy/dx = 0$ and $y = 0$ at $x = x_e$ gives

$$P = 1.57e_b^{1/4}\, b\beta^{-1}\left(\frac{E_a}{\bar{\sigma}_f}\right)^{1/4}\bar{\sigma}_f \tag{18}$$

B. Peeling Adhesives from Rigid Substrates

Frequently, films, coatings, or adhesives are applied alone to rigid substrates rather than being used as interlayers or adhesives for another adherend. In peeling such self-adhered films, application of Equation (2), (4), (7), (9), or (18) requires knowing what fraction acts as adhesive.

The following estimate is based on the concept that peeling occurs in a manner requiring minimum work. If this is so, the peeling moment will be as large as possible for a given force and, using Equation (2),

$$Px_0 = \frac{0.5ebE_a}{\beta x_0^{-1}(1 + \beta x_0)} = \text{maximum} \tag{19}$$

For Equation (19) to be maximum

$$\beta x_0^{-1}(1 + \beta x_0) = \text{minimum} \tag{20}$$

If the effective thicknesses of "adhesive" and "ribbon" are $ft_a$ and $(1 - f)t_a$, respectively, the value of $\beta$ is

$$\beta = 1.316t_a^{-1}f^{-1/4}(1 - f)^{3/4} \tag{21}$$

The derivative of Equation (20) is 0 when $f = 0.25$ and $x_0^{-1} = \beta$. Substituting these values into Equation (2) gives

$$P = 0.108ebt_aE_a \tag{22}$$

For the case of large plastic flow the final result is

$$P = 0.678be^{1/4}t_a\left(\frac{E_a}{\bar{\sigma}_f}\right)^{1/4}\bar{\sigma}_f \tag{23}$$

The utility of Equation (23) will be shown later.

## C. Cases Involving Viscous Responses Within the Adherends

Gent and Hamed [16,17] studied the influence of plastic yielding within the adherend. The contribution of this additional mechanism for energy dissipation to peel force is complex and is influenced by the interplay of thickness, modulus, yield stress, and the strength of the adhesion.

A schematic diagram of peeling an adherend at 180° is given in Figure 1.5. The maximum strain e at a distance x from the neutral axis is given by

$$e = \frac{x}{R} \qquad (24)$$

When the strain exceeds the yield strain $e_y$ there is viscous dissipation of energy.

The total energy dissipated in plastic deformation is

$$W_p = 2 \int_{e_y R}^{t/2} \sigma_y (e - e_y)\, dx \qquad (25)$$

This energy is supplied by the component $P_y$ of the peel force arising from plastic work:

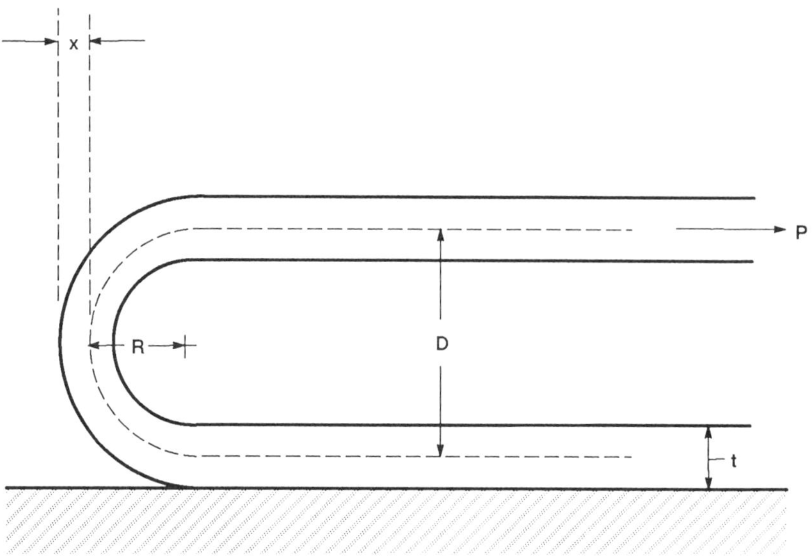


Figure 1.5 Schematic for analysis of yielding in 180°

$$W_p = 2P_y \tag{26}$$

From Equations (24), (25), and (26):

$$P_y = \frac{1}{4}(\sigma_y e_y t)\left[\left(\frac{t}{2Re_y}\right) + \left(\frac{2Re_y}{t}\right) - 2\right] \tag{27}$$

The maximum possible contribution of $P_y$ to the peel force occurs at the minimum radius $\sim t/2$ or $D/2$ when the ribbon folds back on itself. When this occurs the maximum value is

$$P_{y,max} = \frac{1}{4}\sigma_y t \tag{28}$$

When the radius of curvature exceeds a critical value $R_c = t/2e_y$ there is no plastic deformation and $P_y = 0$.

It can be shown that no contribution to the peel force from plastic flow occurs for adhering ribbons having a thickness greater than $t_c$:

$$t_c = \frac{12E_r P}{\sigma_y^2} \tag{29}$$

If the adhesion is low or the adhering ribbon is thick or strong, no plastic deformation occurs. With strong adhesion or relatively thin, ductile adherends, yielding occurs readily and contributes to the total work of detachment. These considerations account for peel forces passing through maxima as adherend thicknesses increase.

## D. Influence of Thickness of Adhesive Layer

Examination of the equations for peel shows that the influence of adhesive thickness ranges from $t_a^{1/4}$ to $t_a$. In several expressions this is complex and involves not only $t_a$ directly but also its effect on values for $\beta$, K, or the slope of the flexible ribbon at the point of detachment [12]. In Equations (14) and (15) the influence of adhesive thickness is clouded by its effect on the work of separation $W_{SEP}$. When peeling adhering films (without attached flexible ribbons) at high angles, both elastic and highly plastic films require forces directly proportional to thickness. Experiments with polyethylene showed linearity over a range of thickness from $\sim 0.02$ to 2.0 mm.

## E. Influence of Peel Rate

Peel forces are determined by the moduli of adherends $E_r$ and adhesives $E_a$, the tensile strengths of adhesives $\sigma_b$, and extension ratios

$\lambda_b$ or strains at break $e_b$. Work over the past 35 years has produced a relatively sophisticated knowledge of the influence of time (rate) and temperature on mechanical responses of viscoelastic materials including all of the above properties [18-22].

Rate effects were recognized long ago and were examined by several workers [3,4,10,11,23]. Relationships between peel forces and time-temperature influences on viscoelastic and tensile properties were noted during the 1960s [12,24-26].

Peeling is inherently a high-strain-rate process. Estimates of effective rates can be obtained using

$$\dot{e} \sim e_b t_a^{-1} x_s^{-1} \frac{dx}{dt} \tag{30}$$

where $\dot{e}$ is the effective strain rate and $x_s$ is the distance along the x axis over which the strain in the adhesive goes from 0 to $e_b$. It is evident that for thin adhesive layers even low peel rates lead to high values of $\dot{e}$.

Utilization of equations for peel force requires values for moduli, $\sigma_b$, and $e_b$ measured at appropriate rates. (Also, recall that over the temperature interval from $T_g$ to $\sim T_g + 100^oC$ the influence of temperature may be very profound, and relatively small changes in both temperature and peel rate may produce dramatic effects.)

Experience has shown that the locus of failure may change from cohesive to interfacial [12,27] and the influence of thickness of the adhesive layer may change with increasing peel rates [12].

At low peel rates many adhesives exhibit low values of rate-dependent modulus and yield. Stresses are distributed broadly within the strained adhesive and are not localized at the point of detachment. Both modulus and yield values increase with increasing strain rates (peel rates), and often increasing localization of stress at the point of debonding leads to exceeding the critical stress for debonding, changing failure from cohesive fracture within the adhesive to separation at the interface. When this occurs, peeling becomes relatively insensitive to any further increase in the rates and the thickness of the adhesive layer.

Hamed [28] has measured stress relaxation during peeling and found that the data could be used to predict rate dependence.

## F. Summary of Factors Influencing Peel Data

Analysis of peeling and practical experience both show that the data are influenced strongly by the physics or mechanics of the test and the viscoelastic behavior of all materials deformed during peeling. Although no fixed relationships exist for many of the important properties, they can be estimated for circumstances of interest if the parameters involved are determined independently at appropriate rates and temperatures and the interplay among thicknesses, moduli, and tensile behavior is considered in the light of known or recognized influences.

For example, when adherends are attached to substrates with adhesives, the peel force is nearly proportional to the adhesive thickness for thin layers, for adhesives with high relaxation moduli, and for high peel rates (until a critical value is reached). Conversely, thick layers, low moduli, and low peel rates will lead to forces proportional to $t_a^{1/4}$.

Since strain rates are inversely proportional to adhesive thickness, it is clear that critical values for altering failure modes will be reached at relatively low peel rates for thin layers and at high rates for thick layers.

Thick adherends or elastic adherends may not contribute to the nonrecoverable work of separation, but thin adherends or those with low yield stress may contribute significantly, increasing the work of separation. It is often easy to recognize when adherends have yielded by their persistent deformation after peeling.

## III. ANALYSIS OF THE ADHESIVE BOND

A little reflection on the preceding analysis of peel testing shows that a given set of materials—adherend, adhesive, and substrate—may give peel forces differing by orders of magnitude depending on the physical character of the bonded structure and the way it is peeled.

Unfortunately, it is often unrecognized, overlooked, or dismissed as being either "too theoretical" or trivial that a given set of materials exhibiting identical physical character and tested in the same way may also give peel forces that differ by orders of magnitude, depending on how the bonds were formed or on changes wrought by time and/or environmental influences. Perhaps the most insidious of these factors is simply how the bond was formed.

### A. Influence of the Thermodynamic State of the Interface

Bonds formed under arbitrary conditions may have effective densities of atoms close to the interface much less than, or effective mean interfacial separations between phases much greater than, their equilibrium values. These differences exist on too small a scale to be detected optically at visible wavelengths. As a direct result of low interfacial "contact density," increased mean separation of phases, or stress concentration in regions on the order of tens of nonometers, the effective critical interfacial stress is greatly diminished. Such bonds, when peeled, give data related mainly to how the sample was made rather than how strongly the materials are capable of adhering. This is a commonly occurring situation and is the basic cause of many practical problems. This was demonstrated experimentally [24] in 1963 and its importance has been emphasized over the past 20 years [25,29-31].

## B. The Bonding Process

Bonding is virtually synonymous with wetting. A liquid adhesive layer must flow to conform to the topography of the substrate, first on a macroscopic scale, then on a microscopic scale, and finally accommodating features on the atomic level. Equilibrium is reached when the minimum free energy of the system is reached. For nearly all practical cases this corresponds to a completely wetted interface [29]. The adhesion or attraction between phases at equilibrium is determined mainly by the atomic density and mean separation of the atoms or molecules across the interface.

## C. Hot Melts

Adhesives or polymer films are often bonded by melting and applying pressure for some arbitrary time. The viscosity is normally high at bonding temperatures, frequently ranging up to $10^5$ Pa·s. Also, flow is very highly restricted in constrained thin adhesive layers. (The entire volume of the adhesive is involved in flow into an interfacial interstice, and the thinner the layer the greater the constraint.) These constraints also retard the flow involved simply in the molecular rearrangements required to achieve the orientations and configurations that minimize the free energy, and often long times are needed at bonding temperatures to reach equilibrium. In practice, limits on allowable temperatures and restrictions on bonding times are set by materials or production requirements. Frequently, equilibrium is not reached, and bonds perform at levels far below the capabilities of the pair of materials involved. Examples of this are given later.

## D. Application from Solution

It might be inferred that application from solution, because of the low viscosity, would obviate the wetting problem described above. This is seldom the case, however, for the final state after removal of the solvent still involves wetting the substrate by polymer, not solution. Solutions comprising, at a minimum, two components offer the possibility of selective adsorption of solvent or some component of the solvent, which may remain at the interface until the polymer solution is very viscous. Often the solution may even drop below its glass transition temperature while solvent is still at the interface, leading to a system very far from equilibrium when the solvent is finally removed.

Conformations of polymer molecules near the interface when solvent is present are significantly different from the conformations they would ultimately assume at equilibrium after the solvent evaporates. Unless such systems are held for relatively long times at temperatures well above $T_g$ after removal of solvent, they also can be far from equilibrium [32,33].

### E. Influence of Third Phases

Practical adhesives often are multicomponent systems, either by design or as a result of impurities. Even when present in small amounts, impurities can influence performance if they are selectively adsorbed at the interface.

If the adhesive layer comprises a two-phase system in which both phases are liquid during bonding, at equilibrium wetting may be complete but the interface may be heterogeneous, or a separate thin interphase layer may be interposed between the bulk adhesive and the substrate [31].

When insoluble fluid components exist in the adhesive and droplets reach the interface they will remain there, exhibiting finite contact angles with the substrate when the interfacial tensions of the three phases are such that

$$\frac{(\gamma_{CS} - \gamma_{LS})}{\gamma_{CL}} < 1 \qquad (31)$$

where C stands for contaminant, L for adhesive, and S for solid. Such trapped droplets act as flaws or cracks and cause stress concentration at their boundaries related to droplet size and contact angles.

When $(\gamma_{CS} - \gamma_{LS})/\gamma_{CL} = (-1)$ the contaminating liquid collects as a stable separate interphase. Bond strength is set by the thickness and viscosity of the layer if it remains fluid at the use temperature. (Examples will be given later.) If such a layer solidifies on cooling, it may influence performance by changing the stress distribution through the influence of its thickness and modulus, or it may fracture at a low applied load if its critical stress is less than that of the adhesive or either new interface.

Interfacial phases can form slowly over relatively long time periods by diffusion or through chemical changes occurring during aging. Also, intrusion of third phases from environmental exposure is commonly observed. Water is a frequent culprit.

It is clear that third phases produce significantly altered peel values and must be considered in interpreting data.

## IV. ANALYSIS OF PRACTICAL EXAMPLES

The following analyses of actual problems involving adhesion demonstrate the utility of applying fundamental principles. These problems include a broad range of materials and provide examples of each of the four basic causes of poor performance. These are:

1. Failure to produce bonds approaching interfacial equilibrium
2. Mismatch of molecular composition and structure between adhesive and adherend
3. Mechanical properties unsuited for the design or intended use of the bonded structures

4. Presence of interfacial phases which dominate the performance

The results of this work show that none of these can be dismissed as "theoretical" or trivial.

## A. Peel Tests Used

All of the experimental values were obtained using an apparatus permitting either T-peel tests or tests at fixed angles from 10 to 90°. For tests at fixed angles the concept was that described by

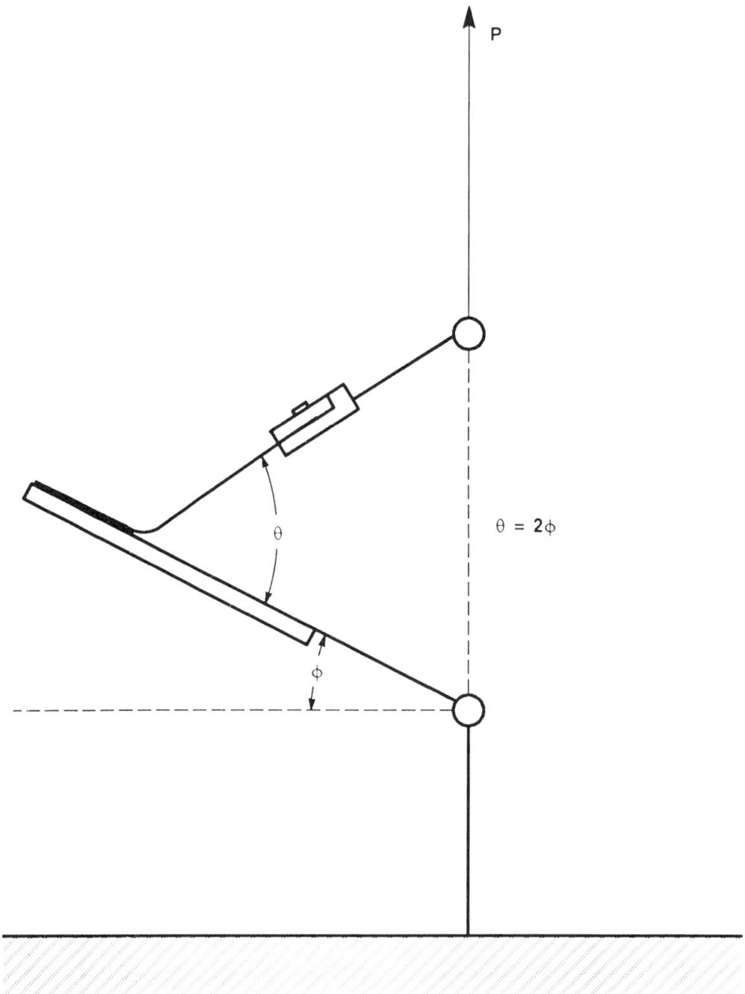


Figure 1.6 Schematic diagram illustrating the principle for constant peel-angle apparatus.

Packham [34]. Figure 1.6 is a schematic representation showing the principle involved in maintaining a constant peel angle. When peeled adhesive or adhesive plus adhering ribbon is inextensible, the peel angle $\theta$ is constant and equal to $2\phi$. If the peeled portion is strained, the peel angle increases, diminishing the force. Force was applied using a constant-speed motor which could be adjusted to give peel rates of $\sim 2 \times 10^{-3}$ to 2 m/sec. A take-up device was incorporated in the load line, delaying application of force until the motor attained full speed. Data were obtained using a transient recorder to store, smooth, and display the input from a load cell.

### B. Performance of Film/Adhesive/Film and Adhesive/Film Bonds

This problem comprised an effort to explain the performance of 25 combinations of dissimilar polymer/adhesive pairs in order to provide a basis for development of superior adhesives. Some of the materials used were proprietary, but the analysis and conclusions drawn are not related directly to the composition and the use of arbitrary designations for the samples does not restrict the utility of the study.

Samples are identified by the letters A to W. They represent a range of surface free energies from $\gamma$ = 34.5 to 71.5 mN/m, fractional polarities from p = 0.07 to 0.58, moduli from E = <0.5 to 348 MPa, stress at fracture from $\sigma_b$ = 3.5 to 75.5 MPa, strain at break from 0.5 to 11.6, yield stress from 0.8 MPa to no yield at 348 MPa, and melt indices ranging from 35 to values too low to be measured—essentially no flow.

The letter designations of the samples are roughly in order of increasing dissimilarity; A represents polyethylene, B to O are copolymers and terpolymers containing ethylene, P and Q are Surlyn* ionomer resins, R is polypropylene, S is polystyrene/butadiene (81/19), T is acrylonitrile, U is Mylar polyester resin,† V is nylon, and W is poly(vinyl alcohol).

The broad spectrum of chemical and physical characteristics of these samples provided an opportunity to develop some basic information which would show correlation with performance. Consequently, all of the properties likely to be important were measured.

Surface energies were measured using the method of Owens and Wendt [35,31]. Table 1.1 lists the surface free energy $\gamma_{SO}$, the fractional contribution of dispersion forces to the surface energy, and the melt index.

Tensile data were measured at $\dot{e}$ = 16 $sec^{-1}$ to obtain strain rates as close as possible to those associated with peeling. Values for stress at break ($\sigma_b$), strain at break ($e_b$), yield stress ($\sigma_y$), and mean stress in the flow region ($\bar{\sigma}_{\hat{f}}$) are given in Table 1.2.

---

*Surlyn is du Pont's trade name for its ionomer resins.

†Mylar is du Pont's trade name for its polyester resins.

Table 1.1 Surface Energies, Dispersion Fractions, and Melt Indices of Polymers Used

| Sample | $\gamma_{SO}$ (mN/m) | d | Melt index (220°C) |
|---|---|---|---|
| A | 39.2 | 0.93 | 2.8 |
| B | 39.5 | 0.93 | 3.0 |
| C | 40.2 | 0.91 | 0.44 |
| D | 40.7 | 0.91 | 13.0 |
| E | 38.7 | 0.89 | 2.0 |
| F | 40.6 | 0.87 | 20.0 |
| G | 39.5 | 0.87 | 2.5 |
| H | 45.8 | 0.88 | 35.0 |
| I | 39.9 | 0.87 | 2.7 |
| J | 38.6 | 0.87 | 3.26 |
| K | 38.1 | 0.85 | 0.44 |
| L | 43.6 | 0.87 | 6.0 |
| M | 38.3 | 0.81 | 0.9 |
| N | 42.6 | 0.78 | 3.0 |
| O | 47.5 | 0.75 | 6.0 |
| P | 36.6 | 0.90 | 1.6 |
| Q | 35.5 | 0.88 | 0.9 |
| R | 34.5 | 0.90 | 0.9 (230°C) |
| S | 41.0 | 0.92 | 2.0 |
| T | 49.1 | 0.80 | 1.0 |
| U | 52.5 | 0.79 | No flow |
| V | 58.0 | 0.65 | 2.0 (230°C) |
| W | 71.3 | 0.42 | No flow |

The elastic moduli could not be obtained at the rates used for the tensile data and were measured independently by determining the instantaneous strain on application of loads during creep tests. Since many of the samples showed nonlinear responses, the moduli were measured at two different loads for each sample. Data are given in Table 1.3.

Samples originally received for peel tests comprised two similar films bonded by an adhesive. They had been prepared using a Sentinel heat sealer and were bonded using 6-sec dwells at 20 psi at various temperatures ranging from 150 to 200°C. They were

Table 1.2 Tensile Data at $\dot{e}$ = 16 $sec^{-1}$

| Sample | $\sigma_b$ (MPa) | $e_b$ | $\sigma_y$ (MPa) | $\bar{\sigma}_f$[a] (MPa) |
|---|---|---|---|---|
| A | 40.2 | 0.5 | | 19.5 |
| B | 16.5 | 3.2 | 12.6 | 8.1 |
| C | 17.5 | 7.75 | 6.2 | 8.7 |
| D | 37.9 | 6.7 | | 18.9 |
| E | 15.1 | 7.3 | 8.7 | 7.5 |
| F | 11.3 | 7.7 | 7.5 | 5.9 |
| G | 12.2 | 9.1 | 6.9 | 6.0 |
| H | 3.5 | 7.8 | 0.8 | 1.7 |
| I | 11.2 | 9.2 | 5.0 | 5.6 |
| J | 11.8 | 7.8 | 6.8 | 6.0 |
| K | 13.2 | 8.7 | 6.0 | 6.6 |
| L | 5.45 | 9.6 | 2.8 | 2.7 |
| M | 14.3 | 10.4 | 8.1 | 7.2 |
| N | 15.4 | 11.6 | 8.2 | 7.7 |
| O | 5.5 | 9.5 | 2.8 | 2.7 |
| P | 38.4 | 3.75 | 20.5 | 19.3 |
| Q | 27.6 | 5.3 | | 14.3 |
| R | 36.9 | 1.75 | 3.2 | 18.3 |
| S | 20.7 | 0.59 | | 10.5 |
| T | 75.5 | 4.1 | 5.4 | 38.0 |
| U | | | | |
| V | | | | |
| W | 67.7 | 1.65 | 30.8 | 33.9 |

[a] $\bar{\sigma}_f$, Mean stress during plastic flow.

tested using the T-peel mode at 0.04 m/sec. Data are given in groups of two temperature ranges to indicate the influence of bonding conditions. These results are given in Table 1.4.

These data show no readily discernible correlations between peel force and either surface energy or mechanical properties. Estimates of work of adhesion including acid-base interactions also failed to correlate. Perhaps the most surprising aspect of the peel

Table 1.3 Young's Modulus for the Samples Used

| Sample | Low stress (MPa) | E (MPa) | High stress (MPa) | E (MPa) |
|---|---|---|---|---|
| A | 6.84 | 316 | 15.6 | 197 |
| B | 2.93 | 95.8 | 4.97 | 106 |
| C | 1.38 | 19.2 | 2.73 | 15.2 |
| D | 3.54 | 221 | 14.2 | 246 |
| E | 2.3 | 58 | 4.96 | 39.4 |
| F | 0.365 | 41 | 2.48 | 25 |
| G | 0.78 | 23 | 2.62 | 12 |
| H | 0.12 | < 5 | | |
| I | 1.02 | 71 | 3.30 | 75 |
| J | 1.8 | 22.7 | 4.07 | 24.5 |
| K | 0.70 | 44 | 2.83 | 24 |
| L | 0.634 | 7.34 | 1.27 | 5.4 |
| M | 0.725 | 27 | 2.44 | 23 |
| N | 1.58 | 30.4 | 2.67 | 15.0 |
| O | 0.634 | 7.4 | 1.27 | 5.4 |
| P | 5.14 | 89 | 8.68 | 76 |
| Q | | | | |
| R | 4.08 | 113 | 10.0 | 127 |
| S | | | | |
| T | 2.89 | 134 | 11.5 | 178 |
| U | | | | |
| V | 13.8 | 348 | 33.9 | 313 |
| W | 1.10 | 204 | 3.76 | 180 |

data is the relatively high force for some films bonded with adhesives H and L, which exhibit very low values for $E_a$, $\sigma_b$, and $\bar{\sigma}_f$.

The data of Table 1.4 show only 2 cases of cohesive failure and 17 cases of interfacial separation. There is also an average increase of 60% in peel force for the bonds formed at 190 to 200°C. Since many of the pairs show high values for $\phi$ and work of adhesion and very low interfacial free energies, it appears that the majority of the samples were not at equilibrium.

To provide a more closely defined basis for comparing peel force with the various surface energy and mechanical property data,

Table 1.4 Peel Forces for Samples Evaluated

| Sample | Peel force (kN/m) (bonded at 150-175°C) | | Peel force (kN/m) (bonded at 190-200°C) | |
|---|---|---|---|---|
| P/N/P | 1.35 | I[a] | 1.70 | I |
| A/L/A | 1.16 | C[a] | 1.16 | C |
| V/G/V | 0.42 | I | 1.08 | I |
| S/L/S | 0.93 | I | 0.85 | I |
| W/F/W | 0.69 | I | 0.85 | I |
| Q/H/Q | 0.62 | I | 0.77 | I |
| U/H/U | 0.46 | I | 0.77 | I |
| P/I/P | 0.54 | I | 0.54 | I |
| V/K/V | 0.46 | I | 0.54 | I |
| P/J/P | 0.58 | I | 0.58 | I |
| V/M/V | 0.19 | I | 0.54 | I |
| R/E/R | 0.15 | I | 0.54 | I |
| V/B/V | 0.27 | I | 0.46 | I |
| P/E/P | 0.46 | I | 0.39 | I |
| R/N/R | 0.31 | C | 0.31 | C |
| P/D/P | 0.19 | I | 0.23 | I |
| W/E/W | 0.19 | I | | |
| W/H/W | 0.08 | I | | |
| W/L/W | 0.04 | I | | |

[a]I, Interfacial failure; C, cohesive failure.

bonded pairs of materials were prepared. Time-temperature conditions leading to maximum strength were established for each pair of materials, and each sample was bonded using optimum conditions. Bonding schedules ranged from 30 sec at 190°C to 120 sec at 220°C. The pairs comprised the 19 represented by the original film/adhesive/film samples plus 6 included to broaden the experimental base.

Since bonding and wetting are virtually synonymous, correlation of bonding conditions directly with spreading pressures and inversely with melt viscosities would be expected. The spreading pressure of a liquid on a solid ($\pi_{LS}$) is equal to the surface energy of the solid ($\gamma_{SO}$) minus the interfacial free energy ($\gamma_{INT}$). Using the data of Table 1.1, values for the interaction parameter $\phi$, $\gamma_{INT}$, $W_{ADH}$, and the thermodynamic failure locus were calculated for each pair of materials. These values are given in Table 1.5.

Table 1.5 Values Calculated from Data of Table 1.1

| Sample | $\phi$ | $\gamma_{INT}$ | $W_{ADH}$ | Thermodynamic failure locus |
|---|---|---|---|---|
| O/W | 0.94 | 9.2 | 109.6 | O |
| H/W | 0.87 | 17.5 | 99.6 | H |
| L/W | 0.88 | 16.9 | 98.0 | L |
| H/U | 0.99 | 0.9 | 97.3 | H |
| F/W | 0.88 | 17.3 | 94.6 | F |
| M/V | 0.98 | 3.6 | 92.7 | M |
| G/V | 0.97 | 5.1 | 92.4 | G |
| D/V | 0.95 | 7.3 | 92.0 | D |
| K/V | 0.97 | 4.7 | 91.4 | K |
| C/W | 0.85 | 20.8 | 90.8 | C |
| E/W | 0.86 | 19.2 | 90.8 | E |
| B/V | 0.93 | 8.1 | 89.4 | B |
| H/C | 0.998 | 0.4 | 85.9 | H |
| L/S | 1.0 | 0.3 | 84.3 | S |
| L/A | 0.99 | 0.5 | 82.3 | A |
| H/Q | 1.0 | 0.7 | 80.6 | Q |
| C/E | 0.999 | 0.06 | 78.9 | E |
| L/R | 1.0 | 0.6 | 77.5 | R |
| P/D | 1.0 | 0.1 | 77.2 | P |
| N/P | 0.995 | 0.5 | 77.1 | P |
| I/P | 1.0 | 0.2 | 76.3 | P |
| N/R | 0.99 | 1.5 | 75.6 | R |
| E/P | 1.0 | 0.04 | 75.3 | P |
| J/P | 1.0 | 0.1 | 75.1 | P |
| E/R | 1.0 | 0.1 | 73.1 | R |

At bonding temperatures each polymer can be considered as a viscous liquid and consequently each would have a spreading pressure with respect to the other in the pair. The polymer exhibiting the highest value for the product of spreading pressure × melt index, $\pi_{12} \times \mathrm{M.I.}_{(1)}$ or $\pi_{21} \times \mathrm{M.I.}_{(2)}$, would be the effective spreading liquid or adhesive and would establish the time-temperature requirements for bonding. Maximum values of $\pi_{LS} \times \mathrm{M.I.}_{(L)}$ for each

Table 1.6 Maximum Values for $\pi_{LS} \times$ M.I. for the Polymer Pairs Used

| Sample | $\pi_{LS} \times$ M.I.(L) |
|---|---|
| H/U | 1883 |
| H/W | 1821 |
| H/C | 1379 |
| H/Q | 1218 |
| F/W | 1082 |
| D/V | 718 |
| P/D | 474 |
| O/W | 372 |
| L/W | 326 |
| L/S | 244 |
| L/A | 232 |
| L/R | 203 |
| B/V | 143 |
| G/V | 130 |
| J/P | 119 |
| N/P | 112 |
| E/W | 104 |
| N/R | 99 |
| I/P | 98 |
| C/E | 80 |
| E/P | 73 |
| E/R | 69 |
| M/V | 48 |
| K/V | 23 |
| C/W | 22 |

pair of materials are given in Table 1.6. There is good correlation between these values and the bonding conditions required.

Pairs were tested in the T-peel mode at 0.04 m/sec at 23°C and 25% relative humidity. Results are given in Table 1.7. These data differ greatly from those of Table 1.4, due in part to the different character of the samples and in part to changes in bonding.

Table 1.7 Measured Peel Force and Observed Locus of Failure

| Sample[a] | Peel force (kN/m) | Failure locus |
|---|---|---|
| M/V | 1.0 | C |
| K/V | 0.99 | C and I |
| G/V | 0.69 | C and I |
| F/W | 0.54 | C |
| B/V | 0.52 | C and I |
| I/P | 0.43 | C and I |
| C/W | 0.41 | I |
| L/A | 0.36 | C |
| E/W | 0.35 | C |
| E/P | 0.33 | C |
| E/R | 0.31 | C |
| J/P | 0.28 | C |
| L/R | 0.27 | C and I |
| C/E | 0.25 | C |
| N/P | 0.22 | C |
| N/R | 0.21 | C |
| H/U | 0.19 | D |
| L/S | 0.19 | C |
| P/D | 0.19 | I |
| H/Q | 0.17 | C |
| H/C | 0.11 | C |
| H/W | 0.11 | C and I |
| O/W | 0.10 | I |
| L/W | 0.05 | I |
| D/V | 0.03 | I |

[a]The first listed is the effective adhesive, the second the substrate.

A major difference is that only 4 of 25 pairs failed distinctly at the interface, whereas 17 of the 19 samples bonded using arbitrary conditions failed at the interface. Also very striking is the performance of the pair M/V, which failed cohesively at the highest peel force when at or near equilibrium but failed interfacially at very low loads when bonded at 150 to 175°C and at moderate loads when bonded at 190 to 200°C. Note that the pair M/V had a low value for $\pi_{LS} \times$ M.I.

These samples, moreover, show correlation with mechanical properties. Eight of the 10 lowest peel forces were for pairs containing polymers that exhibited the lowest mechanical strengths.

The other two samples that failed at low peel forces and showed apparent interfacial failure both contain polymer D. These are especially interesting because one would expect on the basis of all likely correlations that these samples would fail cohesively at high peel forces. Polymer D has high modulus, high elongation at break, high tensile strength, and a high value for $\bar{\sigma}_f$. Both D/V and P/D show relatively high $W_{ADH}$, donor-acceptor interactions, and values for $\phi$, and low or moderate interfacial energies. Both pairs also have moderately high values for $\pi_{LS} \times$ M.I.$_{(L)}$, suggesting that they should be at or close to equilibrium. It seems reasonable to conclude that some component of polymer D separates to form a thin interphase which leads to poor performance.

The other pairs showing interfacial separation contain polymer W [poly(vinyl alcohol)]. This is certainly the most dissimilar with respect to the adhesives, which is evident from the low values for $\phi$ and high interfacial free energies. Pairs containing polymer W are also those for which $\phi_r$ is less than unity. ($\phi = \phi_a \phi_r$, where $\phi_a$ is a measure of dissimilarity in attractive forces due to differences in polar contributions and $\phi_r$ is a measure of effective interphase separation or atomic density at the interface. In the values of $\phi$ used for these polymer pairs, $\phi_r$ has been assumed to be unity.) If the $\phi_r$ values for the pairs containing polymer W were ~0.85 to 0.87, the thermodynamic failure locus would be at the interface. Also, because of the nature of polymer viscoelastic responses at usual test rates, failure is probably frequently biased to occur at the interface.

## C. Comparison of Experimental Data with Calculated Values

Because all of the data required for use of Equations (22) and (23) were measured, it was possible to calculate values for peel forces for each pair. Some refinements were needed to compensate for the departures from simple T-peel tests introduced by the different thicknesses and moduli of the polymers.

In a symmetrical T-peel the effective adhesive thickness applicable to the equations is half the thickness of the layer. To compensate for the asymmetry of these practical cases the effective thickness was estimated using

$$t_{a(eff)} = t_a \left(1 + \frac{R_a}{R_r}\right)^{-1} \tag{32}$$

where $R_a$ and $R_r$ are the radii of curvature of the bent adhesive film and the flexible ribbon or substrate.

$$\frac{R_a}{R_r} = E_a E_r^{-1} t_a^3 t_r^{-3} \tag{33}$$

The asymmetric bending also changes the effective peel angle from $90^o$ to values dependent on the nature of the materials. The effective peel angle was estimated using

$$\theta_{eff} = 180 - \psi \tag{34}$$

where $\psi = \cos^{-1}$ of

$$\cos\psi = (1 - R_a/R_r)(1 + R_a/R_r)^{-1} \tag{35}$$

The factor compensating for the peel angle is (from Equation 4) $(1 - \cos\theta)^{-1}$. Since $\cos\theta = -\cos\psi$ the angle factor can be written as

$$(1 - \cos\theta)^{-1} = (1 + \cos\psi)^{-1} = \frac{1 + R_a/R_r}{2} \tag{36}$$

Multiplying Equation (32) by Equation (36) gives $t_a/2$, and the expression for the force required to peel an adhering film from a flexible substrate using a T-peel test is simply

$$P = 0.34bte_b^{0.25}E^{0.25}\bar{\sigma}_f^{0.75} \tag{37}$$

where all values are those for the polymeric adhesive film exhibiting large plastic deformation.

Equation (37) was used to calculate the peel force for all pairs except P/D and D/V, whose performance was believed to be dominated by an interphase. Despite the arbitrary relationships used to compensate for differing moduli and thicknesses, the agreement of the calculated values given in Table 1.8 with experimental data is satisfying. Calculated and measured forces cover the same range. Pairs O/W and L/W failed at the interface at forces less than the calculated maximum values. There is an average shift of only about two places in the ordering of performance from highest to lowest forces when the calculated and measured values are compared.

I conclude that these polymer pairs, when bonded in a manner allowing close approach to equilibrium, exhibit peel strengths established by mechanical properties.

The experimental peel forces for polymer L were 0.36, 0.27, and 0.19 kN/m for substrates A, R, and S, respectively. Calculated values were identical at 0.28 kN/m. This discrepancy is probably

Table 1.8 Peel Force Calculated Using Equation (37)

| Sample[a] | $t_a$ (mm) | Calc. peel force (kN/m) |
|---|---|---|
| M/V | 0.168 | 1.01 |
| K/V | 0.136 | 0.80 |
| G/V | 0.130 | 0.62 |
| F/W | 0.097 | 0.50 |
| B/V | 0.076 | 0.52 |
| I/P | 0.089 | 0.56 |
| C/W | 0.079 | 0.46 |
| L/A | 0.138 | 0.28 |
| E/W | 0.067 | 0.45 |
| E/P | 0.067 | 0.45 |
| E/R | 0.067 | 0.45 |
| J/P | 0.087 | 0.42 |
| L/R | 0.138 | 0.28 |
| C/E | 0.079 | 0.46 |
| N/P | 0.074 | 0.47 |
| N/R | 0.074 | 0.47 |
| H/U | 0.069 | 0.08 |
| L/S | 0.138 | 0.28 |
| P/D[b] | 0.045 | |
| H/Q | 0.069 | 0.08 |
| H/C | 0.069 | 0.08 |
| H/W | 0.069 | 0.08 |
| O/W | 0.069 | 0.20 |
| L/W | 0.138 | 0.28 |
| D/V[b] | 0.088 | |

[a]The first sample polymer is the effective adhesive.
[b]Performance probably dominated by an interphase.

due to differences in plastic deformation of the substrate polymers unaccounted for by Equation (37).

The rather good correlation of peel force with mechanical properties appeared to be somewhat at odds with the peel forces

Table 1.9 Comparison of Calculated Peel Force for Samples Comprising Pairs of Ribbons Adhered with an Adhesive Layer with Force for Samples Comprising Only Adhesive and Ribbon

| Sample | Calc. peel force (kN/m) |
|---|---|
| A/L/A | 0.95 |
| L/A | 0.28 |
| S/L/S | 1.40 |
| L/S | 0.28 |
| Q/H/Q | 0.53 |
| H/Q | 0.08 |
| U/H/U | 0.42 |
| H/U | 0.08 |

exhibited by the pairs of films bonded with adhesive layers (Table 1.4). Four of the highest seven values were for adhesives L and H, which have the lowest values for $E_a$, $\bar{\sigma}_f$, $\sigma_y$, and $\sigma_b$. Peel forces calculated using Equation (18) showed, however, that peel forces in T-peel tests are very strongly influenced by the mechanical properties of the flexible ribbons. (This dependence comes through the value for $\beta$.) Peel forces calculated for adhesively bonded pairs of ribbons are compared with those for single ribbon/adhesive pairs in Table 1.9.

## D. Performance of Poly(vinylbutyral) as Safety Glass Interlayers

Safety glass functions by dissipating kinetic energy through plastic deformation of unbonded areas of the tough plasticized polymer. For optimum performance the amount of debonding when the glass is fractured must be controlled closely. Debonding occurs as high-rate, low-angle peeling. One method of effecting control is to incorporate $\sim 0.1\%$ by weight of potassium acetate into the polymer sheeting and control the amount of water in the sheeting at the time of lamination to the glass. The peel strength is very sensitive to the amounts of both.

A comprehensive study of this problem [36] showed that a thin interfacial layer of aqueous salt solution controlled the peel strength, which depended strongly on the thickness of the aqueous layer. Optimum thickness (for best performance of the safety glass) was on the order of 10 nm.

For plates separated by a fixed volume of liquid, Healey [37] proposed a modified form of the Stefan equation. The force (F) required to separate the plates depends on the rate of separation:

$$F = 3\eta v^2 (8\pi t d^4)^{-1} \tag{38}$$

where V is the volume of liquid, η the viscosity, and d the initial separation of plates.

For steady-state peeling the effective volume of the liquid layer could be expressed as the product of the width of the specimen, the thickness of the liquid layer, and some effective length $l_e$ extending from the point of attachment through the deformed segment. Making such a substitution in Equation (38) and combining all constants into a single term, where U is the steady peel rate, gives

$$F = KUd^{-2} \tag{39}$$

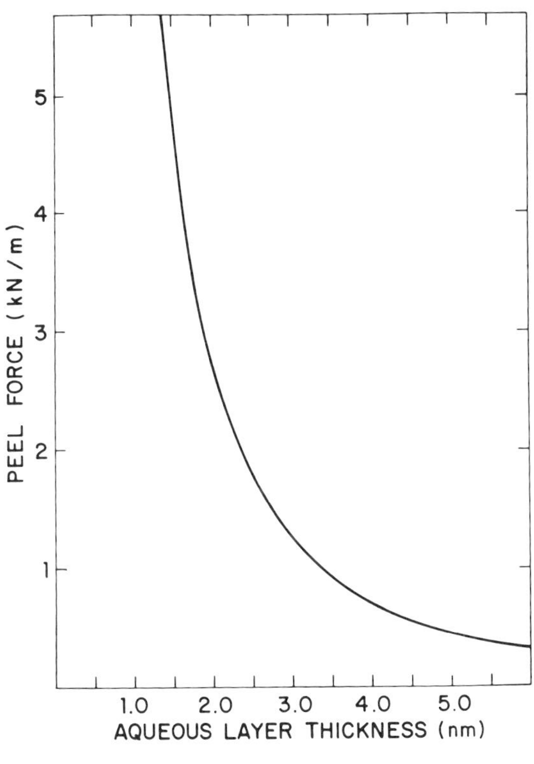


Figure 1.7 Peel force versus aqueous layer thickness.

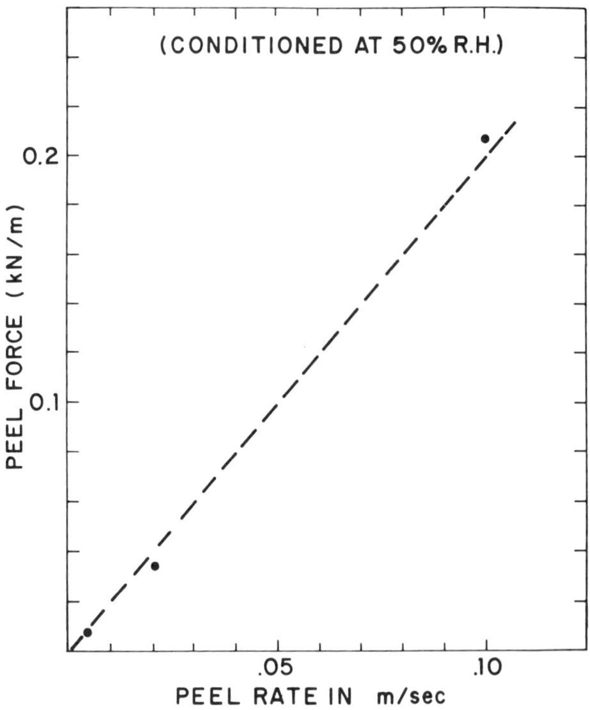


Figure 1.8 Peel force versus peel rate at low rates for plasticized PVB

Estimating d = 1.5 nm for the thickness of the interfacial water in additive-free sheeting equilibrated at 50% relative humidity allowed evaluation of K, giving for a fixed nominal angle

$$F = 27.6Ud^{-2} \tag{40}$$

where F is in kN/m, U is in m/sec, and d is in nm.

A plot of peel force versus aqueous layer thickness is given in Figure 1.7. Equations (39) and (40) also indicate that peel force should be linear with rate and approach zero at low rates. Data at low rates are given in Figure 1.8. Experimental data over a broad range of angles and rates were consistent with Equation (39). This type of response should be expected whenever a liquid interlayer of any kind is present and Newtonian viscosity is displayed at the test rates used.

## V. SUMMARY

Reliable evaluation of adhesives and fruitful approaches to solution of problems involving poor performance require knowledgeable assessment of the character of the bonded structures involved and interpretation of test results. Peel tests are especially useful because they can be interpreted well and data required for comparison of calculated maximum values with test results can be measured independently.

Experimental peel forces significantly less than values calculated using the most appropriate equations indicate unsatisfactory bonding conditions, presence of an interfacial phase, or heterogeneity at the interface, or suggest that the adhesive and adherend are mismatched. It is possible to "zero in" on the most likely cause by measuring melt viscosities and establishing the influence of varying bonding times and temperatures and/or by measuring surface energies and values for $\phi$. If an interfacial phase seems to be a likely cause of poor performance, inventive ways of identifying the source and nature of the material are in order.

## REFERENCES

1. N. A. DeBruyne, in "Adhesion and Adhesives," (N. A. DeBruyne and R. Houwink, eds.), Elsevier, Amsterdam, 1951.
2. J. J. Bikerman, *J. Appl. Phys.*, *28*, 1484 (1957).
3. D. H. Kaelble, *Trans. Soc. Rheol.*, *3*, 161-180 (1959).
4. D. H. Kaelble, *Trans. Soc. Rheol.*, *4*, 45-73 (1960),
5. F. S. Chang, *Trans. Soc. Rheol.*, *4*, 75-89 (1960).
6. C. Jouwersma, *J. Polym. Sci.*, *45*(145), 253-255 (1960).
7. D. H. Kaelble, in "Adhesion and Cohesion" (P. Weiss, ed.), Elsevier, Amsterdam, 1962.
8. G. E. R. Lamb, *J. Polym. Sci.*, *51*(156), S73-76 (1961).
9. N. A. Brunt, *J. Appl. Polym. Sci.* VI *23*, 548-557 (1962).
10. J. L. Gardon, *J. Appl. Polym. Sci.*, *7*, 625-641 (1963).
11. J. L. Gardon, *J. Appl. Polym. Sci.*, *7*, 643-665 (1963).
12. J. L. Gardon, "Treatise on Adhesion and Adhesives" (R. L. Patrick, ed.), Marcel Dekker, New York, 1966.
13. P. B. Lindley, J. Inst. Rubber Ind. 5, 243 (1971).
14. G. P. Anderson, K. L. DeVries, and M. L. Williams, Int. J. Fract. *10*(4), 565-583 (1974).
15. G. P. Anderson, K. L. DeVries, and M. L. Williams, Expmtl. Mechanics *16*(1), 1-15 (1976).

16. A. N. Gent and G. R. Hamed, *Polym. Eng. Sci. 17*(7), 462-466 (1977).

17. A. N. Gent and G. R. Hamed, *J. Appl. Polym. Sci., 21*, 2817-2831 (1977).

18. T. Alfrey, "Mechanical Behavior of High Polymers," Wiley-Interscience, New York, 1948.

19. J. D. Ferry, "Viscoelastic Properties of Polymers," Wiley, New York, 1960.

20. A. V. Tobolsky, "Properties and Structure of Polymers," Wiley, New York, 1960.

21. B. Rosen, ed., "Fracture Processes in Polymeric Solids, Wiley-Interscience, New York, 1964.

22. R. A. Dickie and T. L. Smith, *J. Polym. Sci., A-2*, 687 (1969).

23. W. M. Bright, in "Adhesion and Adhesives" (J. Clark, R. L. Savaage, and J. E. Rutzler, Sr., eds.), Wiley, New York, 1954.

24. J. R. Huntsberger, *J. Polym. Sci., A-1*, 2241 (1963).

25. J. R. Huntsberger, in "Treatise on Adhesion and Adhesives," (R. L. Patrick, ed.), Marcel Dekker, New York, 1966.

26. D. H. Kaelble, in "Treatise on Adhesion and Adhesives" (R. L. Patrick, ed.), Marcel Dekker, New York, 1966.

27. J. R. Huntsberger, *J. Polym. Sci., A-1*, 1339 (1963).

28. G. R. Hamed, *J. Adhes., 13* (2), 101 (1981).

29. J. R. Huntsberger, *Adv. Chem., 43*, 180 (1964).

30. J. R. Huntsberger, Adhesive Restorative Dental Materials II. U.S. DHEW (1965)

31. J. R. Huntsberger, *J. Adhes., 12*, 3-12 (1981).

32. J. R. Huntsberger, *J. Polym. Sci., 43*(142), 581 (1960).

33. J. R. Huntsberger, *J. Paint Technol., 39*(507), 199-211 (1967).

34. D. E. Packham, in "Aspects of Adhesion," Vol. 6 (D. S. Alner, ed.), London Univ. Press, London, 1970.

35. D. K. Owens and R. C. Wendt, *J. Appl. Polym. Sci., 13*, 1241 (1969).

36. J. R. Huntsberger, *J. Adhes., 13*, 107-129 (1981).

37. A. Healey (in R. W. Griffiths), *Trans. Inst. Rubber Ind., 1*, 334 (1926).

# Two

# *Energy Conservation During Peel Testing*

GARY R. HAMED *University of Akron, Akron, Ohio*

## I. INTRODUCTION

When at least one of the bonded members of an adhesive joint is flexible, peel testing is commonly employed to determine joint strength. Three common test geometries, 90°-, 180°-, and T-peeling, are shown in Figure 2.1. In each case, stresses are concentrated at the peel front. Calculations of these stresses were first carried out by Spies [1] and then others [2-8] who assumed linearly elastic behavior while applying small bending theory to a beam (stripping member) on an elastic foundation (adhesive). Later stress analyses have considered plastic yielding in the stripping member [9-13]. However, at the point of failure initiation, the stresses cannot be defined when the stress analysis is completed using a linear elastic continuum theory.

An alternative method of analyzing peel tests is to employ the principle of conservation of energy during peeling [14-16]. This approach is the focus of the present chapter, which summarizes primarily the work of the author and his colleagues over the past 10 years. This is *not* meant to be an exhaustive review of all the literature on this subject.

## II. ENERGY CONSERVATION DURING PEELING

Consider a flexible adherend of width w bonded with an adhesive interlayer to a rigid substrate (Figure 2.2), being pulled at a constant rate R and peel angle $\theta$. During steady peeling, the peel force P and the moment arm between the line of action of P and the line of separation will be constant, i.e., independent of the distance already peeled. Furthermore, the stored strain energy in the adhesive and stripping member is constant. Conservation

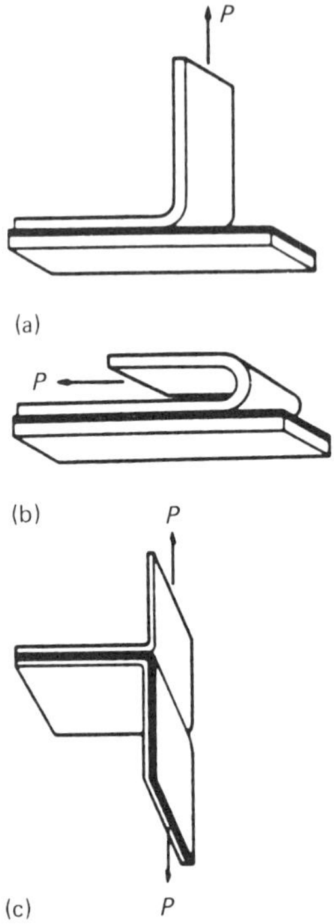


Figure 2.1 Peel test geometries: (a) 90°, (b) 180°, (c) T-peel.

of energy requires that the energy *supplied* to cause fracture (peeling) equal that *expended* in the fracture processes [17]:

Work supplied by peel force P
= surface energy expended to create free fracture surfaces
+ bulk energy dissipated within the adhesive
+ bulk energy dissipated within the stripping member
+ strain energy in the newly detached strip (1)

The last term in this expression is usually small and may be neglected [18], since in most instances the stripping layer is flexible but rather inextensible. The first term on the right-hand side of Equation (1) is the intrinsic work of adhesion or cohesion, $W_i^\circ$ or $W_c^\circ$, depending on whether the failure locus is interfacial or cohesive tearing, respectively. Their values are equal to the energy per unit area required *only* to disrupt the interactions that existed across the free surfaces before fracture. In principle, $W^\circ$ is independent of test rate, peeling angle, and the thickness of the adhesive and stripping member. Values of $W^\circ$ range from about 0.07

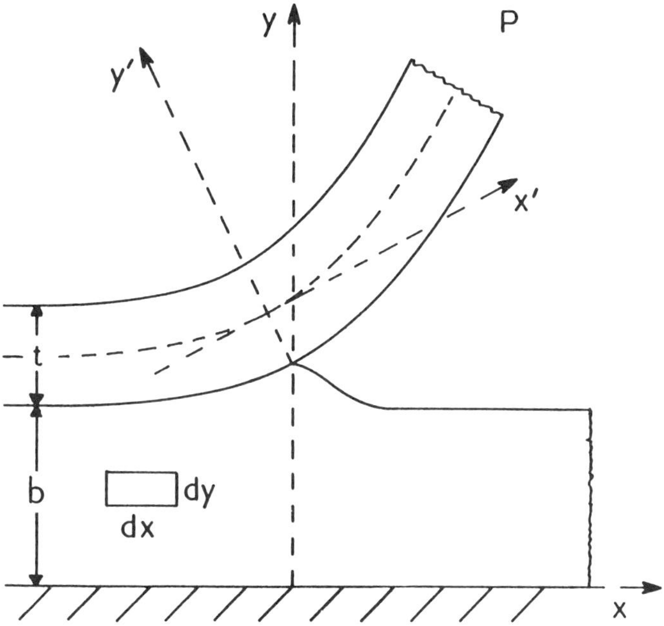


Figure 2.2 Magnified view of peel test specimen near the peel front.

$J/m^2$ for hydrocarbon van der Waals interactions to about 2 $J/m^2$ for carbon-carbon covalent bond strengths in an elastomer network. If an adhesive joint were fractured *reversibly*, without energy being expended away from the fracture plane, then the peel force would reflect only $W^\circ$. However, for normal peel testing of adhesive joints joints, this condition does not hold, and the energy expended in fracture can be several orders of magnitude larger than $W^\circ$ because of energy dissipation within the adhesive and stripping member [second and third terms on the right-hand side of Equation (1)].

Consider an element of the adhesive layer w dx dy as depicted in Figure 2.2. At a sufficient distance ($\pm$ x) from the fracture front, this element is stress-free. As peeling proceeds "through" the element, it will experience a cycle (extension) of stress from zero to some maximum (where either cohesive or interfacial failure commences) and back to zero (retraction) again. This is shown schematically in Figure 2.3, where, in general, the retraction curve lies below the extension curve because of bulk mechanical hysteresis. The energy lost per unit volume is $h_a$. The total energy dissipated, $H_a$, within the adhesive per unit area fractured is obtained by summing the contributions from each element through the adhesive thickness b:

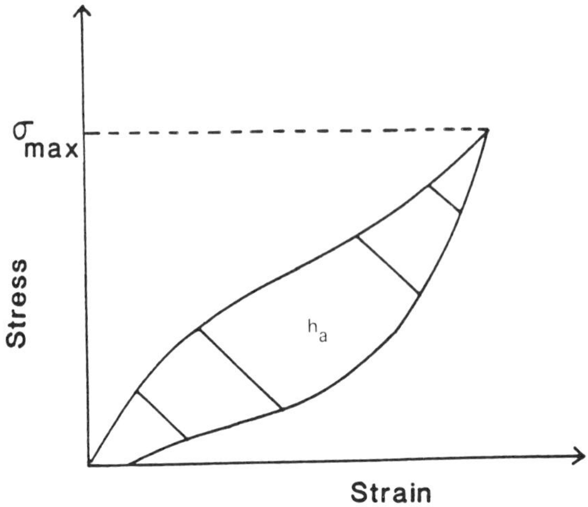


Figure 2.3 Schematic diagram indicating the hysteretic loss within an element of the adhesive as the peel front advances.

$$H_a = \int_0^b h_a(y)\, dy$$

Similarly, the total energy dissipated, $H_s$, in the stripping member during peeling can be approximated by

$$H_s = \int_{-t/2}^{t/2} h_s(y')\, dy'$$

where $h_s$ = energy dissipated per unit volume in the stripping member as it traverses the highly bent region
$y'$ = locus of points at which the radius of curvature is a minimum (stress in strip maximized)
$t$ = thickness of stripping member

Equation (1) may now be expressed as [17]

$$\frac{P}{w}(1 - \cos\theta) = W^o + \int_0^b h_a(y)\, dy + \int_{-t/2}^{t/2} h_s(y')\, dy' \qquad (2)$$

Usually, the three terms on the right-hand side of Equation (2) are lumped together into the fracture energy G:

$$G = \frac{P(1 - \cos\theta)}{w} \qquad (3)$$

Note that the fracture energy can be determined *without* having a constitutive equation for the adhesive or stripping member. This is a direct consequence of the driving force for fracture coming *directly* and *solely* from the test machine without any change in the bonded member's stored strain energy as peeling proceeds. In most fracture geometries (e.g., a blister or butt tensile joint), the energy supplied for fracture is the stored strain energy lost as the crack advances; to calculate this, a stress-strain constitutive law is required. This factor provides an advantage of a peel test for determining G, especially for materials in which constitutive laws are uncertain. Another advantage of peel testing can be gleaned from Equation (3); P does not depend on the size of any precrack length, so bond strengths can be determined at each segment over the entire bonded area; if bonding is uniform, the average P should be constant along the length.

Returning to Equation (2), it is instructive to consider each of the three components of G in more detail. Several investigators [19-27] have shown that, in peel testing, P is generally dependent on test speed, temperature, peel angle, and adhesive and stripping member thickness. The reason for this does *not* lie in the dependence of $W^\circ$ on these parameters but rather in the dependence of the two hysteretic terms on these variables. Bulk energy losses per unit volume in polymeric materials are well known to be rate and temperature sensitive. In addition, the total extent of energy dissipation, since it is *volume* dependent, may change with adhesive and stripping member thicknesses.

Under *normal* test conditions, $H_a + H_s >> W^\circ$, so that $G = H_a + H_s$. [Typically, values of G for rubbery adhesives range from about 0.5 $J/m^2$ (weak physical bonds) to 50 $kJ/m^2$ (chemical linkages)—covering *five* orders of magnitude.] This does *not* imply that $W^\circ$ is unimportant in determining the peel force. Although the terms in Equation (2) are additive, $H_a$ and $H_s$ must depend on $W^\circ$. If $W^\circ$ is zero, then necessarily so are $H_a$ and $H_s$. In some instances [28-29], hysteretic losses have been found to be directly proportional to $W^\circ$, such that

$$G = W^o(1 + k) \tag{4}$$

where k is the proportionality constant indicative of hysteretic losses. However, the very simple Equation (4) has been shown and argued *not* to have general validity in peel testing [22,30,31]. Nonetheless, it is apparent that the peel strength is better viewed as resulting from an intrinsic interaction ($W^\circ$) *multiplied* by hysteretic contributions rather than simply the sum of the two. Again, it is worth emphasizing that, although $W^\circ$ may be negligible in magnitude compared to $H_a + H_s$, this does not diminish the importance of $W^\circ$.

Moreover, without substantial hysteretic losses during peel fracture, it is not possible to have high peel strengths, even if $W^\circ$ is relatively large (say, due to chemical linking). On the other hand, even with relatively low values of $W^\circ$, peel forces can be substantial if there are sufficient hysteretic losses at the peel front. The *capacity* for high energy dissipation is a necessary though insufficient condition for high peel strengths. Adhesives are often modified with additives (e.g., fillers) that result in increased peel strengths, not because of increases in W° but rather because of enhanced dissipation.

Although, with the energy balance approach, the direct contribution of hysteresis to the peel force is established, the importance of energy dissipation can be considered from another viewpoint. When mechanical energy is put into a peel joint, it may be either stored (elastically) or dissipated. The stored energy results in stresses tending to pull bonds apart and thereby cause fracture, whereas energy dissipation is a mechanism for converting the mechanical energy into enhanced molecular motions (thereby resulting in heat)—and making it unavailable to promote fracture. This is particularly important right at the advancing front, where hysteresis is manifested by crack tip blunting and the reduction of locally high stresses. This is not possible with purely elastic materials, which are necessarily weak and brittle.

## III. DETERMINATION OF W°

Often it is desirable to determine W° for an adhesive or cohesive bond. Can this be accomplished with a peel test? Equation (2) indicates that P will be a direct measure of W° only if $H_a + H_s \to 0$. This condition has been approached for a lightly cross-linked polybutadiene (BR) layer bonded directly to a rigid glass substrate [30]. At elevated temperatures and low rates of peeling (denoted threshold conditions), the cross-linked BR deforms nearly elastically (without dissipation) and P is quite small (and independent of reduced test rate) as expected (Figure 2.4). However, the inferred value of W° was about 25 times greater than expected based on simple surface energies.

This suggests that for mechanical measures of joint strength, it may not be possible to *only* disrupt the bonding across the fracture plane, without some (although in some circumstances small) energy dissipation away from this plane. Indeed, Lake and Thomas [32], in their theory of the threshold strength of elastomers, assume that energy dissipation is involved to a dimension of the chain length between cross-link points, even under equilibrium tearing.

## IV. BOND THICKNESS EFFECTS

In principle, another way to eliminate hysteretic contributions to the peel force is to determine joint strength as bond thickness $b \to 0$. (There can be no bulk energy dissipation if there is no volume in which it can occur.) An attempt to do this has been carried out [17]. Figure 2.5 shows the fracture energy as a function of bond thickness for an SBS elastomer (Kraton 3202) bonded

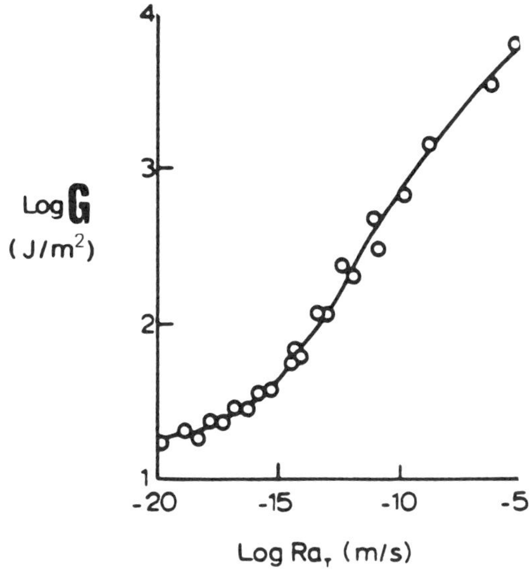


Figure 2.4 Fracture energy versus reduced test rate for a cross-linked polybutadiene layer peeled from glass [30].

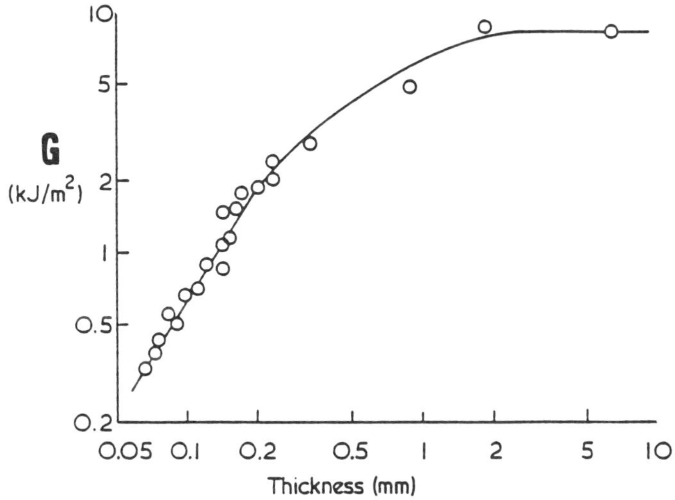


Figure 2.5 Dependence of fracture energy on bond thickness for Mylar stripped from a layer of Kraton 3202 [17].

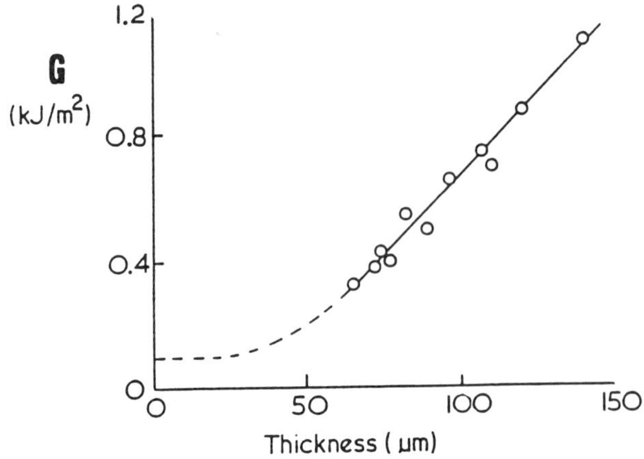


Figure 2.6 As in Figure 2.5, but for low bond thickness only [17].

to a polyester substrate. The fracture energy increases with bond thickness (because of increased hysteretic contributions) before reaching a plateau at higher thicknesses. The limitation occurs when the region of high stresses and hence high dissipative losses does not extend throughout the thickness of the adhesive. The critical thickness will be comparable to the size of the highly deformed zone, which will be greater for stronger joints and softer adhesives. The volume-dependent dissipative losses decrease as the thickness is decreased, while $W^\circ$ remains constant. Hence, at sufficiently low thicknesses, it is expected that $G \rightarrow W^\circ$. Extrapolation is attempted in Figure 2.6, which shows the low-thickness region of Figure 2.5. The experimental data are linear, but, if extrapolately linearly, the plot would intersect the thickness axis. This is clearly not possible. Rather, in the low-thickness region the behavior must curve in the manner indicated (although the specific extrapolation drawn in Figure 2.6 is arbitrary). In principle, G will become independent of bond thickness when bulk energy losses become small relative to the intrinsic strength.

The thickness dependence of peel strength depends on the stress-strain response of the adhesive [33]. This is demonstrated in Figure 2.7, where the results of Figure 2.5 are replotted along with data for another, much stiffer, SBS adhesive (Kraton 1101). Over the region tested, the peel joint with the Kraton 1101 exhibits a fracture energy which is independent of thickness. It has been found, although it is not shown in Figure 2.7, that G does not start to diminish in the case of Kraton 1101 until the bond thickness is below approximately 0.05 mm. The stress-strain curves of the two SBS elastomers are given in Figure 2.8. The Kraton 1101 is initially much stiffer than the Kraton 3202 and has an "elastic-plastic" character at low strain. When the adhesive layer is

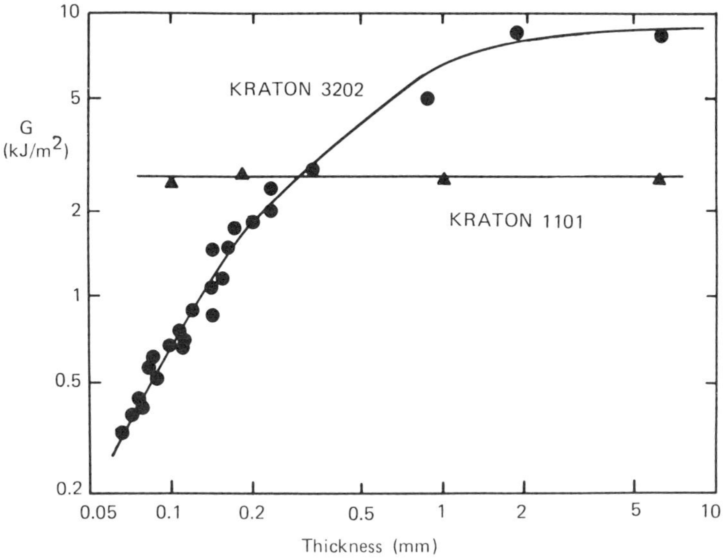


Figure 2.7 Comparison of the thickness dependence of the peel fracture energy for Mylar stripped from Kraton 1101 and Kraton 3202 [33].

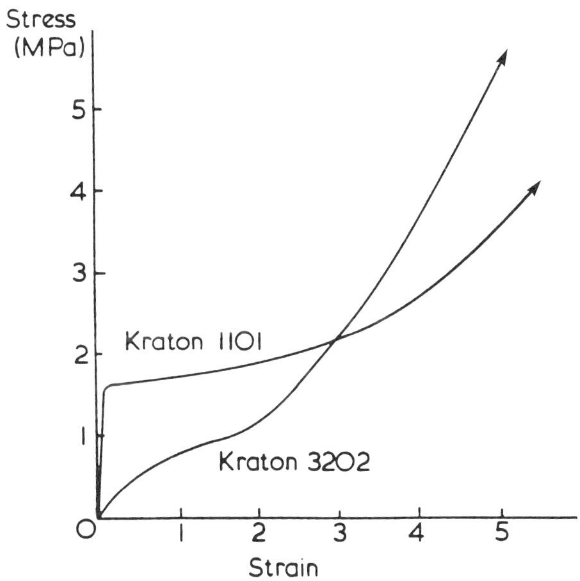


Figure 2.8 Stress-strain curves of Kraton 1101 and Kraton 3202 [33].

relatively soft (Kraton 3202), the peel strength increases with thickness up to high thicknesses. However, for an elastic-plastic material (Kraton 1101), the stresses at some distance below the line of detachment may never rise to the yield stress, so the bulk of the adhesive will not deform plastically on peeling. In this case the bulk of the material will be relatively nondissipative; only a region near the surface (~0.05 mm thick) will deform plastically on peeling and contribute the major part of the dissipative term. These results have a practical implication. It is clearly important to specify and control the thickness of the various adhesive layers employed in peel tests, because the order of merit can be reversed when different thicknesses are used. Thus, it is advisable to examine potential adhesives over a wide range of thicknesses.

To this point, discussion of Equation (2) has been qualitative. Let's now consider its quantification, starting with $H_a$. By appropriate selection of test system and conditions, it is possible to cause $H_a$ to greatly exceed both $H_s$ and $W°$. Equation (2) then becomes

$$\frac{P(1 - \cos\theta)}{w} = \int_0^b h_a \, dy \tag{5}$$

If the interlayer is thin, so that the stress through the thickness (and hence $h_a$) can be considered approximately constant, Equation (5) simplifies to

$$\frac{P(1 - \cos\theta)}{w} = h_a b \tag{6}$$

The peel force can be predicted if the amount of bulk energy expended per unit volume ($h_a$) in the fracture process can be determined. The validity of Equation (6) has been studied [34,35]. To make $H_s$ negligible, a high-modulus, but quite thin and flexible, nonyielding cloth was used as a backing layer for T-peel specimens. The adhesive interlayers were thin and consisted of various uncross-linked elastomeric compositions. During the constant-rate peeling of these joints, the locus of fracture was tearing within the elastomer. Experimentally determined peel tear forces per unit width are given in Table 2.1 for natural rubber samples masticated for various times [35]. To assess $h_a$ to be used in Equation (6) to predict P, it is necessary to determine the *effective* strain rate $\dot{\varepsilon}$ experienced by the adhesive layer during peeling. This is given approximately by [36]

$$\dot{\varepsilon} = \frac{R}{b} \tag{7}$$

At the $\dot{\varepsilon}$ calculated from Equation (7), a strip of elastomer was elongated in tension; just prior to breaking (predetermined from another test sample), the sample was retracted at the same rate. The area enclosed by the stress-strain loop was taken as $h_a$, and

Table 2.1 T-Peeling Tear Strength of Green Natural Rubber [35]

| Mastication time (min) | Experimental P (kN/m) | Calculated P (kN/m) (Equation 6) |
|---|---|---|
| 1 | 2.62 ± 0.05 | 3.0 |
| 5 | 2.00 ± 0.06 | 2.0 |
| 10 | 1.50 ± 0.04 | 1.3 |
| 30 | 0.88 ± 0.02 | 0.8 |

Equation (6) was used to calculate values of P. These are included along with directly measured results in Table 2.1, where agreement is reasonable. Data from a similar study [34] are given in Table 2.2, where the agreement between experimental values of P and those calculated from Equation (6) is somewhat better. It is concluded that the energy expended per unit volume during peel tearing of a thin uncross-linked rubbery layer is similar to that experienced when a tensile strip is cycled up to near fracture. This is somewhat surprising, since the detailed stresses in the material for peeling and cyclic tensile testing are not expected to be the same. Nonetheless, the bulk energy expended for both cases must be similar.

Table 2.2 T-Peeling Green Strengths for SBR, NR, and Their Blends [34]

| Ratio NR/SBR | Experimental P (kN/m) | Calculated P (kN/m) (Equation 6) |
|---|---|---|
| 0/100 | 1.70 ± 0.25 | 1.6 |
| 20/80 | 1.77 ± 0.18 | 1.7 |
| 40/60 | 2.60 ± 0.23 | 2.7 |
| 60/40 | 2.21 ± 0.25 | 1.9 |
| 80/20 | 3.15 ± 0.30 | 3.2 |
| 100/0 | 2.24 ± 0.16 | 2.1 |

## V. ADHEREND THICKNESS EFFECTS

An experimental arrangement for determining $H_S$ is shown in Figure 2.9. Two Mylar strips were detached simultaneously by peeling them away from two thinly rubber-coated metal plates [12]. Adhesion

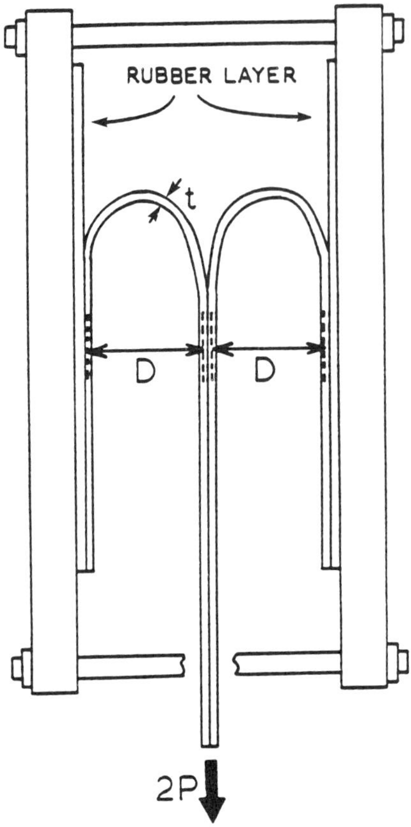


Figure 2.9 Experimental arrangement for determining the contribution of energy dissipation due to plastic yielding of the adherend during peeling [12].

to the rubber was very low, virtually eliminating $H_a$, so that Equation (6) becomes

$$\frac{P}{w}(1 - \cos\theta) = \int_{-t/2}^{t/2} h_s \, dy' \qquad (8)$$

The rubber layer served merely to prevent slipping and buckling of the Mylar strips under the force P. The degree of bending of the Mylar strips was controlled by the distance D between the central planes of the adhering and fully removed portions of the film. Various values of D were obtained by altering the separation of the

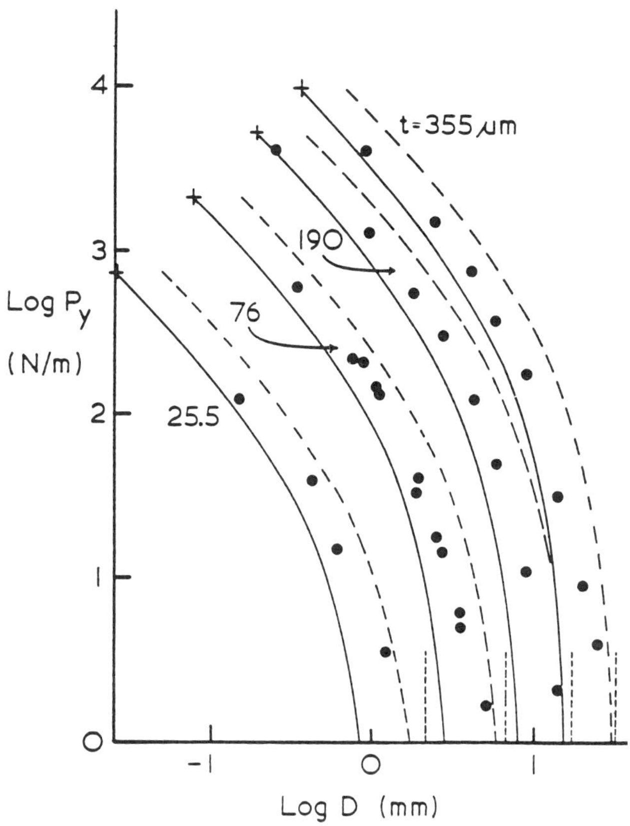


Figure 2.10 Experimental and theoretical values of the peel force caused by plastic yielding of Mylar of various thicknesses [12].

two steel backing plates. When the spacing D was reduced below a critical value for plastic deformation, the force P was found to increase greatly due to energy dissipated in plastic yielding, $H_S$. Equation (8) has been solved to allow prediction of P from the stress-strain and hysteretic properties of an elastic-plastic material such as Mylar; the results are [12]

$$\frac{P(1 - \cos \theta)}{w} = \frac{\sigma_y \varepsilon_y t}{2} \left( \frac{t}{2R\varepsilon_y} + \frac{2R\varepsilon_y}{t} - 2 \right) \tag{9}$$

where $\sigma_y$ = yield stress
$\varepsilon_y$ = yield strain
R = minimum radius of curvature of neutral axis of strip during peeling

There are two limiting conditions relating R to the experimentally controlled variable D. When the strip deforms elastically or is in the earlier stages of plastic yielding [37], $R \approx D/4$. For more severe bending deformations approaching a plastic hinge, $R \approx D/2$. In Figure 2.10, theoretical results calculated from Equation (9) are shown for the two extremes $D = 4R$ (dashes) and $D = 2R$ (solid) and four different film thicknesses [12]. Also plotted are data directly determined with the setup shown in Figure 2.9. The experimental data fall well within the expected theoretical range and behave as expected (approach dotted line at high D and solid line at low D) in the limiting cases. The crosses represent the theoretical upper limit of the peel force due to energy-dissipative yielding when D attains a lower limit value of t. The vertical dotted lines denote critical values of the spacing, $D_0 = 2t/\varepsilon_y$, above which bending deformations are insufficient to cause any yielding and hence P due to this cause must vanish. The predictions are in accord with the experimental findings.

Even for strongly bonded adherends in which $H_a$ is substantial, Equation (9) can be used to determine the fraction of the measured peel force that is due to dissipation within the stripping member.

Considerations of energy dissipation due to plastic yielding can also account for published reports that the 180° peel force passes through a maximum as the stripping member thickness is increased (shown in Figure 2.11) [25,26]. (This is true for situations in which the force that is required to hold the stripping member back before peeling is small compared to the peeling force.) For a given

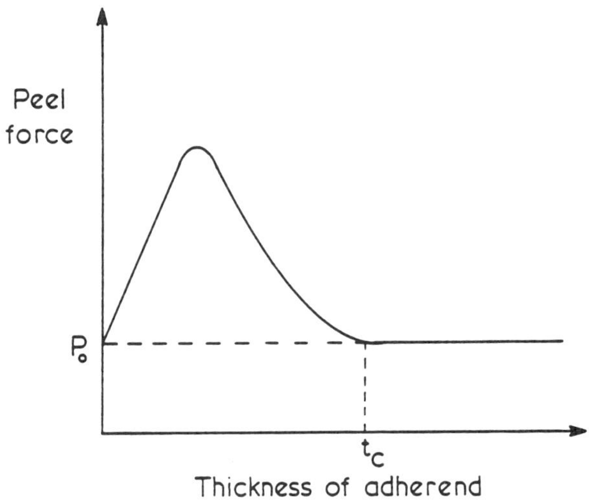


Figure 2.11 Schematic showing the peel force as a function of adherend thickness.

level of adhesion a very thin adherend will yield during peeling, but the total energy dissipated will be small because t is small, and the contribution of plastic yielding to the peel force will be neglibible. As the thickness is increased, more energy will be dissipated and the peel force will increase. However, at sufficiently large thicknesses, the detaching layer will become too stiff to undergo yielding throughout its thickness and the peel force will begin to decrease as the thickness is increased further. Eventually, when the thickness exceeds $t_c$, the detaching layer will not undergo plastic yielding at all and the peel force will return to a value reflecting fracture in the absence of plastic yielding.

## VI. EFFECT OF PEEL ANGLE

If, as the peeling angle $\theta$ is changed, the extent of bulk energy dissipation during peeling is unaltered, the peel force should be inversely proportional to $1 - \cos\theta$ (again see Equation 2). This has been found to be true for rather weak brittle joints, in which dissipative processes are small [38]. However, this is not generally true for strongly bonded joints, for example, a Mylar strip bonded to Kraton 1101. Here, the peel force at $\theta = 180^c$ has been found to be substantially greater than twice that for 90° peeling [12]. Moreover, after detachment at 180°, the Mylar strip exhibited a high degree of permanent curvature, indicating that extensive plastic yielding had occurred during the peeling process, whereas after peeling at 90 the residual curvature was much less pronounced. There has clearly been more plastic work at 180° than at 90°. When plastic yielding was prevented by pulling the detaching strip around a freely rotating roller (to prevent severe bending), then not only did the peel force decrease in both cases, but now $2P_{180°} = P_{90°}$.

## VII. RATE AND TEMPERATURE EFFECTS

The dissipative properties of polymers are well known to be rate and temperature sensitive. Indeed, as pointed out previously, this is thought to be the primary factor in the dependence of peel force on rate and temperature. In this section, this dependence is examined for several peel joints.

Several authors [36,39,40] have found that peel results are often time-temperature superposable for joints containing simple rubbery adhesives. (One example has already been given in Figure 2.4.) WLF-type shift factors for peel data are similar to those applicable for other viscoelastic properties, such as dynamic modulus. Peel master curves have been constructed, as shown in Figure 2.12 [36]. These data were obtained from T-peel specimens in which a soft, uncross-linked SBR layer was bonded to two Mylar adherends. At low reduced rates, the peel force increases with rate and failure takes place entirely within the adhesive layer, which fails by

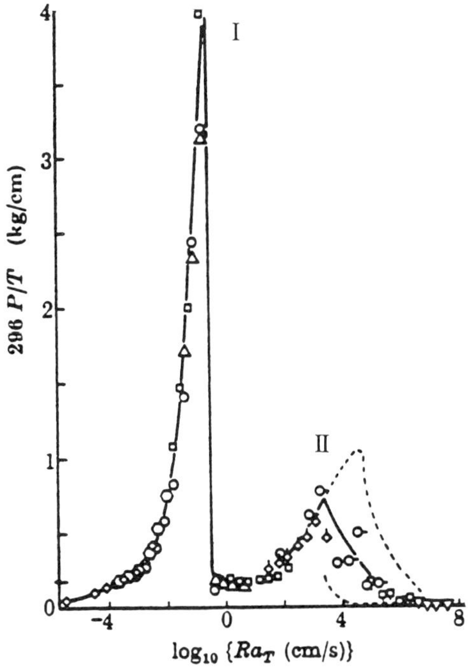


Figure 2.12 Peel force master curve for the T-peeling of an un-cross-linked SBR bonded between Mylar films. For rates to peak I, failure is cohesive; thereafter, failure is interfacial [36].

flowing apart. At a critical rate of peel, an abrupt transition takes place to interfacial fracture, i.e., clean separation of the adhesive from the adherend, at much smaller peel forces. This transition occurs when the rate of deformation of the adhesive layer at the peeling front becomes so high that the adhesive molecules are unable to disentangle and flow apart like a liquid but remain intertwined as a coherent elastic solid. Although the local stress required to disentangle the molecules at low rates is relatively small, the work expended in ductile flow is large and the peel force, which measures the energy expended in separation, is correspondingly high. In the elastic state, the work of separation is expended mainly near the interface and is then relatively small. As rate is further increased (or temperature decreased), the peel force passes through a second maximum as the adhesive becomes glassy. The origin of the two peaks was confirmed in separate dynamic mechanical studies: peak I, flow region → rubbery plateau; peak II, transition zone → glassy. It should be pointed out, however, that the cohesive-to-interfacial transition can exist even

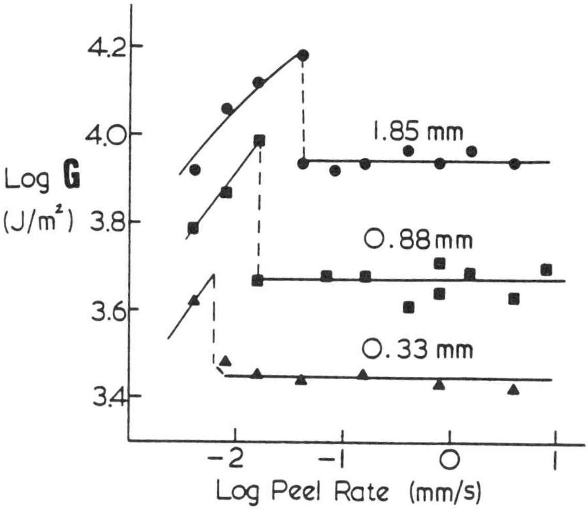


Figure 2.13 Peel fracture energy versus peel rate for three thicknesses of Kraton 3202. Vertical dotted line in each case separates (low-rate) cohesive tearing from interfacial failure [33].

if the adhesive is cross-linked and hence incapable of viscous-type flow. This has been found for a joint in which a cross-linked polychloroprene layer was tenaciously bonded to brass shim [22].

Figure 2.13 [33] shows the effect of rubber thickness and peel rate on the 180°-peel tearing energy for Mylar stripped from a layer of Kraton 3202. As the rate of peel was increased, there was an abrupt decrease in the peel strength and change in the mode of failure from cohesive to interfacial at a critical rate. The peel strength and the critical rate both increased with the thickness of the rubber layer. For a given rate of peel, the rubber in a thinner layer experiences a higher rate of strain than that in a thicker layer (see Equation 7). Hence, if the cohesive-interfacial failure transition is related to a critical rate of straining of the rubber, it should be possible to shift the transition rate for each of the adhesive thicknesses to a single value by dividing the critical rate by the bond thickness. Indeed, when this was done, a single value of about 0.2 $sec^{-1}$ was found. When $\dot{\varepsilon}$ exceeds 0.2 $sec^{-1}$ at room temperature, the rubber is sufficiently strong to resist tearing and failure is interfacial; below this strain rate, the rubber is too weak to resist tearing [33].

A further consequence of the peel force being a direct measure of the energy expended in fracture is given below for a Mylar film adhered (to varying degrees) to a Kraton 1101 layer [33]. Figure 2.14 shows the effect of rate of peeling on G for specimens with different values of W°. By varying the temperature at which the thermoplastic elastomer was pressed into contact with the Mylar

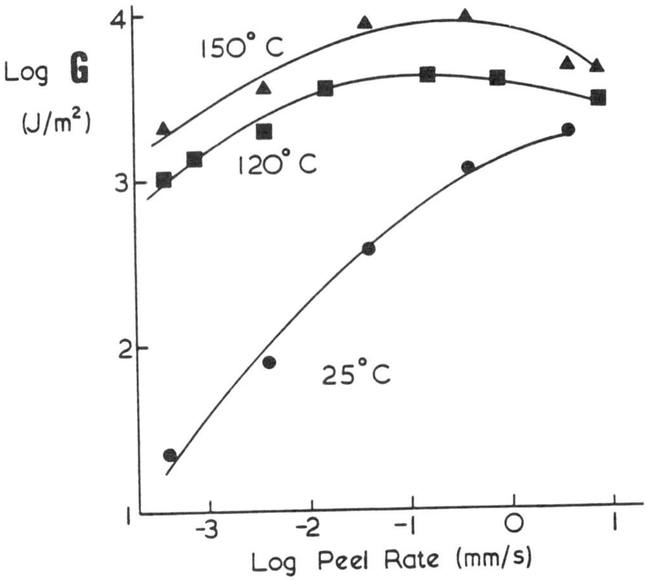


Figure 2.14 Peeling fracture energy as a function of test rate and degree of interfacial wetting for Mylar stripped from Kraton 1101. Molding temperatures are indicated. All specimens were tested at 23°C [33].

film, it was found possible to vary the degree of interfacial interaction without changing the bulk properties of the rubbery layer. The higher the application temperature, the higher W°, probably as a result of better wetting. Failure appeared to be interfacial in all cases. However, there was a marked difference in the dependence of the measured peel strength on the rate of the peeling when different molding conditions were employed. Test pieces prepared at 150 and 120°C showed peel strengths that were only slightly dependent on peel rate, in comparison with bonds formed at room temperature, which showed nearly a 100-fold increase in peel strength as the peel rate was increased from $10^{-4}$ to 10 mm/sec. Indeed, at high rates the bond formed at room temperature was nearly as strong as that formed when the adhesive was applied as a low-viscosity melt.

To understand this difference in behavior of samples bonded at high and low temperatures, it is helpful to examine again the stress-strain curve of Kraton 1101 (Figure 2.8). Suppose that during a peel experiment the maximum stress experienced by the rubber layer during detachment was a relatively low value, below the yield stress. Little energy would then be dissipated as peeling proceeded, and the peel force would be small. Now consider a situation, as a result of more complete wetting or a higher rate of peeling, when the maximum stress is increased to a value above the yield stress. The hysteretic losses per unit volume given by the area under the stress-strain curve will now be many times as large as before. Because of this added sissipation, there will be a manyfold increase in the peel force compared to the previous case, even though the maximum stress before detachment may have increased by only a small amount. Moreover, if the interfacial strength and rate of detachment are increased still further, there will be only a moderate further increase in the energy dissipated and hence in fracture energy compared to the case when the maximum stress first exceeds the yield stress.

The maximum load experienced by the rubber layer during detachment will depend on the degree of interaction between the adhering materials. It will also depend on the rate of detachment, in a way which can be inferred from the observed dependence of peel strength on peel rate. For example, for the sample bonded at room temperature, which shows a large increase in peel force with rate, there *must* be a corresponding increase in energy dissipation, such as would occur in increasing the detachment stress in the neighborhood of the yield stress. Moreover, the peel force would be expected to be quite variable if the maximum stress before detachment was near the plateau of the stress-strain curve, since, in this region, small variations in interfacial interaction would lead to large differences in energy dissipation. In accordance with this, the peel results were found to be highly variable in the intermediate range of peel rates for test pieces bonded at room temperature. At higher rates, when the detachment stress was apparently well above the yield stress, there was much less variability in peel strength.

Peel forces for specimens bonded at high temperatures, 120 and 150°C, did not show much change with rate because the detachment stress probably exceeded the yield stress even at low rates. When the detachment stress lies in the strain-hardening region of the stress-strain curve there will be relatively little change in energy dissipation, and hence peel strength, even for relatively large changes in the magnitude of the maximum stress.

## ACKNOWLEDGMENT

The author is grateful to the National Science Foundation under DMR-83-06533 for partial support of the research program for this review.

## REFERENCES

1. G. J. Spies, *Aircr. Eng.*, *25*, 64 (1953).
2. J. J. Bikerman, *J. Appl. Phys.*, *28*, 1484 (1957).

3. J. J. Bikerman, *J. Appl. Polym. Sci.*, *2*, 216 (1959).
4. D. H. Kaelble, *Trans. Soc. Rheol.*, *3*, 161 (1959).
5. D. H. Kaelble, *Trans. Soc. Rheol.*, *4*, 45 (1960).
6. D. L. Gardon, *J. Appl. Polym. Sci.*, *7*, 643 (1963).
7. S. Yurenka, *J. Appl. Polym. Sci.*, *6*, 136 (1962).
8. A. N. Gent and G. R. Hamed, *J. Adhes.*, *7*, 91 (1975).
9. A. J. Duke and R. P. Stanbridge, *J. Appl. Polym. Sci.*, *12*, 1487 (1968).
10. W. T. Chen and T. F. Flavin, *IBM J. Res. Dev.*, *16*, 203 (1972).
11. A. J. Duke, *J. Appl. Polym. Sci.*, *18*, 3019 (1974).
12. A. N. Gent and G. R. Hamed, *J. Appl. Polym. Sci.*, *21*, 203 (1977).
13. A. D. Crocombe and R. D. Adams, *J. Adhes.*, *13*, 241 (1982).
14. P. B. Lindley, *J. Inst. Rubber Ind.*, *5*, 243 (1971).
15. K. Kendall, *J. Phys. D Appl. Phys.*, *6*, 1782 (1973).
16. R. S. Rivlin, *Paint Technol.*, *9*, 215 (1944).
17. A. N. Gent and G. R. Hamed, *Plast. Rubber Mater. Appl.*, *3*(1), 17 (1978).
18. K. Kendall, *J. Adhes.*, *5*, 179 (1973).
19. A. N. Gent and G. R. Hamed, *Polym. Eng. Sci.*, *17*, 462 (1977).
20. J. C. Chen and G. R. Hamed, *Rubber Chem. Technol.*, *60*, 319 (1987).
21. G. R. Hamed and C.-H. Shieh, *J. Polym. Sci. Polym. Phys. Ed.*, *21*, 1415 (1983).
22. G. R. Hamed and F.-C. Liu, *Rubber Chem. Technol.*, *57*, 1036 (1984).
23. D. W. Aubrey, G. N. Welding, and T. Wong, *J. Appl. Polym. Sci.*, *13*, 2193 (1969).
24. T. Hata, M. Gamo, and Y. Doi, *Kobunshi Kagaku*, *22*, 152 (1965).
25. D. Satas and F. Egan, *Adhes. Age*, *9*, 22 (1966).
26. J. Johnston, *Adhes. Age*, *11*, 20 (1968).
27. G. R. Hamed and C.-H. Shieh, *Rubber Chem. Technol.*, *57*, 227 (1984).
28. E. H. Andrews and A. J. Kinloch, *Proc. R. Soc. London Ser. A*, *332*, 385 (1973).

29. E. H. Andrews and A. J. Kinloch, *Proc. R. Soc. London Ser. A*, *332*, 401 (1973).

30. A. Ahagon and A. N. Gent, *J. Polym. Sci. Polym. Phys. Ed.*, *13*, 1903 (1975).

31. G. R. Hamed and C.-H. Shieh, *Rubber Chem. Technol.*, *58*, 1038 (1985).

32. G. J. Lake and A. G. Thomas, *Proc. R. Soc. London Ser. A*, *300*, 108 (1967).

33. A. N. Gent and G. R. Hamed, *Rubber Chem. Technol.*, *55*, 483 (1982).

34. G. R. Hamed, *Rubber Chem. Technol.*, *54*, 403 (1981).

35. G. R. Hamed and C.-H. Shieh, *Rubber Chem. Technol.*, *55*, 1469 (1982).

36. A. N. Gent and R. P. Petrich, *Proc. R. Soc. London Ser. A*, *310*, 433 (1969).

37. R. V. Southwell, "An Introduction to the Theory of Elasticity," 2nd ed., Oxford Univ. Press, London, 1941, reprinted by Dover, New York, 1969, Chap. 13.

38. K. Kendall, *J. Mater. Sci.*, *11*, 638 (1976).

39. D. H. Kaelble, *J. Colloid Sci.*, *19*, 413 (1964).

40. D. W. Aubrey and M. Sherriff, *J. Polym. Sci. Chem. Ed.*, *16*, 2361 (1978).

# Three

# Analysis of Standard Bond-Strength Tests

GARRON P. ANDERSON
*Morton Thiokol, Inc., Brigham City, Utah*
K. L. DEVRIES
*University of Utah, Salt Lake City, Utah*

## I. INTRODUCTION

Testing of adhesive joints is required for a number of reasons, including

1. Quality control—to ensure that changes such as adhesive age, mix method, and surface preparation have not been introduced that will affect bond quality
2. Comparative analyses—to determine which of a series of adhesives, primers, surface preparations, cure methods, etc. is best suited for a given bond application
3. Engineering data—to predict the load-carrying capability of a given bonded joint

The geometry of the test specimen selected is very important even for quality control tests. The data in Figure 3.1 provide a very good illustration of this point. These data were obtained for a bond between a solid propellant and a rubber insulator. The tensile adhesion test specimen consisted of a 2.5-cm (1-in.) cube of solid propellant bonded to a thin sheet of rubber insulation. The insulation and the propellant side opposite the insulation were then bonded to steel end tabs. The peel test data were obtained by peeling a strip of insulation 2.5 cm (1 in.) wide at an angle of 90° with respect to a block of solid propellant.

The bond strengths of the two specimen types are plotted as a function of a manufacturing variable, in this case the time delay between applying the adhesive (liner) to the insulation and casting the solid propellant against the adhesive. The important point of this example is that if the tensile adhesion test was the only source

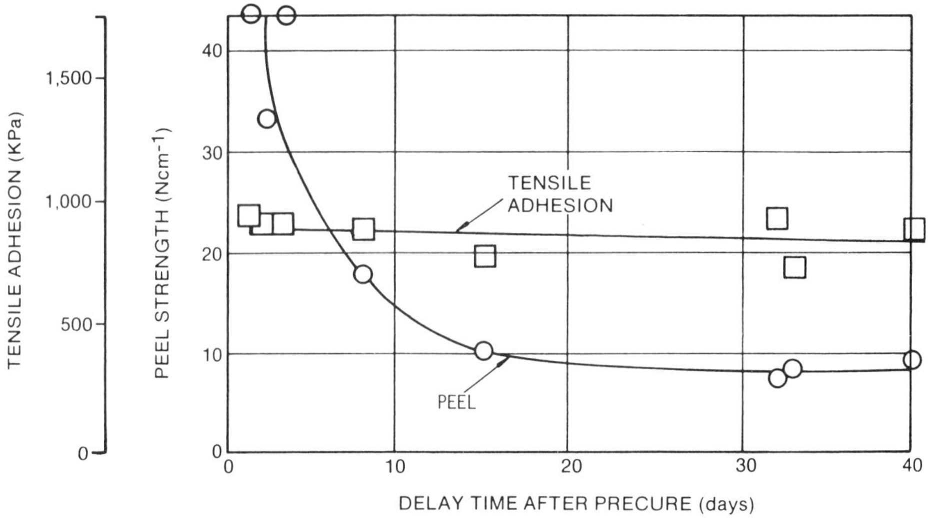


Figure 3.1 Solid propellant bond strength versus state of precure.

of data for monitoring the consistency of this solid propellant bond strength, changes in the manufacturing procedure (casting delay time) could not be detected. If peel strength is important for the bonded joint, poor bonds could be made and not detected by the tensile adhesion tests.

In fact, many variables may affect bond strength. Each variable may affect the bond strength in a different manner, and the manner in which the bond strength is affected may be different for each test geometry. This was clearly demonstrated in the example just cited; casting delay time may affect peel strength in a different manner than it affects tensile strength. Another example is that bond thickness often affects lap shear strength in a different way than it affects tensile button strength. If a series of lap shear joint tests are completed for various adhesive thicknesses and an optimum adhesive thickness is determined, it is not safe to assume that this thickness is optimum for a given joint unless the geometry, adhesive, and loading are very nearly the same as for the lap shear joints tested.

Thus, even for comparative analyses, a great deal of care must be used in applying laboratory test data to joints whose geometry or load is different from that of the laboratory test specimen.

One of the purposes of obtaining engineering data is to ensure that a bonded joint can withstand the loads for which it was designed. One approach is to apply a load to each joint that is bonded, i.e., proof-test each joint, as is presently the procedure for critical bonds in the Space Shuttle system. The load applied is at least as large as the maximum expected during the service life of the joint. Straightforward application of this approach is not 100% safe. The proof load is not expected to fail the joint, but it could damage the joint and initiate flaws that subsequently grow as the joint is subjected to fatigue loading, aging, and environmental factors. The proof load should be larger than the service loads to allow a decrease in joint strength with service life without precipitating a joint failure. In addition to a proof test, it is normally desirable to test a statistically significant number of the joints to failure to determine the margin of safety of the bond system.

The above two approaches are not practical for all bonded joints. An engineer is often required to obtain bond strength data from laboratory-size samples and infer the strength of a given bonded joint from these data. A straightforward approach would be to prepare and test several "standard" laboratory specimens using the adhesive type and surface preparation technique to be evaluated. The strength of these standard bond tests is usually reported as an average failure stress, which is defined as the breaking load per unit of bond area. This stress is then compared with the average stress that exists in the joint being evaluated when its maximum load is applied. However, if the joint geometry, loading time, and other conditions are not identical to the laboratory test conditions, the direct comparison can lead to unsafe joint designs.

Direct comparison of average stress values is not valid because there are many factors that affect the failure load (averaged bond stress) in bonded joints. These factors must be accounted for by completing additional laboratory tests and/or by analysis of the test data and bonded joint. These factors include loading mode (direction of load application with respect to a crack or debond termination line; see Figure 3.2), loading rate, joint temperature, adhesive thickness, residual stresses, moisture content and moisture

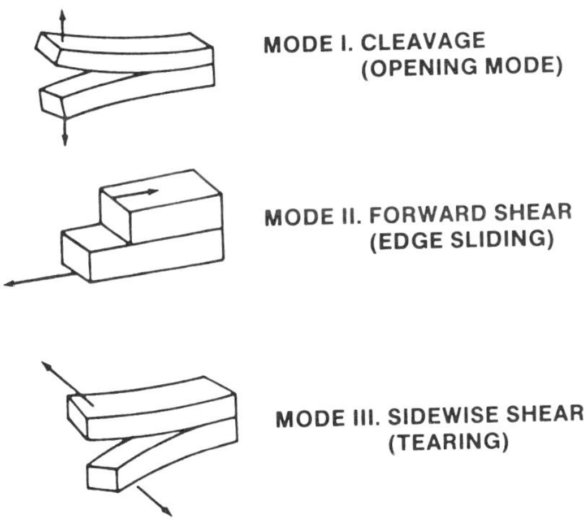


Figure 3.2 Basic loading modes.

distribution within the adhesive, adherend stiffness, and adhesive stiffness and compressibility (Poisson's ratio). Each one of these factors can change the load-carrying capability of a joint by more than a factor of 2. Improper testing and data interpretation can, therefore, lead to predictions of load-carrying capability that are in error by more than an order of magnitude.

As the average stress in a bond line is generally not a reliable tool for predicting failure in a bonded joint, an alternative approach is required. If it is hypothesized that failure of a bonded joint occurs when the stresses (or some functional of the stresses) reach a critical value, one might evaluate the stress at each point in a test specimen bond line. The value of the maximum stress(es) at which the bond broke would be termed the bond stress capability. The next step would be to evaluate the stresses at each point in the joint bond when it is subjected to its maximum expected load. The highest resulting stress (or stress functional) at any location in the bond would then be termed the joint requirement. The amount by which the bond capability exceeded the joint requirement would provide a measure of the margin of safety of the joint. In practice, both the capability and requirement are evaluated in terms of combinations of stresses, such as a maximum shear stress or octahedral shear stress, or simply the maximum principal stress.

This procedure has been very useful for homogeneous materials except when a notch or crack (360° notch) is present in the material. In such cases, the stresses are not defined at the notch tip when linear elastic analyses are employed; i.e., the notch tip is a point of stress singularity [1-3]. One normally relies on a fracture mechanics analysis to predict load-carrying capability for such geometries.

The fracture mechanics procedure is very similar to that for a material without a notch. Simplistically, the procedure is to determine the material capability using a notched test specimen. The capability of the notched material is quantified in terms of a fracture mechanics parameter. A comparison of this capability parameter to that required in the notched body being analyzed again provides a measure of the margin of safety. Details of this procedure are given in Ref. 1.

Notches in an adhesive and initial debonds are obvious points of bond line stress singularity. In addition, many bond termination geometries are points of singular stress [1,4-6]. Thus, for debonds initiating at bond edges even in the absence of voids or initial debonds, the joint requirement and bond capability must both be quantified in terms of fracture mechanics parameters. Bond failure does not always initiate at a point of stress singularity. Thus a failure criterion which is applicable to both singular and nonsingular points is highly desirable. Our approach is to determine an "inherent" flaw size for which energy release rates can be calculated whether failure initiates within a bonded joint or at the bond edge.

For many adhesives such as rubber-modified epoxies, stresses are not linear functions of strains. For these materials it may be necessary to use nonlinear analysis techniques which are currently in use for homogeneous materials. Bond joint stresses can be determined using nonlinear numerical techniques, which are available in many finite-element formulations.

## II. ANALYSIS OF BONDED JOINTS

Notches in homogeneous materials have been recognized for many years as points from which fracture propagates. As demonstrated in two individual examples by Sneddon [2] in 1946 and later generalized by Williams [3], linear elastic analysis of plane problems yield series solutions for the stresses near the tip of a crack in a homogeneous material. The form of the series can be written

$$\sigma_{ij} = \frac{K}{r^{1/2}} f(\theta) + r^{1/2} f_1(\theta) + \cdots \tag{1}$$

where r, $\theta$ are polar coordinates with origin at the crack tip. The parameters K and $f(\theta)$ vary with applied load and geometric factors. The first term in the series becomes dominant near the crack tip. For a general enclosed material angle of $\beta$, Equation (1) may be written

$$\sigma_{ij} = \frac{K}{r^{\lambda}} f(\theta) + r^{1-\lambda} f_2(\theta) + \cdots \tag{2}$$

where $\lambda$ is a function of $\beta$ as plotted in Figure 3.3. For material angles less than 180° no singular term is present ($\lambda$ = 0), but for 360° the square root of r singularity of Equation (1) is retained.

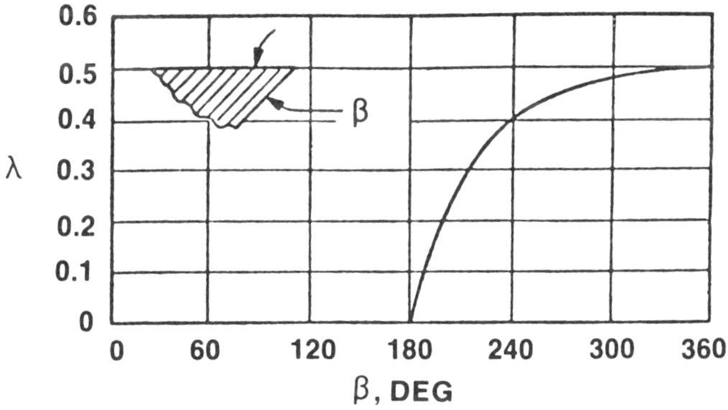


Figure 3.3 Singularity exponent for a sharp corner in a homogeneous material.

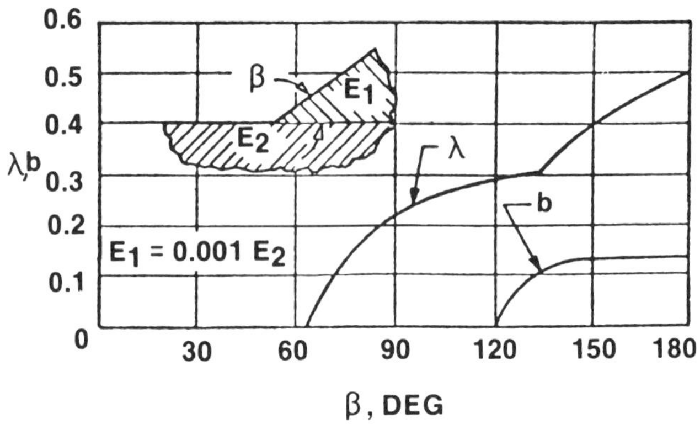


Figure 3.4 Singularity parameters for a half-plane to wedge bimaterial, $\nu_1 = \nu_2 = 0.2$.

When two materials are bonded, the stresses near the bond line terminus may be written

$$\sigma_{ij} = \frac{K}{r^{\lambda}} f(r,\theta,b) + r^{1-\lambda} f_3(r,\theta,b) + \cdots \tag{3}$$

where $f(r,\theta,b)$ represents a geometry and material property function of the variables, r, $\theta$, and b. The function b typically appears in the form sin(b ln r) or cos(b ln r). The resulting oscillatory characteristics of the solution are discussed in Refs. 1, 4, and 5.

For plane problems in which one material is bonded at an angle $\beta$ to a semi-infinite sheet of a second, more rigid material [4] ($E_1 = 0.001E_2$, $\nu_1 = \nu_2 = 0.2$), values of $\lambda$ and b are plotted as in Figure 3.4. The function b in Equation (3) obviously depends on material properties, since for a homogeneous material b is zero for all material angles. For the nonhomogeneous case, no singularity exists ($\lambda = 0$) until a material angle of over 60° is obtained. Thus, fillet angles less than 60° are advisable for such systems.

A more general analysis of this bimaterial wedge configuration was completed by Dill et al. [6]. If one material is relatively rigid and a wedge of a second (more flexible) material with included angle $\beta$ is bonded to the rigid material, the exponent $\lambda$ in Equation (3) is as plotted in Figure 3.5 as a function of Poisson's ratio for the flexible material. It is important to note that the singularity exponent $\lambda$ is zero for all material angles less than 45° independent of the value of Poisson's ratio. Thus, in general, fillet angles $\beta$ less than 45° are required in order to avoid a stress singularity at the bond termination point.

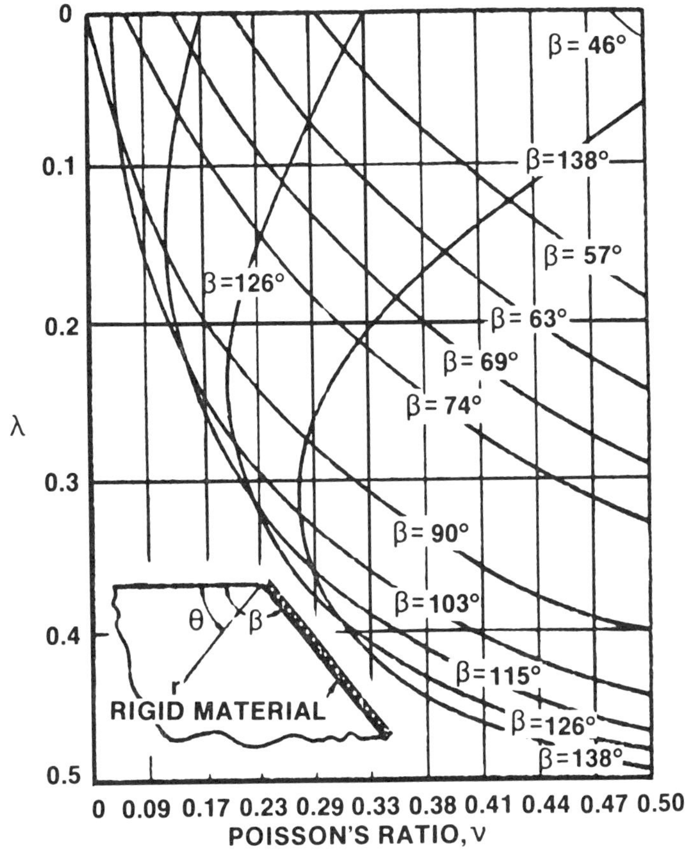


Figure 3.5 Singularity exponent for a free-fixed wedge. (From Ref. 6.)

Another geometry pertinent to bond terminations is shown in Figure 3.6. The plot of $\lambda$ in Figure 3.6 indicates that singular stresses are expected for all bond systems in which two materials terminate at a 90° angle. However, the order of the singularity is smaller than that found in sharp notches.

Thus, most bond lines in practice have singularity points, and stress failure criteria are not applicable at such points when linear elastic analyses are used. As will be shown in the discussion of tensile adhesion tests, debonds do not always initiate at bond edges in spite of the stress singularity at this point.

Up to this point we have discussed the need to use fracture mechanics parameters when analyzing bonded joints without describing

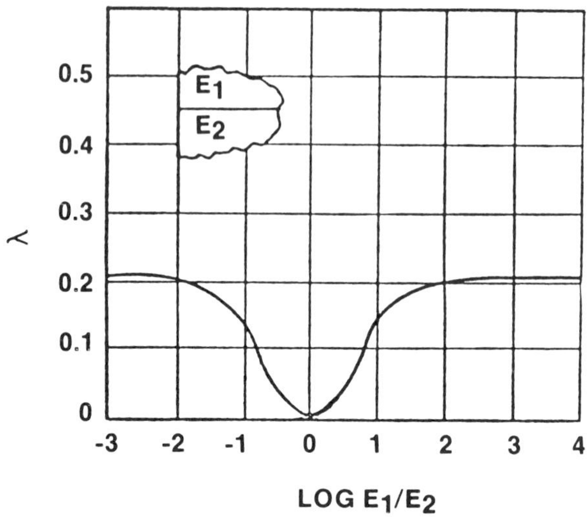


Figure 3.6 Singularity exponent for two quarter-planes, $\nu_1 = \nu_2 = 0.2$.

specific parameters. Two sets of parameters have been used in the analysis of homogeneous materials; one is the strain energy release rate G or $\partial U/\partial A$, and the other is a stress intensity factor K. Both G and K may be subscripted with Roman numerals to denote the respective values contributed by mode I, II, and III loadings (see Figure 3.2). The same parameters can be extended for use in fracture mechanics analyses of bond lines as described below.

The strain energy release rate approach was originally proposed by Griffith [7] in 1924. This approach is based on the first law of thermodynamics and predicts that a crack in a homogeneous linear elastic system can propagate only when the difference between the increment of work and the increment of strain energy in a body caused by propagation of a crack becomes greater than the increment of energy required to create new crack surface area. The restriction of a "linear elastic system" can be removed if one accounts properly for the various dissipative mechanisms introduced, for instance, in plastic flow or viscous dissipation. The validity of removing the restriction is evidenced by the research of Orowan [8] and Irwin [9] for ductile materials and of Williams [5] for viscoelastic materials.

The restriction of a homogeneous system is certainly not necessary as long as one can properly account for energy dissipation through the entire system as fracture propagates. Thus, the application to adhesively bonded systems is direct. For debond to propagate over an area $\Delta A$ for a given applied load, the following energy balance must be satisfied:

Change in work input = change in stored energy
+ energy dissipated
+ energy to debond an area $\Delta A$

or

$$\frac{\Delta W}{\Delta A} = \frac{\Delta U}{\Delta A} + \frac{\Delta E_d}{\Delta A} + G \tag{4}$$

where $\Delta A$ = change in crack surface area of one side (projected)
$\Delta W$ = change in work input
$\Delta U$ = change in strain energy stored
$\Delta E_d$ = change in dissipated energy
$G$ = energy release rate

At the condition of incipient crack propagation G has the value $G_c$, called the critical energy release rate. At the critical load $P_{cr}$, Equation (4) becomes

$$\left.\frac{\Delta W}{\Delta A}\right|_{P_{cr}} = \left.\frac{\Delta U}{\Delta A}\right|_{P_{cr}} + \left.\frac{\Delta E_d}{\Delta A}\right|_{P_{cr}} + G_c \tag{5}$$

where $G_c$ is the energy in a system that is required to form a new fracture surface in a homogeneous material or to separate two materials.

If $G_c$ is interpreted as the energy per unit area required to break bonds, its magnitude will depend only on the number of bonds existing per unit of fracture area or debond surface area and the average energy required to break a single bond. Thus, $G_c$ for a linear elastic system with a planar crack should be a property of the adhesive bond strength only, independent of loading mode. However, it may be very difficult to determine (and even harder to predict in advance) the actual fracture area existing on debonding of two surfaces. In addition, the amount of energy dissipated ($\Delta E_d/\Delta A$) in the immediate vicinity of the debond terminus through such effects as localized plastic flow and localized viscous dissipation is normally neglected when the bulk behavior of the materials is linearly elastic. The energy required for these mechanisms ($\Delta E_d/\Delta A$) is therefore added to that required to break bonds and lumped in $G_c$ as shown:

$$\left.\frac{\Delta W}{\Delta A}\right|_{P_{cr}} = \left.\frac{\Delta U}{\Delta A}\right|_{P_{cr}} + G_c \tag{6}$$

By adding all such localized energy loss sources and assuming the debond surface is planar, the $G_c$ values obtained from laboratory testing may be orders of magnitude greater than the energy required to break bonds. The $G_c$ values now represent *all* the energy per unit of planar (projected) debond surface area lost in the process of separating the two materials.

When dissipated energy is neglected and the crack area becomes infinitesimally small, Equation (4) can be written

$$\frac{dW}{dA} = \frac{dU}{dA} + G \qquad (7)$$

In Ref. 1 it is shown that if a body is loaded only by body forces and/or surface tractions (external forces and/or pressures), the change in work input as a crack grows is equal to twice the change in strain energy. Thus Equation (7) becomes

$$G = \frac{dU}{dA} \qquad (8)$$

If loads are induced only by temperature changes or fixed displacements, there is no change in work input as the flaw propagates and Equation (7) becomes

$$G = -\frac{dU}{dA} \qquad (9)$$

Thus, for most structural systems, the strain energy release rate is all that is required in order to calculate the total energy release rate G. However, when a body is simultaneously loaded by fixed displacements or temperature changes combined with tractions or body forces, the generalized expression, Equation (7), must be used. Since plane strain analyses assume a constant out-of-plane displacement, such geometries must be treated as having a fixed-displacement boundary condition.

In a body which is subjected to constant forces $P_{0i}$, applied at particular points, and fixed nodal displacements $u_{0j}$ at other locations, the total energy release rate can be rewritten

$$G = \sum_{\sigma} \frac{P_{0i} \cdot du_i}{dA} - \frac{dU}{dA} \qquad (10)$$

where the $u_i$ are displacements at the points of the applied constant forces and the summation index is used to designate the points where the constant forces are applied.

There is an alternative method of obtaining the total energy release rate in terms of work on displacement boundary areas (complementary work). We first note that for a given geometry (initial crack area $A_0$), the work applied to a body must be stored as strain energy (again neglecting heat generation etc.). In equation form,

$$U_0 = \sum_{\sigma} \frac{P_{0i} \cdot u_i}{2} + \sum_{u} \frac{P_j \cdot u_{0j}}{2} \qquad (11)$$

where $u_i$ represents the displacements at each point where a force $P_{0i}$ is applied and $P_j$ represents the forces required to apply each displacement $u_{0j}$.

For the same boundary conditions but with a second crack area $A_1$, the stored strain energy is

$$U_1 = \sum_{\sigma} \frac{P_{0i} \cdot u_i^*}{2} + \sum_{u} \frac{P_j^* \cdot u_{0j}}{2} \tag{12}$$

where $u_i^*$ and $P_j^*$ are the displacements and forces at the applied boundary conditions $P_{0i}$ and $u_{0j}$ when the crack area has increased to $A_1$. The change in strain energy for the two different crack depths per unit of crack area change is therefore

$$\frac{\Delta U}{\Delta A} = \sum_{\sigma} \frac{P_{0i} \cdot (u_i^* - u_i)}{2\Delta A} + \sum_{u} \frac{(P_j^* - P_j) \cdot u_{0j}}{2\Delta A} \tag{13}$$

In the limit as $\Delta A$ approaches zero

$$\frac{dU}{dA} = \sum_{\sigma} \frac{P_{0i}}{2} \cdot \frac{du_i}{dA} + \sum_{u} \frac{u_{oj}}{2} \cdot \frac{dP_j}{dA} \tag{14}$$

This is a rather interesting equation as it provides a means of calculating the strain energy release rate directly from boundary conditions. Equation (14) may also be written

$$\sum_{\sigma} P_{0i} \cdot \frac{du_i}{dA} = 2\,\frac{dU}{dA} - \sum_{u} u_{0j} \cdot \frac{dP_j}{dA} \tag{15}$$

Substituting this expression into Equation (10) gives

$$G = \frac{dU}{dA} - \sum_{u} u_{0j} \cdot \frac{dP_j}{dA} \tag{16}$$

Depending on their ease of application in a given situation, either (or both) of Equations (10) and (16) can be used. Except for a possible small numerical error, both methods will give the same value for the total energy release rate.

Energy release rate can be obtained using numerical techniques such as finite-element analyses. Several approaches have been used successfully.

### A. Numerical Differentiation ($\Delta U/\Delta A$)

This method has been widely used in the past. Two computer runs are required for a calculation. The strain energy release rate, $\Delta U/\Delta A$, is the difference in strain energies for two crack lengths divided by the corresponding change in the crack surface area A. The crack extension can be modeled either by extending a row of void elements or by releasing nodal constraints (as explained below). Strain energy can be accurately calculated in finite-element computer programs using the stiffness matrix and force and displacement vectors as discussed in Ref. 1.

As previously explained, $\Delta U/\Delta A$ provides an accurate value for G for simple loading conditions, but not for certain combined loading situations. To properly consider all possible loading conditions, Equation (7) should be used.

## B. Crack Closure Integral (CCI)

This less well known method for finding the energy release rate is based on the work theoretically required to reverse flaw propagation by an amount $\Delta A$. Figure 3.7 shows a finite-element grid in the vicinity of a crack. In the first run, nodes B and E are constrained (no relative movement allowed); the output indicates the reaction forces $P_{xBE}$ and $P_{yBE}$ required to constrain the node pair. The second run, with nodes B and E released, gives their relative displacements.

The mode I and mode II components of energy release rate may be obtained from the following equations:

$$G_I \sim \frac{P_{yBE}^{(1)} \left| u_{yB} - u_{yE} \right|^{(2)}}{2\Delta A} \tag{17}$$

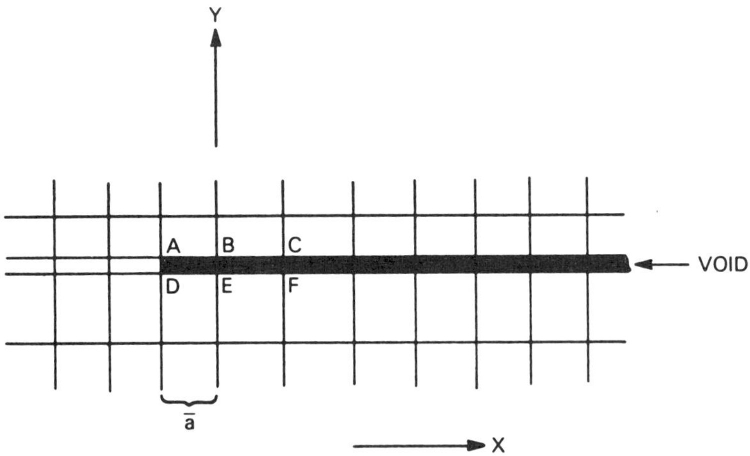


Figure 3.7 Crack closure integral method.

$$G_{II} \sim \frac{P_{xBE}^{(1)} \left| u_{xB} - u_{xE} \right|^{(2)}}{2\Delta A} \tag{18}$$

where the superscripts indicate the first and second computer runs.

In the above equations, P is divided by 2 to represent the average force during the nodal displacement. Depending on the orientation of the coordinate axes, subscripts x and y may be transposed or changed to r and z for an axisymmetric geometry.

An important advantage of this method is the separation of G into modal components $G_I$ and $G_{II}$, which correspond to crack surface opening and sliding, respectively. The total energy release rate G is the summation of the two components.

It is usually not necessary to worry about the sign of G. As long as the displacements required to close a crack are in the same directions as the crack closure forces, G is positive. We have seen cases where one loading mode is dominant and the secondary loading mode produced a small negative G. Thus the total G can be smaller than the value for one particular loading mode.

## C. Modified CCI [10]

In the modified crack closure integral approach, it is assumed that the relative crack opening displacements at points C and F (Figure 3.7) with nodes B and E constrained are approximately the same as the relative displacements of nodes B and E after their constraints have been removed. Thus, the components of energy release rate may be approximated using the equations

$$G_I \sim \frac{P_{yBE} \left| u_{yC} - u_{yF} \right|}{2\Delta A} \tag{19}$$

$$G_{II} \sim \frac{P_{xBE} \left| u_{xC} - u_{xF} \right|}{2\Delta A} \tag{20}$$

Using this approximation will reduce computer time, since nodal forces and displacements are taken from the same run. The modified CCI is satisfactory if node C and F relative displacements are nearly equal to what the node B and E displacements would have been if their constraint had been removed.

## D. Crack Opening Displacement (COD)

For a sharp crack in a homogeneous material with a plane stress geometry, the equation for crack opening displacements can be written in a series expansion [3], the first term of which is dominant near the crack tip

$$u_y = \frac{4K_I}{\sqrt{2\pi}E} \sqrt{r} + \cdots \tag{21}$$

where $u_y$ = crack opening displacement (COD)
$K_I$ = stress intensity factor
E = material modulus
r = distance from crack tip

In logarithmic form the first term of the series is

$$\log u_y = \log \frac{4K_I}{\sqrt{2\pi}E} + \frac{1}{2} \log r \tag{22}$$

As the equation is of the form y = C + Dx, we can see that a log u versus log r plot is linear with a slope of 1/2. Finding the y intercept, C, will allow us to solve for $K_I$. Then $G_I$ can be obtained from the plane stress relationship

$$G_I = K_I^2/E \tag{23}$$

For bimaterial geometries, the crack opening displacement has the series form [5]

$$u_y = Kr^n\{A \cos[f_1(r)] + B \sin[f_2(r)]\} + \cdots \tag{24}$$

where n depends on geometry and material properties and the sine and cosine terms contain functions of r, the distance from the crack tip. These terms are relatively constant except for a very small distance from the crack tip. Thus, Equation (24) can be approximated in logarithmic form by the equation $\log u_y = K' + n \log r$, where n is the slope of the "linear" portion of the curve and K' is the intercept. For some systems, the "linear" portion of the log $u_y$ versus log r curve may be influenced by the sine and cosine terms and therefore may not be a straight line.

When the COD approach is implemented using a finite-element computer program, a very fine grid is required near the crack tip.

## E. Summary of Numerical Techniques

When calculating the energy release rate with numerical differentiation, the type of loading is very important. Numerical differentiation of strain energy is valid when all loads are thermal changes and/or nodal displacements (there is no change in boundary work as the crack propagates). It is also valid when all loads are body forces, point loads, and/or pressures. Strain energy numerical differentiation is not valid if loads include a temperature change and/or applied displacement along with body force, external force, and/or pressure loads. For these situations, one must subtract the strain energy release rate ($\Delta U/\Delta A$) from the boundary work change ($\Delta W/\Delta A$) or use a crack closure integral in order to get the total energy release rate.

In general, *the crack closure integral method is preferred over numerical differentiation.* The CCI method depends directly on nodal forces and displacements at a crack tip. Consideration of input loading conditions is not required. In addition, the crack closure integral method separates the energy release rate into individual components of loading mode.

Combinations of mode I and mode II loading are very common in adhesive systems, since fracture normally follows bond lines or takes place within the adhesive but propagates parallel to the bond surface. Several authors including Anderson et al. [11], Trantina [12], and Johnson and Manzalgire [13] have noted a dependence of critical strain energy release rate on loading mode.

What analytical techniques can be applied to a point of stress singularity if no initial debond exists and the inherent flaw size is not known? Can the strain energy release rate at a singular bond termination be obtained by extrapolating the strain energy release rate for a small initial debond to the point where the initial debond length is zero? To answer these questions, we first note that according to the data in Figure 3.4 the stresses near the bond termination with an initial debond ($\beta = 180°$) are of the form $\sigma = K/r^{0.5}$. If no initial debond is present and the adhesive (material 1 in Figure 3.4) terminates at a right angle, the bond termination stresses have the form $\sigma = K/r^{0.22}$. Because these values differ, a discontinuity in stresses exists at an initial flaw depth of zero. Thus the strain energy release rate G cannot be defined when no debond exists at the bond line. In such cases, the constant K in Equation (3) could be used to quantify the singular stresses. This is very similar to using K in Equation (1) to quantify the stress singularity behavior in cracked homogeneous materials. In the latter case, K is referred to as a stress intensity factor, and the value of the stress intensity factor that causes fracture to propagate is called either the critical stress intensity factor or the fracture toughness.

As discussed in Ref. 1, a relationship between stress intensity factors and energy release rates has been demonstrated for homogeneous materials. It is noted that the units of K in Equation (3) are no longer those of pressure times square root of length, $Pa\sqrt{m}$ (psi $\sqrt{in.}$). The units are now pressure times length to the $\lambda$ power. An example using stress intensity factors is presented in our discussion of lap shear testing. Details of numerical calculation procedures for stress intensity factors are given in Ref. 1.

## III. STANDARD BOND-STRENGTH TESTS

Many different techniques exist for determining the strength of an adhesive bond. Over 50 bond strength test methods have been accepted by the American Society for Testing and Materials as standard test procedures. In 1983 one of the authors conducted a poll of 43 individuals who attended short courses in adhesion sponsored by the Adhesion Society and the American Chemical Society. These individuals, representing 38 companies, were asked what testing

Table 3.1 Adhesion Tests in Use (1983)

| Test | Percent of respondents[a] Primary test method | Some degree of use |
|---|---|---|
| Peel | 46 | 93 |
| Lap shear | 22 | 74 |
| Tensile adhesion | 24 | 63 |
| Other | 8 | 33 |

[a]Forty-three total respondents representing 38 companies.

techniques their company was using. In some instances, different branches within a company employed different test methods. In these cases, the results of both branches were included. Thus the results from a total of 43 respondents were evaluated. These data are summarized in Table 3.1. Although the survey was not conducted on a scientific basis, it provides some insight into the types of individual tests which are currently in use.

Lap shear tests, tensile adhesion tests, or peel tests are used as the primary test technique by 92% of the respondents. Some amount of testing with each of these three techniques is completed by over 60% of the respondents. Since these are the most popular bond strength tests, the remainder of this chapter will be devoted to discussions of the interpretation of data from the tests. Additional analytical efforts are continuing; however, the results to date provide a good basic understanding of the failure mechanism involved.

## IV. LAP SHEAR TESTS

One of the most widely used groups of test methods can be generally classified as lap shear tests. Typical lap shear test specimens for which ASTM standards have been written are presented in Figure 3.8. It will be shown that the stresses along the bond line of these specimens are not constant. The bond stress distribution and failure load are highly dependent on the thickness of the adherends. Failure load also varies with bond thickness, increases very nearly linearly with width of the overlap, and increases in a nonlinear manner with length of the overlap. Debonds generally initiate at the bond terminations, where the stresses are singular. Debond initiation in lap shear specimens must therefore be characterized in terms of fracture mechanics parameters. In addition, it is shown in the following sections that debonds initiate and are propagated by opening mode displacements (mode I).

## A. Structural Analysis—Lap Shear Tests

Many analyses have been completed for lap shear geometries. As early as 1938, Volkersen [14] obtained expressions for the stresses in a lap shear joint by considering the differential displacements of the adherends and neglecting bending. This was followed in 1944 by the now classical treatment of Goland and Reissner [15]. They used standard beam theory and other strength of materials concepts to obtain expressions for the joint stresses. Plantema [16] combined the results of these two earlier investigations to include shear effects in the system.

Since the stress state of the lap shear joint is so complex and does not lend itself to closed-form solutions, it is only logical that as numerical methods became available, researchers would apply them to analyze adhesive joints.

Wooley and Carver [17], for example, used finite-element methods to calculate the joint stresses. They compared their results with the results obtained by Goland and Reissner and reported very good agreement. Adams and Peppiatt [18] used a two-dimensional finite-element code to analyze the stresses in a standard lap shear joint. They too reported good agreement with Goland and Reissner. These authors also investigated the effect of a spew (triangular adhesive fillet) on the calculated stresses. A nonlinear finite-element analysis of the single lap joint was completed by Cooper and Sawyer [19] in 1979. They compared their results with those of Goland and Reissner and with experimental results from photoelastic models. They concluded that although nonlinear behavior can

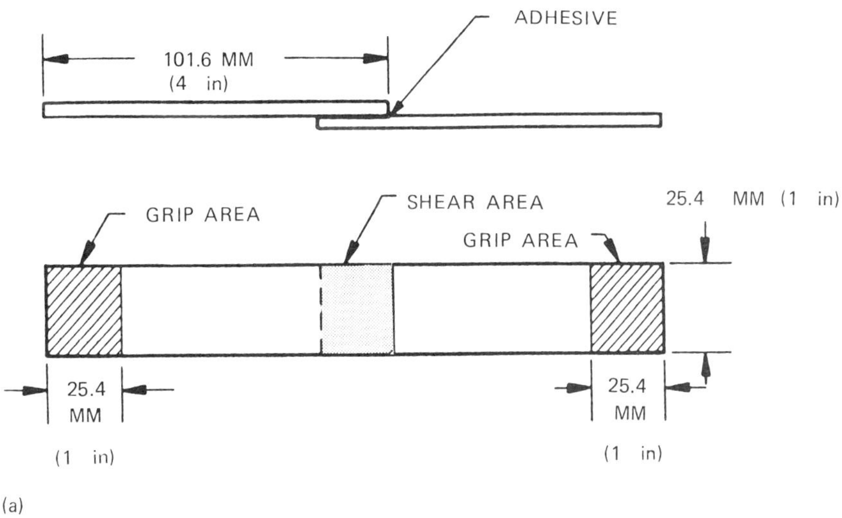


Figure 3.8 (a) Single lap shear test, ASTM D1002.

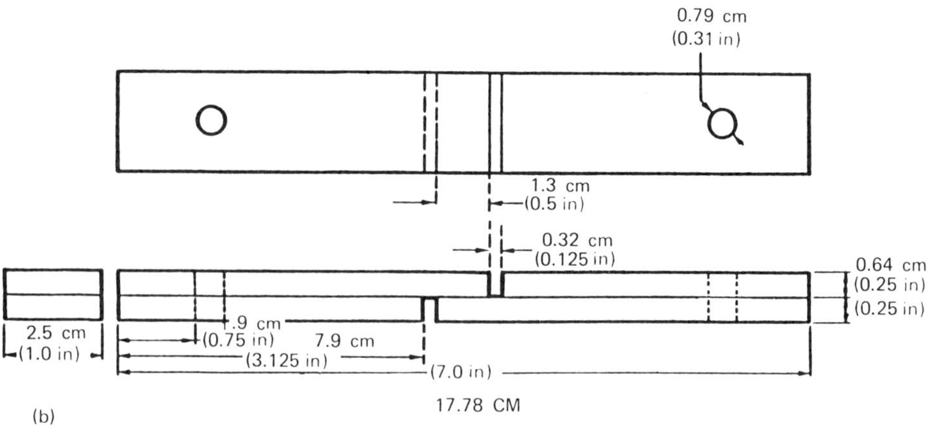


Figure 3.8 (b) Thick adherend single lap shear joint, ASTM D3165.

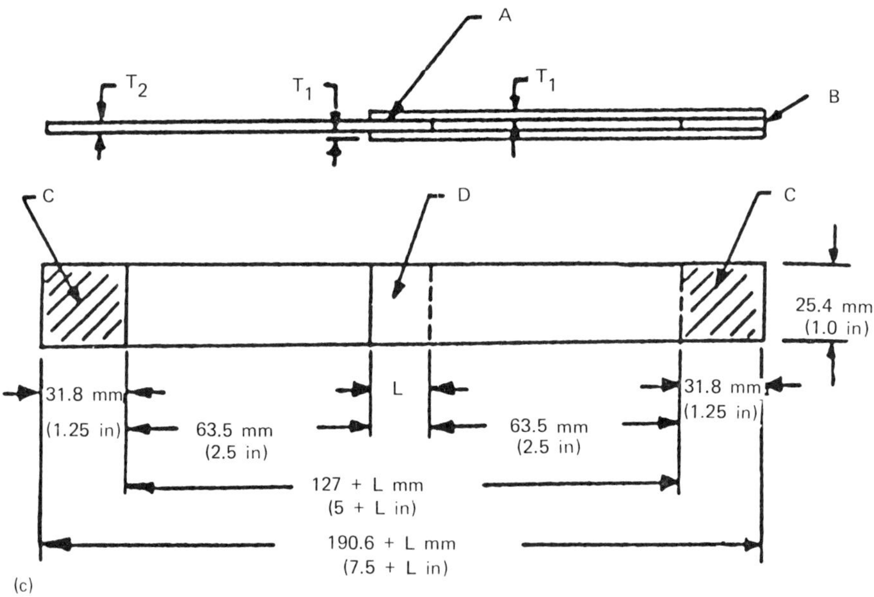


Figure 3.8 (c) Double lap shear test, ASTM D3528,

at times have a sizable effect on joint stresses, the Goland and Reissner method is usually sufficiently accurate for prediction of midsurface stresses.

The double lap shear joint (Figure 3.8c) has the advantage of greatly reducing the bending stresses. This joint has also received analytical attention. Amijima et al. [20] used finite-element methods to calculate the stresses. They were able to claim some agreement between predictions of joint stresses obtained in their numerical analysis and experimentally observed strength.

Sen and Jones [21,22] also used finite-element techniques for the analysis of double lap joints. They considered a viscoelastic adhesive. They reported that the magnitude of the maximum adhesive stress decreases as the ratio of the adhesive modulus to adherend modulus decreases from one. They also investigated the effect of adhesive thickness and length of overlap on the stress distribution along the joint and reported that decreasing the adhesive thickness results in a more uniform stress distribution.

*Measurement* of the exact stress state in an adhesive joint is also a difficult task. There have, however, been several intensive efforts in this respect. Sharpe and Muha [23] used a laser technique to measure the adhesive shear strain. Their technique is suitable only for transparent linearly elastic adhesives. They reported stresses in good agreement with those of Goland and Reissner.

Several investigators have suggested modifications to the standard ASTM lap shear test geometries. Leaving an adhesive spew and rounding adherend corners as analyzed by Adams and Peppiatt [18] (noted previously) can eliminate the mathematical singularity at the end of the overlap, thereby reducing the maximum stress, increasing the joint strength, and reducing test variability. Renton [24] suggested thickening the adherends, over those recommended in ASTM D3165, thereby reducing bending and hence reducing cleavage stress in the lap joint. He described a technique of determining adhesive strain by measuring changes in a magnetic field resulting from the displacement of the steel parts of the lap joint and suggested optimum specimen designs for various adhesive-adherend combinations. In the most thorough study of this type with which the authors are familiar, Guess et al. [25] presented finite-element and experimental evidence that failure in standard lap shear specimens is often controlled by the tensile strength of the adhesive rather than its shear strength. As will be discussed in more detail later, they reported an apparent increase in shear strength with increasing adherend thickness. River [26] also described methods, and test results, for reducing the stress concentration at the end of the overlap and obtaining relatively accurate measurements of the shear properties of an adhesive.

Because of the singularity and other problems associated with stress criteria of failure, several groups have applied the principles of fracture mechanics to adhesive joints [1,27-29]. Of particular interest for the lap shear joint is the attempt by DeVries et al. [30] to apply a Griffith-type energy balance. They were able to claim reasonable agreement with experimental data and noted an apparent dependence of the adhesive fracture energy on mode of stress at the crack tip.

Specifications published by the ASTM and other sources describe a variety of different tests, but the most commonly used lap shear bond strength tests are ASTM D1002 and ASTM D3165 (Figure 3.8). The adherend thickness recommended in these standards for most metal sheets is 0.16 cm (0.0625 in.) with 1.3 cm (0.5 in.) of overlap. Generally, the quantity reported as the "lap shear strength" or "tensile shear strength" is the breaking load per unit bond area. However, failure initiation in materials is usually a localized phenomenon that is more dependent on maximum stresses (or energy) at a point reaching some critical value than on the average induced values. As mentioned previously and reinforced by plane strain finite-element stress analyses in this study, the standard lap shear specimen shows a variation in shear stress within the adhesive layer and, in fact, can exhibit a singular behavior at the bond termination.

The authors have completed linear elastic stress analyses of lap shear joints using a plane strain finite-element computer program. These analyses have concentrated particularly on the stresses near the bond termination. Lap shear specimens with ASTM D3165 geometry, adherend thicknesses of 0.16 cm (0.0625 in.) to 1.3 cm (0.50 in.), and adhesive thicknesses of 0.025 cm (0.01 in.) and 0.16 cm (0.06 in.) were analyzed. The specimens were 16.5 cm (6.5 in.) long with 1.3 cm (0.5 in.) of overlap. The plane strain approximation is justified since Adams and Peppiatt [31] have shown that the tensile and shear stresses in the direction of the applied load are not significantly influenced by transverse stresses caused by Poisson's ratio strains in the adherends.

The finite-element code used quadrilateral elements which were internally divided into four linear displacement triangular elements. A grid network consisting of 2016 quadrilateral elements was used in each analysis with elements concentrated near the adhesive-adherend interface termination as shown in Figure 3.9. The smallest element width was 0.00025 cm (0.00001 in.) for stress and stress intensity factor analyses and 0.048 cm (0.019 in.) for energy release rate analyses. Axial displacements were applied and transverse displacements set equal to zero at the specimen ends to stimulate clamped ends. The following material properties were used:

| | Modulus, MPA (ksi) | Poisson's ratio |
|---|---|---|
| Adherend (steel) | 207,000 (30,000) | 0.30 |
| Adhesive (epoxy) | 2,760 (400) | 0.34 |

The adhesive ends were terminated at right angles to the adherends. Often a fillet is formed at bond edges by adhesive squeeze-out. However, the right-angle geometry is obtained when the lap

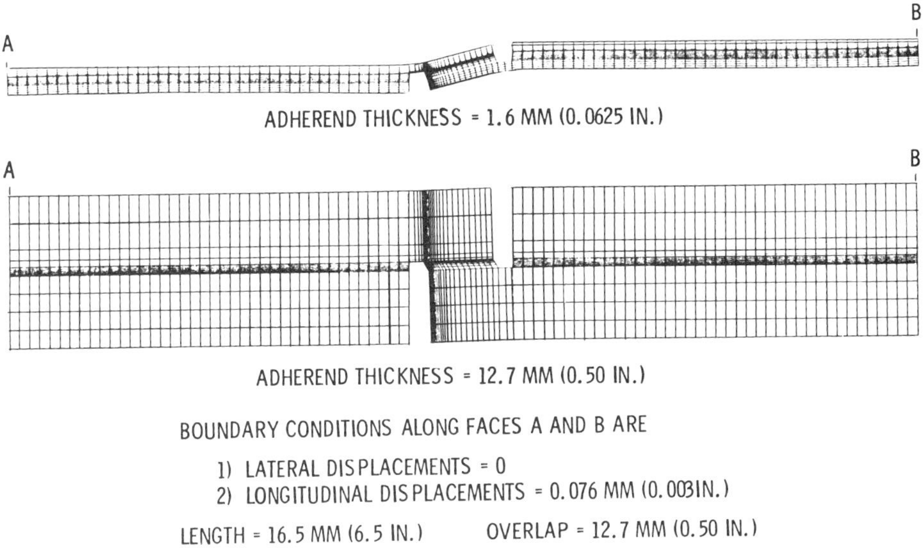


Figure 3.9 Lap shear deformed grids, displacements magnified 10 times.

shear specimens are cut and notched from bonded plates as described in ASTM D3165. The effect of adhesive squeeze-out is discussed in Refs. 18 and 32.

The computed deformed shapes for lap shear specimens with adherend thicknesses of 0.16 cm (0.062 in.) and 1.3 cm (0.50 in.) are presented in Figure 3.9. An axial elongation of 0.016 cm (0.006 in.) was applied, and the resulting displacements were multiplied by a factor of 10 to obtain the deformed grids. The amplified displacements provide a visual indication of the bending which takes place near the bond termination for thin adherend specimens. One would thus expect relatively large bond stresses (cleavage) at these points induced by adherend bending in thin adherend specimens. In addition, for linear elastic stress analyses, both the normal and shear stresses become singular as the bond termination is approached [4,6].

The adhesive stresses adjacent to the nonterminating adherends are plotted in Figures 3.10 and 3.11 as a function of distance from the bond termination. The distances were normalized by dividing by the bond length, a, and the stresses by dividing by the average applied shear stress, $\bar{\tau} = P/A$. Adhesive thickness was 0.152 cm (0.06 in.).

As expected, the stresses change radically as the bond termination is approached. The shear and tensile stresses are nearly constant and equal to the average values over 80% of the bond area

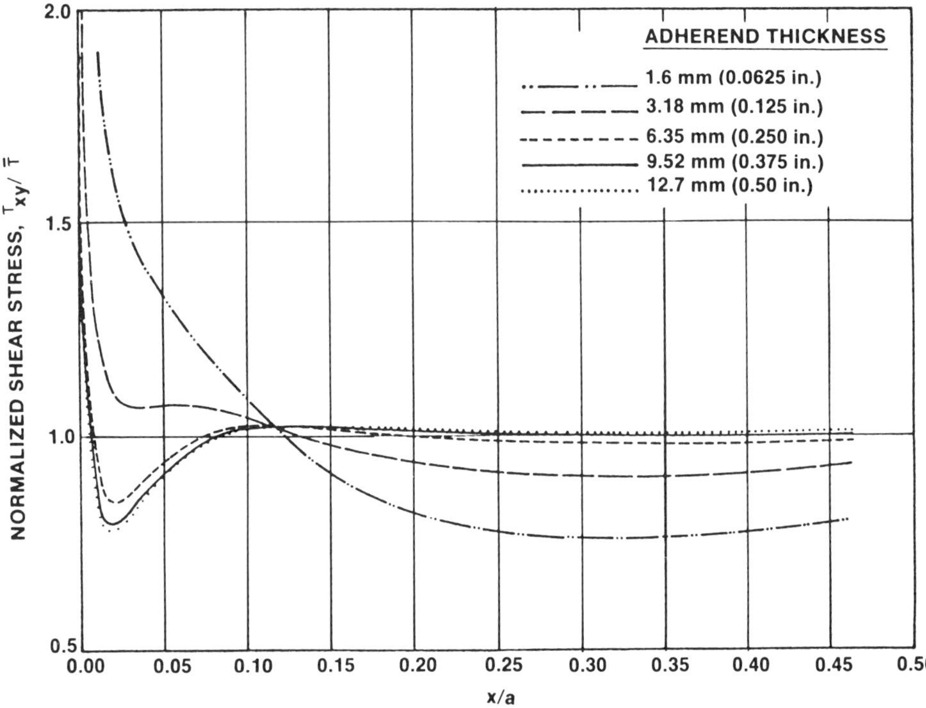


Figure 3.10 Bond line shear stresses in lap shear specimens.

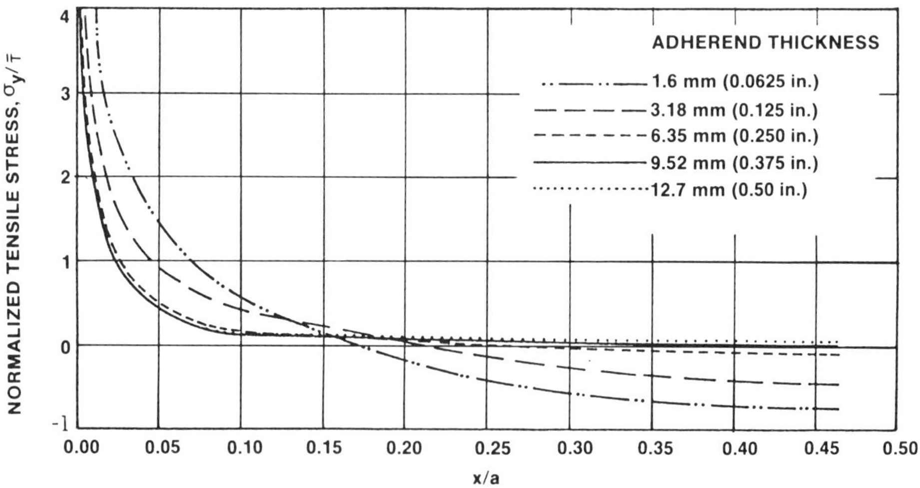


Figure 3.11 Bond line tensile stresses in lap shear specimens.

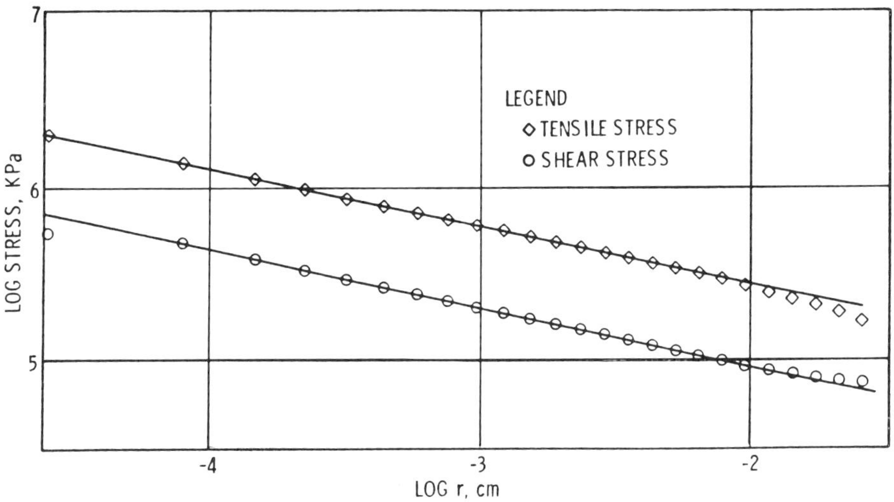


Figure 3.12 Log stress plot, lap shear test.

for adherend thicknesses greater than 0.64 cm (0.25 in.). For thin adherends, the stresses vary continuously along the bond line.

The character of the stress singularity was examined by plotting the tensile and shear stresses in log-log form (Figure 3.12). The tensile and shear stresses exhibit a $1/r^{\lambda}$ singularity behavior (log linear) for all elements within 0.0076 cm (0.003 in.) of the adhesive edge except for the final element. This element typically varies somewhat from the singularity pattern [1] because of the coarseness of the grid network compared to the changes in this area. As the distance from the bond termination increases beyond 0.076 cm (0.003 in.), the stress behavior deviates from the singular (log linear) form as expected.

Stress intensity factors and the singularity exponents are obtained from the straight-line portion of the log-log plots. The "singularity exponent" $\lambda$ in the equation stress = $K/r^{\lambda}$ is the slope of the straight-line portion of the curves in Figure 3.12. This value was approximately 0.34 for tensile as well as shear stresses and does not depend on adherend thickness. The stress intensity factors (intercepts of the curves in Figure 3.12), however, were found to depend on adherend thickness as shown in Table 3.2.

Both the mode I stress intensity factor $K_I$ and the mode II stress intensity factor $K_{II}$, per unit average shear stress, increase with decreasing adherend thickness; see Figure 3.13. Thus one would expect thick adherend specimens to resist larger loads. The decrease in $K_{II}$ with increasing adherend thickness is greater than that in $K_I$. For all adherend thicknesses, the mode I stress intensity factor is greater than that for mode II. As shown in the last column of Table 3.2, the tensile component is nearly three times

Table 3.2 Stress Intensity Factors for Lap Shear Tests

| Adherend thickness cm (in.) | Shear: $K_I / \overline{\tau}$, $cm^{0.34}$ ($in.^{0.34}$) | Shear: $K_{II}/K_{II\ 0.5}$[a] | Tensile: $K_I / \overline{\tau}$, $cm^{0.34}$ ($in.^{0.34}$) | Tensile: $K_I/K_{I\ 0.5}$ | $K_I/K_{II}$ |
|---|---|---|---|---|---|
| 0.16 (0.0625) | 0.448 (0.326) | 2.59 | 0.914 (0.666) | 1.77 | 2.04 |
| 0.32 (0.125) | 0.272 (0.198) | 1.57 | 0.660 (0.481) | 1.28 | 2.43 |
| 0.64 (0.250) | 0.189 (0.138) | 1.10 | 0.560 (0.408) | 1.09 | 2.96 |
| 0.95 (0.375) | 0.178 (0.130) | 1.03 | 0.526 (0.383) | 1.02 | 2.95 |
| 1.27 (0.500) | 0.173 (0.126) | 1.00 | 0.516 (0.376) | 1.00 | 2.98 |

[a] $K_{II\ 0.5}$ is the value of $K_\sigma$ for an adherend thickness of 0.5 in.

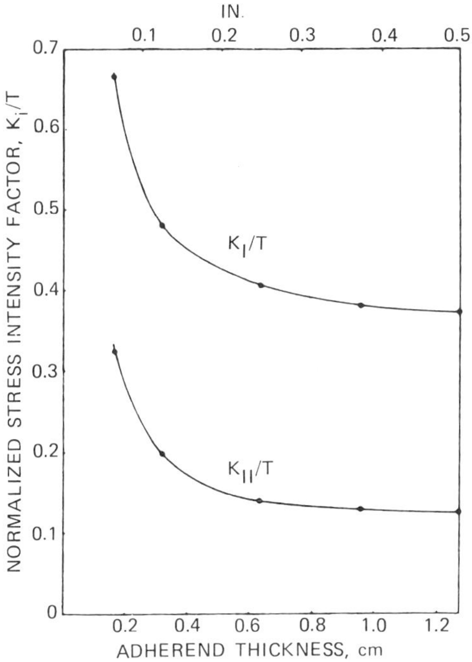


Figure 3.13 Lap shear stress intensity factors.

the shear component for large adherend thicknesses. This tendency was not expected but is consistent with the work of Wang and Yau [33]. Their study included the effects of relative stiffness of the adherends and adhesive (Figure 3.14) and included debonds of various lengths (Figure 3.15). In all cases, the component of stress intensity factor in the mode I direction is greater than the mode II component except when the adherend and adhesive moduli are nearly equal.

As shown in Figure 3.5 for a Poisson's ratio of 0.34 and a 90° bond termination angle, the singularity exponent $\lambda$ has a value of 0.32. Thus, the stresses near the bond termination are singular and proportional to $1/r^{0.32}$. This is in essential agreement with the exponent of 0.34 obtained numerically during the present analyses.

An energy release rate G was determined for each of five different adherend thicknesses using debond depths of 0.048 cm (0.019 in.) and 0.10 cm (0.038 in.) (average x/a = 0.057). As shown in Figure 3.16, the square root of the energy release rate is essentially constant for adherend thicknesses greater than 0.64 cm (0.25 in.) and approximately one-third of the value at the ASTM recommended thickness, 0.16 cm (0.0625 in.). Thus the change in

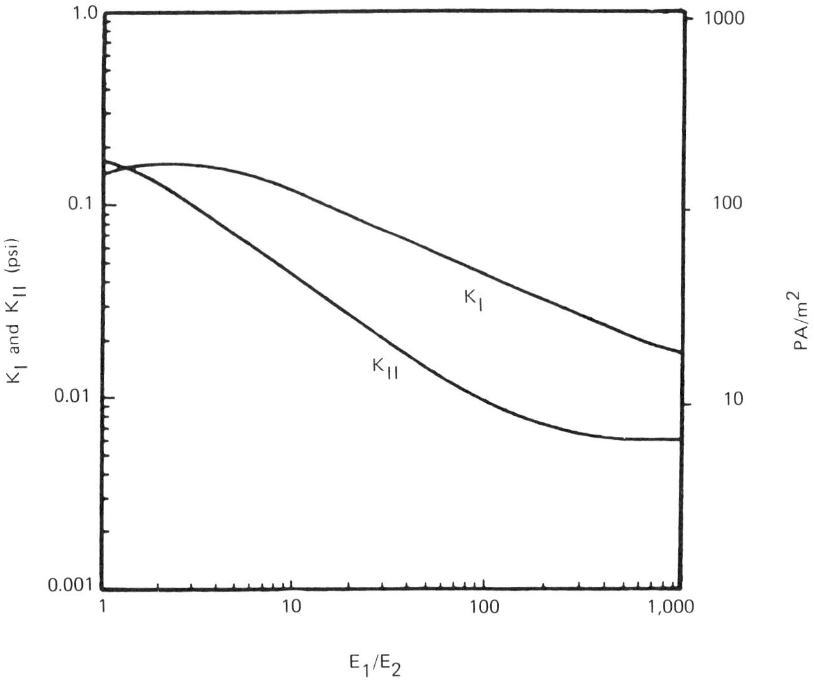


Figure 3.14 Stress intensity factors for interface crack in lap-shear joints with various adhesive moduli. Adherends: E = 69 GPa ($10^7$ psi), $\nu = 0.33$; adhesive: $\nu = 0.35$. (From Ref. 33.)

energy release rate with adherend thickness for a small initial debond has the same basic characteristics as observed for stress intensity factors, i.e., nearly a constant value for adherend thicknesses greater than 0.64 cm (0.25 in.) and increasing by a factor of approximately 3 as adherend thickness is reduced to 0.16 cm (0.0625 in.).

An analysis for a 0.025 cm (0.01 in.) thick adhesive (1.3 cm, 0.5 in. adherend) showed stresses of basically the same form as those for thicker adhesives. However, the stress singularity is shifted closer to the bond termination.

## B. Test Results—Lap Shear Tests

Testing was completed by the authors using single lap shear specimens. Steel adherends with thicknesses ranging from 0.16 cm (0.0625 in.) to 1.3 cm (0.5 in.) were used with a 1.3-cm (0.5-in.) overlap. The adhesive was a fairly rigid two-part epoxy (Hysol EA 934). All testing was completed at 72°F and a displacement

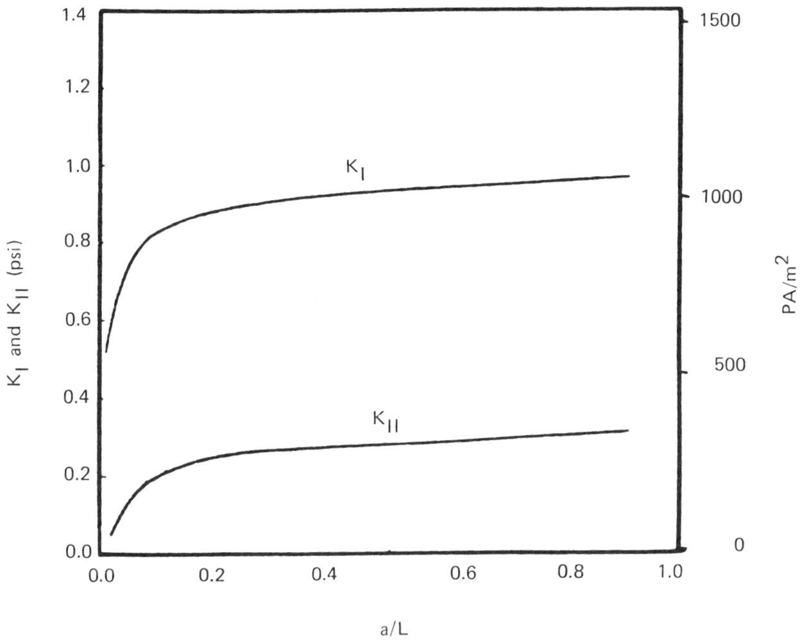


Figure 3.15 Stress intensity factor solutions versus interface crack length in lap-shear adhesive joints. L = 0.5 in.; adherends: E = 69 GPa ($10^7$ psi), $\nu$ = 0.33; adhesive: E = 3.4 GPa ($0.5 \times 10^6$ psi), $\nu$ = 0.35. (From Ref. 33.)

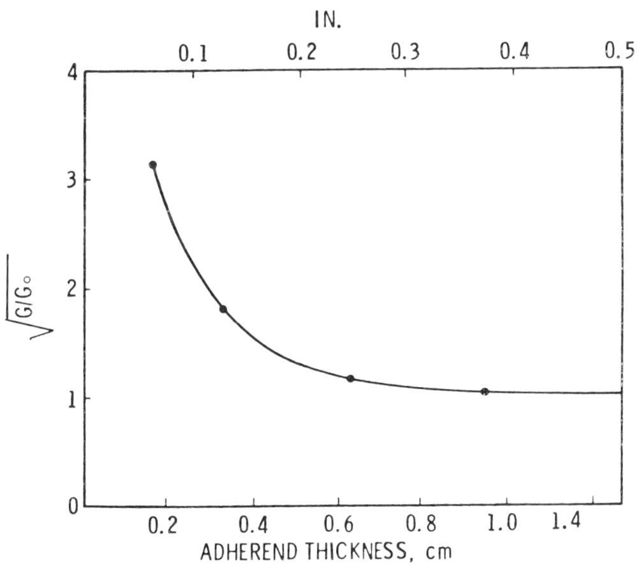


Figure 3.16 Effect of adherend thickness on energy release rate, G, for lap shear tests.

Table 3.3 Lap Shear Test Data

| Adherend thickness, cm (in.) | Average stress at debond, N (psi) | Reduction, % |
|---|---|---|
| 0.16 (0.0625) | 10,500 (1530) | 63 |
| 0.32 (0.125) | 16,800 (2440) | 40 |
| 0.64 (0.25) | 24,800 (3600) | 12 |
| 0.95 (0.375) | 29,000 (4200) | 0 |
| 1.27 (0.50) | 27,600 (4010) | 0 |

rate of 0.13 cm/min (0.05 in./min). The resulting data are presented in Table 3.3 and Figure 3.17. These data show a fairly constant breaking force for specimens with adherend thickness greater than 0.64 cm (0.25 in.). At the ASTM D1002 recommended thickness of 0.16 (0.0625 in.), the breaking force drops by 63%.

Similar data were obtained by Guess et al. [25], who also used steel lap shear adherends with 1.3 cm (0.5 in.) overlap. Their adhesives were EC-2114-R structural paste adhesive and FM-123-5 structural tape adhesive. Again, for thick adherends the breaking force was essentially independent of adherend thickness. The

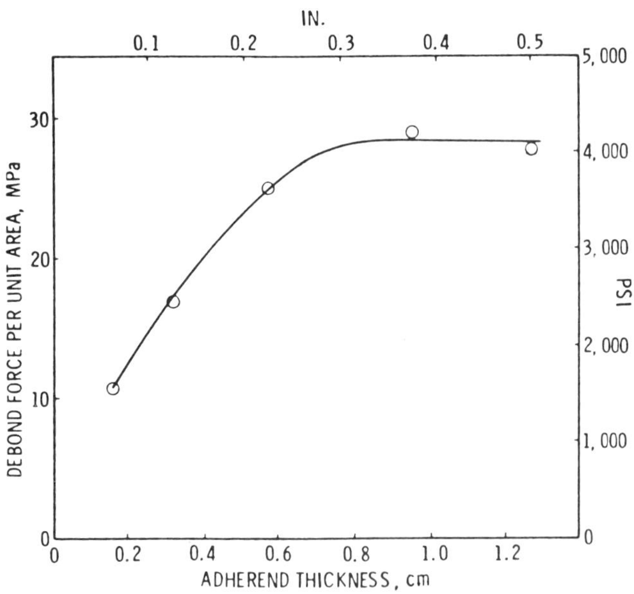


Figure 3.17 Debond force for lap shear tests.

EC-2114-R adhesive apparent strength dropped by 64%, and the corresponding drop for the FM-123-5 film adhesive was 23% for thin adherends. The decrease in bond strength for thin adherend specimens is essentially the same as the increases in stress intensity factor and square root of energy release rate, as would be expected if these parameters controlled debond initiation load.

### C. Discussion and Conclusions—Lap Shear Tests

In the stress analyses discussed here it is assumed that the adhesives as well as the adherends were linear elastic materials. For some structural adhesives, particularly the elastomer-based adhesives and the rubber-modified epoxies, the linearity assumption is invalid and a nonlinear stress analysis is preferred. Such analyses are not difficult to perform, since various nonlinear finite-element computer programs are available. However, the mechanical properties required for input to such programs are not commonly available. Thus most analyses are still performed using the linearity assumption. This is a major weakness in current analytical approaches and may provide misleading data for many bond systems.

In much of the literature on linear elastic analysis of lap shear specimens, the researchers have discussed the stress concentration factor at the bond terminus. When linear elastic analyses are used and the bond terminus is not rounded, the stress concentration factor is infinity because of stress singularity at this point.

Even in the idealized case where the adherends are rigid and restricted to remain parallel, the transverse stress has a singular behavior very near the bond termination. The transverse stress becomes negligibly small (as expected) and the shear stress remains relatively constant (rigid and parallel adherends) for distances greater than one-tenth of the adhesive thickness from the bond termination. Whether stress intensity factors for the initial bond geometry actually control debond initiation is open to question, since inherent flaws (such as bond line voids) that are larger than the entire singularity region could exist in the high-stress area. Thus the most appropriate criteria for debond initiation may be obtained through the use of an energy release rate at the depth of the largest inherent flaw. By what means can this size be best determined? Microscopic examination, extrapolation from finite flaws, and so forth, need further exploration. However, one method for finding the inherent flaw size is discussed in Section VB. It combines adhesion testing with energy release rate analyses. The problem of bond termination singularities could, of course, be avoided entirely by using properly contoured bond termination fillets and rounded adherend edges. This does not, however, circumvent the problem of stress singularities that exists at all inherent flaws, initial debonds, etc.

The numerical study reported here places particular emphasis on the regions very close to the bond terminations. It is in these regions that failure is generally thought to originate. Numerical stress analyses demonstrated that at regions near the bond terminus, the normal and shear stress distributions exhibited the singular behavior reported by others.

It appears fairly certain that for many "standard" lap shear specimens failure initiation is a function not so much of the reported "average" shear stress as of the induced cleavage stresses. Experimentally, higher strengths and less variability result from using thicker specimens that reduce bending in the adherends and hence reduce these cleavage stresses.

## V. TENSILE ADHESION TESTS

Another widely used method for evaluating adhesive strength might generally be classified as tensile adhesion tests. The geometries of several tensile tests for which ASTM test procedures have been written are presented in Figure 3.18. Although the test geometries seem relatively simple, it will be shown that the stresses along the bond line have a rather complex dependence on geometric factors and adhesive compressibility. Debonds can initiate at either the bond edge or bond center depending on these factors. When edge debond initiation is obtained, linear elastic failure stresses are not defined. Failure can then be quantified in terms of fracture mechanics parameters. For the conditions under which debonds initiate at the bond center, stresses are well defined and a stress failure criterion may be applied.

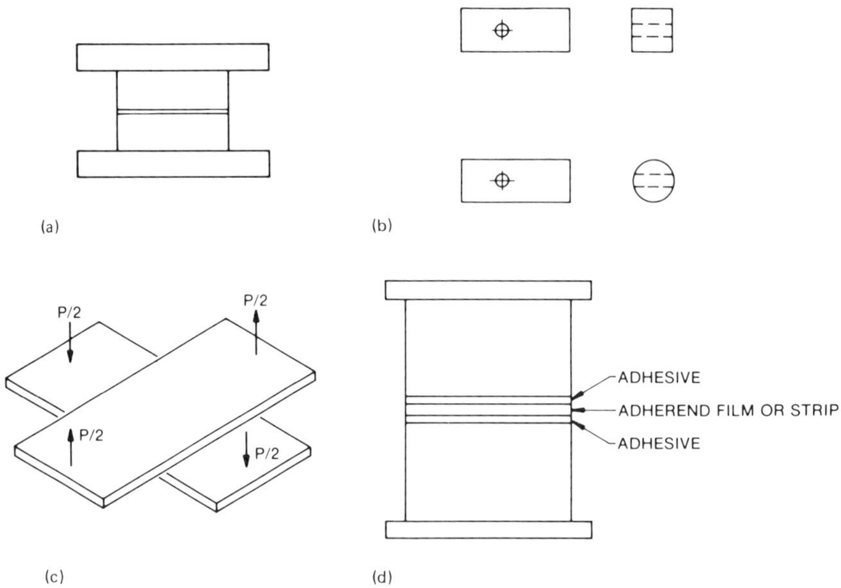


Figure 3.18 Tensile adhesion test geometries. (a) Pi tensile test ASTM D897, (b) bar and rod tensile test, ASTM D2094, (c) cross-lap tensile test, ASTM D1344, (d) sandwich tensile test, ASTM D257.

It is very difficult to load axisymmetric tensile adhesion specimens in an axisymmetric manner. Asymmetric loads are shown to reduce the bond failure load capability and to cause a large scatter in the resulting failure data.

## A. Structural Analysis—Tensile Adhesion Tests

Superficially, the geometry for standard tensile adhesion tests is deceptively simple. The result of the tensile adhesion tests normally reported by experimentalists is simply the failure load divided by the cross-sectional area of the adhesive. It has been known for some time [4-6] that because of the differences in mechanical properties of the adhesive and adherend, the stresses may become singular at the bond edges when analyzed using linear elastic analysis. Even if the edge singularity is neglected, the stress field in the adhesive is very complex. Lindsay [34], for example, analyzed the stress state in what is called a "poker chip" test. These tests, which are essentially identical to standard tensile adhesion tests, were proposed to investigate the dilatational behavior of solid propellants. Lindsay's solution is valid for tensile adhesion tests in which the adherend material can be considered rigid compared to the adhesive and the adhesive Young's modulus/bulk modulus ratio, E/k, is nearly zero ($\sim$ incompressible). However, because singular stresses are neglected, the solution is not valid in an annular region of about 1.5 adhesive thicknesses from the outer boundary. The nonsingular stresses and displacements can be written

$$\frac{\sigma_r}{E\varepsilon} \sim \frac{\sigma_\theta}{E\varepsilon} \sim \frac{3\nu}{1+\nu}\left(\frac{k}{E}\right)\left[1 - \frac{I_0(r\sqrt{E/k})}{I_0(a\sqrt{E/k})}\right] \tag{25}$$

$$\frac{\sigma_z}{E\varepsilon} \sim \frac{\sigma_r}{E\varepsilon} + \frac{1}{1+\nu}\left[1 + \frac{1}{2}\left(\frac{I_0(r\sqrt{E/k})}{I_0(a\sqrt{E/k})}\right)\right] \tag{26}$$

$$\frac{\tau_{rz}}{E\varepsilon} \sim \frac{3\nu}{1+\nu}\sqrt{\frac{k}{E}}\left[\frac{I_1(r\sqrt{E/k})}{I_0(a\sqrt{E/k})}\right] z \tag{27}$$

$$u = -3\nu\varepsilon\sqrt{\frac{k}{E}}\left[\frac{I_1(r\sqrt{E/k})}{I_0(a\sqrt{E/k})}\right](1 - z^2) \tag{28}$$

$$w = \varepsilon z \tag{29}$$

where r, z = radial and axial coordinates, respectively, nondimensionalized by the adhesive disk half-thickness (i.e., at the grip surfaces z = ±1)

$I_0$, $I_1$ = modified Bessel functions of the first kind
$a$ = ratio of disk diameter to thickness (because of the normalization with respect to thickness (a is also equal to the outer radius of the disk)
$\varepsilon$ = applied gross strain (grip separation divided by disk thickness)

Excellent agreement was found between these analytical solutions and corresponding finite-element solutions except near the bond edges, where the stresses become singular, a factor that was not incorporated into Lindsay's original treatment.

Some characteristics of the stress distribution can be deduced from boundary conditions. For example, when a tensile load is applied to rigid adherends, the boundary conditions require that both hoop strain and radial strain be zero at the bond lines. This results in the following relationship between the stress components on the bond plane:

$$\sigma_r = \sigma_\theta = \frac{\nu}{1 - \nu} \sigma_z \tag{30}$$

This equation indicates that pure hydrostatic tension is obtained for incompressible materials. The hydrostatic tension exists completely through the adhesive thickness at the bond center for relatively thin adhesive layers.

In general, complex stress distributions exist on the bond planes of these specimens, but in one limiting instance there is a simple relationship between axial stress and radius. This limiting condition occurs when Poisson's ratio is 0.5 and the diameter-to-thickness ratio is large. Stresses on the bond planes for tensile loads can then be written

$$\sigma_z = \sigma_r = \sigma_\theta = 2\bar{\sigma}\left[1 - \left(\frac{2r}{D}\right)^2\right] \tag{31}$$

where $\bar{\sigma}$ is the average applied stress (P/A), r is the radial dimension from the centerline, and D is the bond diameter. In this case, a stress concentration factor of 2 exists at the bond center. We emphasize that this equation neglects the singularity at the bond termination.

In 1963, numerical analysis techniques were used by Messner [35] to analyze stress distributions in tensile adhesion tests for adhesive Poisson's ratio values of 0.49 and 0.5 and for adhesive diameter-to-thickness (D/h) ratios from 0.5 to 100. In these analyses it was assumed that the adherends were rigid compared to the adhesive. The resulting bond stresses, shown in Figures 3.19 through 3.22, illustrate a strong sensitivity to Poisson's ratio. Since many of the nonelastomeric adhesives have a lower Poisson's ratio, we have completed a similar analysis using a Poisson ratio of 0.34. These results are presented in Figures 3.23 and 3.24.

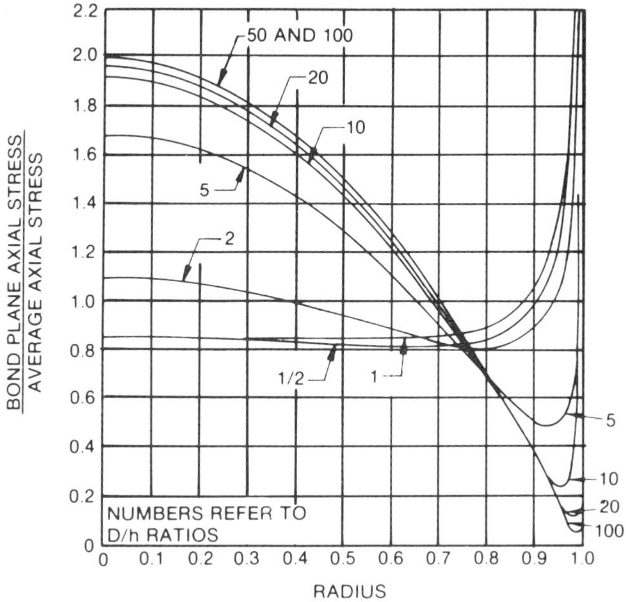


Figure 3.19 Butt-joint axial stress, ν = 0.50.

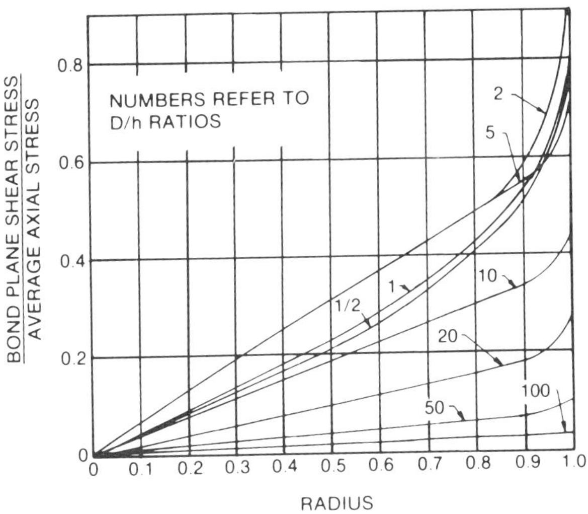


Figure 3.20 Butt-joint shear stress, ν = 0.50.

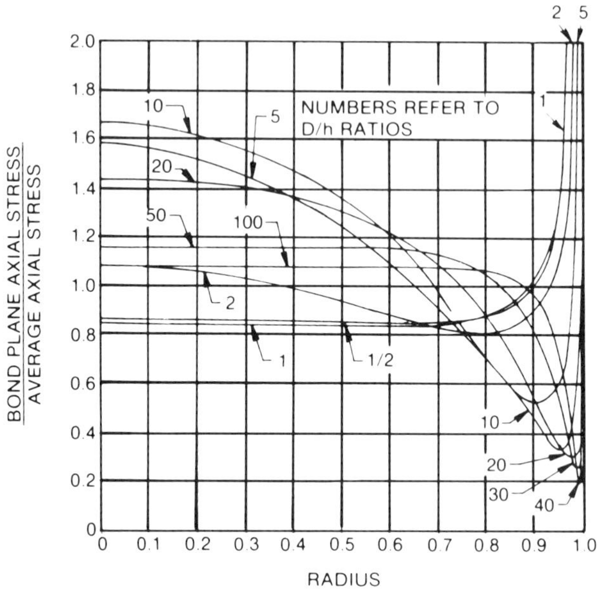


Figure 3.21 Butt-joint axial stress, ν = 0.49.

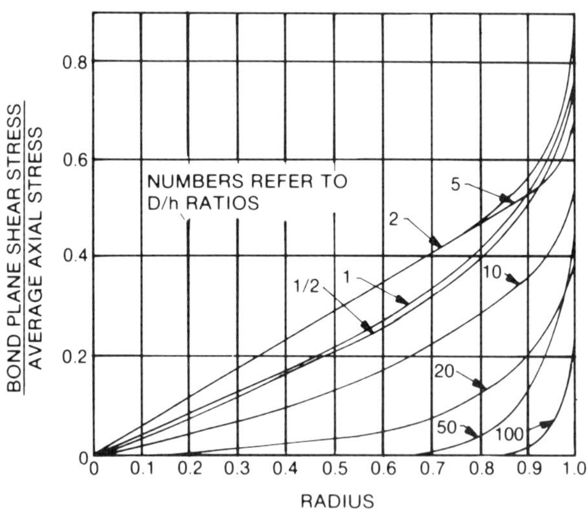


Figure 3.22 Butt-joint shear stress, ν = 0.49.

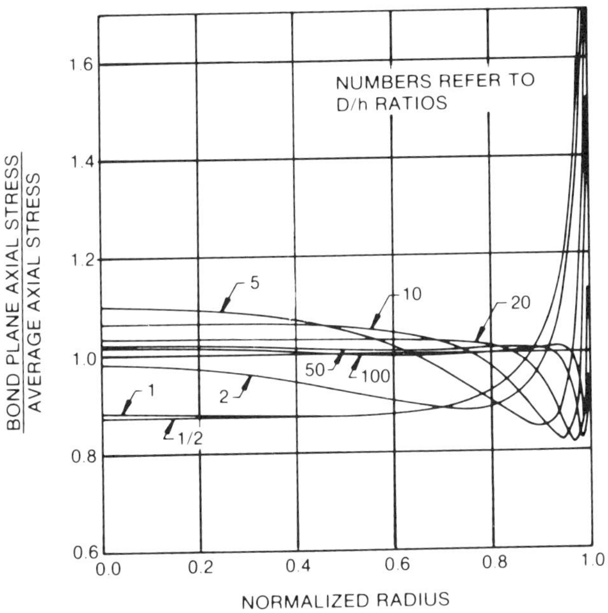


Figure 3.23 Butt-joint axial stress, ν = 0.34.

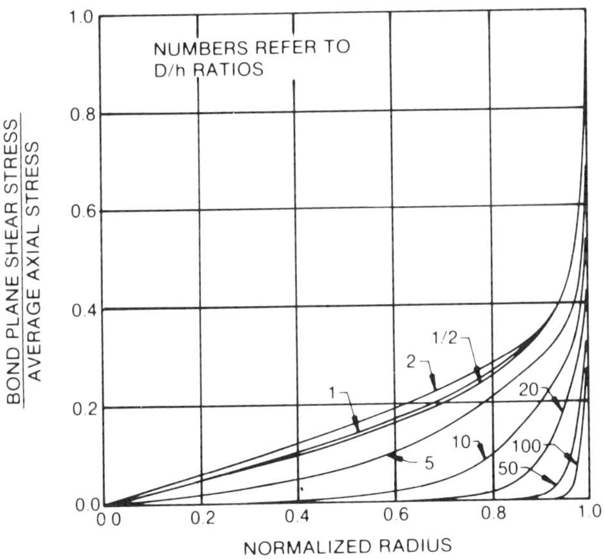


Figure 3.24 Butt-joint shear stress, ν = 0.34.

Several features are apparent from these analyses. For relatively thin bond lines using adhesives which are incompressible ($\nu$ = 0.5), the axial stresses at the bond center are a factor of 2 greater than the average stress (P/A), which agrees with Lindsay's [34] analyses and Equation (31). Furthermore, at the outer edge of the sample, the stresses at the bond line are mathematically singular. This suggests that, based on a maximum stress failure criterion, debonds should initiate at the outer boundary. However, for these thin bonds, the area affected by the mathematical singularity is very small. In a pi tensile test [bond diameter of 2.87 cm (1.13 in.) with an adhesive thickness smaller than 1.4 mm (0.056 in.)], only the outer 0.24 mm (0.010 in.) of the bond is significantly affected by the singularity. One might expect to see inherent flaws (voids, weakly bonded areas, etc.) at any location within the adhesive with dimensions greater than the entire singular region. Thus our early conjecture was that debond initiation would not be controlled by the bond edge singularity. It was hypothesized that debond would initiate at an inherent flaw near the location which produces the greatest energy release rate.

As the adhesive thickness increases compared to the bond diameter, the stress concentration factor at the bond center decreases and for sufficiently thick bonds becomes less than unity; i.e., the local axial stress at the bond center becomes less than average stress (P/A). In addition, the area affected by the edge singularity increases radically with bond thickness. In this case, debonds initiating at the outer edge of the adhesive are anticipated.

An axisymmetric finite-element computer program was used to obtain energy release rate values as a function of adhesive diameter-to-thickness ratio and flaw location for small flaws (unbonded areas). In the analyses for Figures 3.19 through 3.24, the adherends were considered rigid. However, in butt joints bonded with structural adhesives, the stiffness of the adherends may have a significant influence on the stress distribution. Therefore, in the analyses for the remainder of this section, the adherends were included in the structural analyses.

In all of the finite-element analyses presented in this chapter, the stresses are those existing at the bond line and the energy release rate values correspond to unbonded areas. These assumptions were made in order to interpret the bond line failures observed in the polyurethane-to-Plexiglas bond tests discussed later in this chapter. It is recognized that failure is often cohesive within the adhesive layer. Analyses for cohesive failure are discussed later in this section.

In the computer analyses, two sets of computations were completed: one with a small flaw (unbonded area) at the bond edge and one with a center debond. The results of these computations for an incompressible adhesive and a ratio of flaw length to bond diameter of 0.003 are shown in Figure 3.25 in terms of square root of total energy release rate $G_T$ per unit of applied force P. It appears from this figure that edge debonds of a given size become less critical (more force required to initiate debond) as the

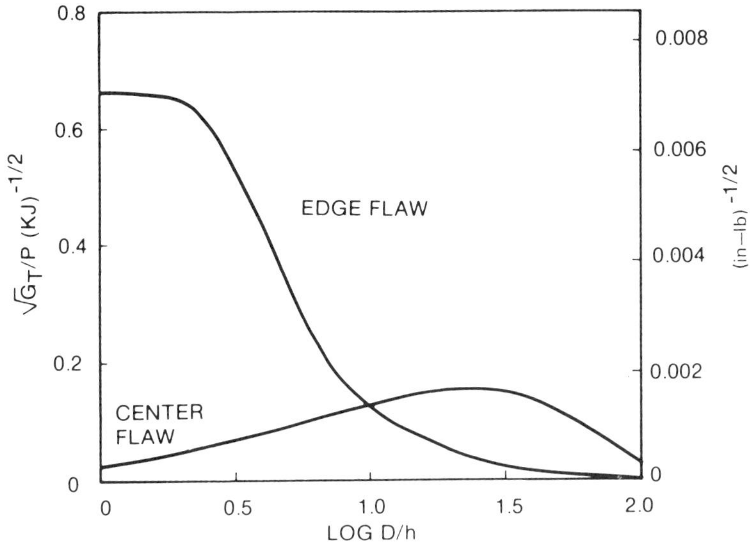


Figure 3.25 Energy release rate for small edge and center flaws in a tensile adhesion test.

diameter-to-thickness ratio increases. The opposite trend is observed for debonds initiating at the bond center. If edge and center inherent flaws are the same size and approximately 0.003 times the bond diameter, this numerical analysis predicts that debonds would propagate from the bond edge for D/h ratios less than 10 and from the bond center for D/h greater than 10. Since energy release rate is decreasing with increasing D/h up to a ratio of 10, this analysis also predicts that bond strength would increase with decreasing bond thickness. As the D/h ratio increases beyond 10, debond load is controlled by the energy release rate of inherent flaws near the bond center and bond strength is predicted to decrease with D/h. When a D/h of 25 is reached, the energy release rate curve reaches a peak and the predicted bond strength again increases with increasing D/h ratio. These trends, of course, assume that the adhesive behaves as a linear elastic material and that the inherent flaw size, adhesive strain rate, etc. are independent of bond thickness.

For adhesives with a Poisson's ratio near 0.34 (typical of epoxies), the calculated stresses are depicted in Figures 3.23 and 3.24. For normal bond lines (bond diameter-to-thickness ratio, D/H, greater than 20), the singular tensile stresses are concentrated in a very narrow region at the outer boundary. The calculated stresses through the major portion of the bond line are very nearly constant and equal to the average stress. Near the boundary there is a small region where the tensile stresses are lower than the average stress (P/A), but as the exterior edge is approached the stresses increase to very high values. As a consequence, one would anticipate that, except for the region influenced by the edge singularity, the probability of debond initiation would be roughly the same at any point along the bond line.

For thicker bonds, the area affected by the edge singularity increases significantly. Thus, at some thickness, failures are expected to initiate at the bond edge, where the linear elastic stresses are not defined, and again the average stress at failure loses significance.

Energy release rate analyses were completed for small edge and center debonds for an adhesive Poisson's ratio of 0.34. Axisymmetric finite-element analysis techniques as described above were used.

For a flaw of length 0.003 times the bond diameter, the energy release rate per unit load is larger for an edge flaw than for a center flaw whenever $1 \leq D/h \leq 320$. At a D/h value of 320, the energy release rate per unit load is approximately the same for center and edge debonds. Thus, except for extremely thin bonds, debonds are expected to initiate from bond edges whenever the adhesive has a low Poisson's ratio (0.34). The energy release rate for an edge debond increases monotonically with bond thickness. Thus bond strength would be expected to decrease monotonically as bond thickness increases.

The plot in Figure 3.26 summarizes predicted bond strength for compressible ($\nu = 0.34$) and incompressible ($\nu = 0.5$) adhesives as a function of bond diameter-to-thickness ratio for inherent flaws 0.003 times the bond diameter. The ordinate in Figure 3.26 is the ratio of the predicted failure load to the failure load for a debond diameter-to-thickness ratio of one. This figure clearly illustrates the difference in failure behavior for compressible and incompressible adhesives. Energy release rates for a bond diameter-to-thickness ratio of 20 are presented in Table 3.4 for compressible and incompressible adhesives.

Two conclusions can be drawn from these computations. First, the bond thickness at which debond initiation changes from the bond edge to the bond center decreases as more compressible adhesives are used. This might have been deduced directly by comparing stress plots of Figures 3.19, 3.21, and 3.23, since the stresses at the bond center decrease with increasing compressibility. The second conclusion is that if inherent flaw sizes and bond strengths for two different adhesives are comparable, the adhesive with the higher Poisson's ratio will have more apparent strength (breaking load) in typical tensile butt joints ($D/h = 20$).

Strain energy release calculations were also completed for center and edge flaws located at the mid-thickness of the adhesive (cohesive failure). For either compressible or incompressible adhesives ($\nu = 0.34$ or $0.5$) and an inherent flaw 0.003 times the bond diameter, the energy release rate was smaller for mid-thickness flaws than for comparable flaws at the bond line. Thus, if the critical

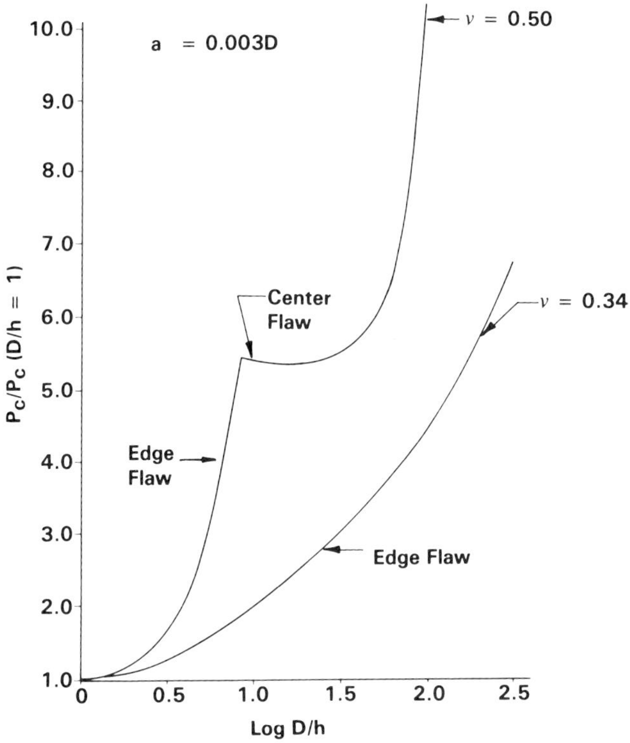


Figure 3.26 Predicted bond strength versus bond diameter-to-thickness ratio.

Table 3.4 Energy Release Rate for Edge and Center Flaws

| | $\sqrt{G_T}/P$, $(kJ)^{-1/2}$ $(in.\text{-}lb)^{-1/2}$ | |
|---|---|---|
| Poisson's ratio | Edge flaw | Center flaw |
| 0.4999 | 0.056 (0.00060) | 0.156 (0.00166) |
| 0.34 | 0.279 (0.00297) | 0.113 (0.00120) |

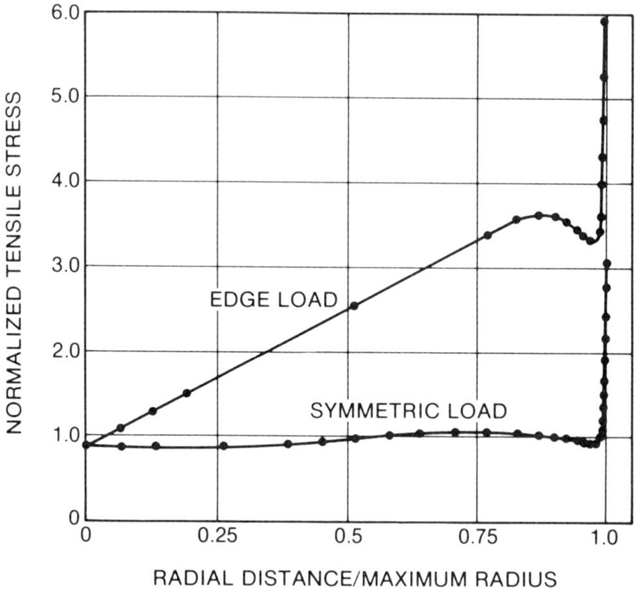


Figure 3.27 Stress in edge-loaded tensile adhesion tests.

energy release rates and inherent flaw sizes for cohesive and adhesive fracture are the same, bond failure would be expected to initiate at the bond line rather than cohesively. Failure often propagates into the adhesive after initiating at the bondline.

We emphasize that all of the above analyses assume that the load is applied in a symmetric manner about the bond center. A stress analysis was completed to evaluate the effects of noncentric loading. In the following discussion, edge loads are defined as those for which the line of action of the load passes through a point on the bond edge. A tensile adhesion specimen consisting of two steel adherends bonded with a 1.6 mm (0.060 in.) thick epoxy ($\nu = 0.34$) was modeled with an axisymmetric finite-element computer program (asymmetric loading permitted). The maximum principal stress distributions are shown in Figure 3.27. The ratio of peak to average stress away from the edge singularity and the ratio of the edge stress intensity factors (using Ref. 1 techniques) to the average applied stress for symmetric and edge loads are given in Table 3.5. Thus, whether stresses or stress intensity factors are considered the load-carrying capability of the bond is reduced by a factor of 3.3 by edge loads. As will be explained in the following section, the case of pure edge loading might not be so unreasonable as it may first appear.

Table 3.5 Effect of Edge Load on Tensile Adhesion Tests

| | Symmetric load | Edge load |
|---|---|---|
| Stress ratio, $\sigma/\bar{\sigma}$ | 1.1 | 3.6 |
| Stress intensity factor ratio, $K_I/\bar{\sigma}$, $cm^{0.30}$ ($in.^{0.30}$) | 0.39 (0.22) | 0.96 (0.73) |

B. Test Results—Tensile Adhesion Tests

To confirm some of the predictions of the analyses discussed above, a series of carefully controlled experiments was conducted. The geometry in Figure 3.28 shows the adhesion butting, "slip" grips, and Teflon spacer. The buttons are bonded together with the adhesive of interest, using the spacer to prevent flowout and to control adhesive thickness. The adhesive is then cured according to

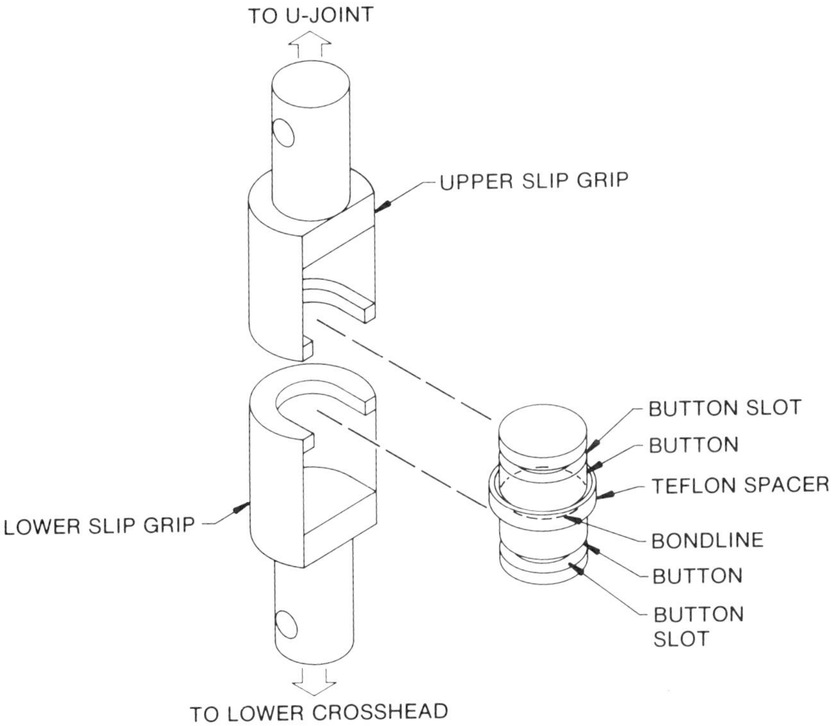


Figure 3.28 Standard tensile adhesion test apparatus.

the procedure to be evaluated. The slip grips are mounted on the testing machine, in our case an Instron machine. One of the grips is secured to the grip holder of the lower crosshead and the other by means of a U-joint to a load cell mounted in the Instron head (Figure 3.29a). The grooves of the bonded buttons are slipped into the slots of the grips for tensile testing.

The design of the grips, careful alignment of the grip holder and load cell, and use of a U-joint are all intended to apply to load in an axisymmetric manner and hence produce a very "uniform stress distribution in the adhesive." It should be noted, however, that the adhesive is generally very thin (0.025 to 1.5 mm, 1 to 60 mils). Since the smaller of these numbers is about the order of machining tolerances, accurate alignment using conventional techniques is not an easy task. For example, assume that a bond nominally 0.12 mm (5 mils) thick has one side 0.10 mm (4 mils) thick and the other 0.14 mm (6 mils) thick. Alternatively, one side of the grooves in the buttons or the slide pieces in the grip may be out of alignment by a similar amount. With such a geometry, one side of the slip grips would make contact with the slots in the tensile buttons well before the other side. As a consequence, one would expect the resulting load to be more "edge" than centric. Our experience has shown that it is difficult to maintain even this close a tolerance. The data in the first two rows of Table 3.6 show the bond strength and variability we were able to obtain using what we considered to be extraordinary care with the standard button-grip combination.

The adhesive used in this study was an unmodified two-part epoxy which was very brittle (less than 1% elongation capability) and had a history of producing a very large bond strength variability. All specimens were cured at ambient temperature for a minimum of 7 days. The reported "bond stress capability" is the peak force per unit of bond area averaged for 10 samples.

Measurements were obtained using a series of bonded buttons to determine if the button slots were in parallel planes. An alignment index was defined as the maximum difference in linear dimension between the two slots in a bonded button pair. An average alignment index in a given sample set was typically 0.076 mm (3 mils) with a range of 0 to 0.13 mm (0 to 5 mils). In a sampling of 200 bonded specimens, the largest alignment index observed was 0.3 mm (11 mils). When standard jaws (Figure 3.28) are used to grip the buttons with nonzero alignment index, the load is essentially applied at only one edge of the specimen. The resulting cleavage loads produce a low debond force and apparently a low bond strength. In addition, the data scatter is large, which we attribute to the variability in the alignment index.

To help alleviate the alignment problems, we felt it would be desirable to have a very compliant link in the load string. Our first attempt to provide such a link is shown in Figure 3.29b. This device consists of two steel disks 20 cm (8 in.) in diameter separated by and bonded to a 1.27 cm (0.5 in.) thick, 20 cm (8 in.) diameter pad of very soft rubber. The lower disk is fitting with a fixture for attachment to the Instron. The upper disk has a groove

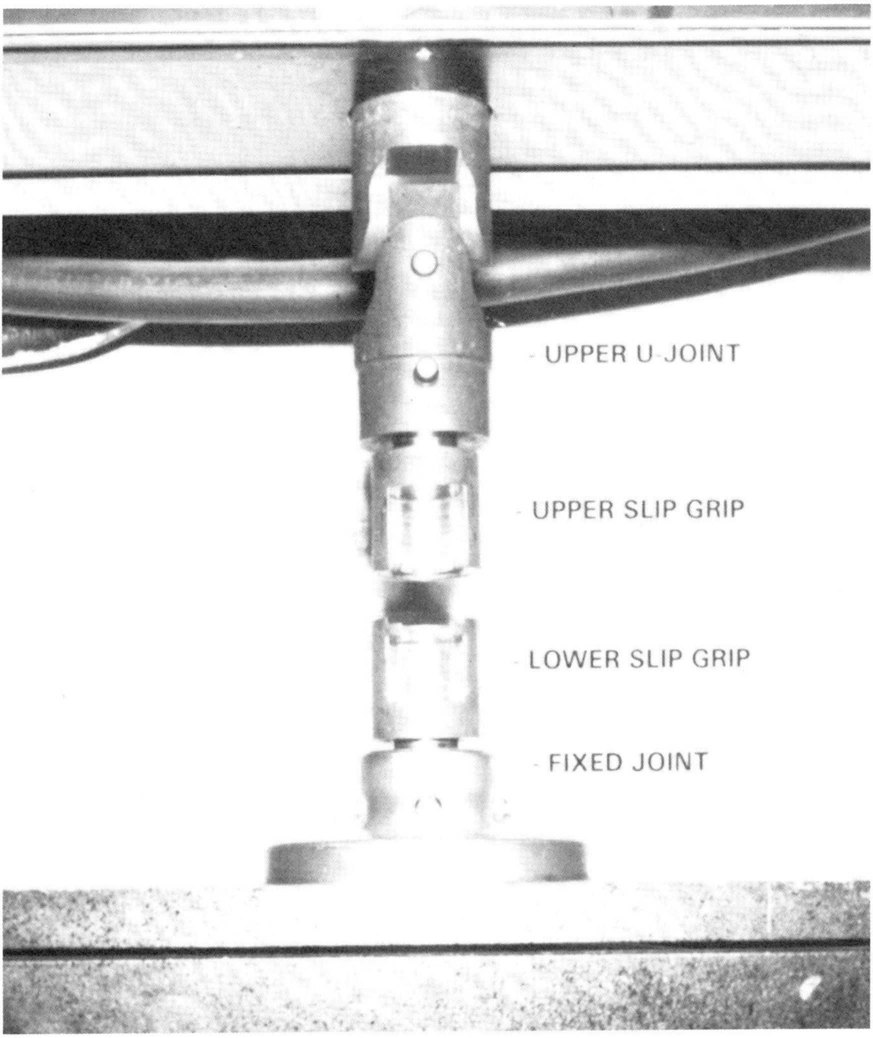


(a)

Figure 3.29 (a) "Standard" tensile button linkages. Over: (b) Modified tensile test grips. (c) Grips with ball joint added.

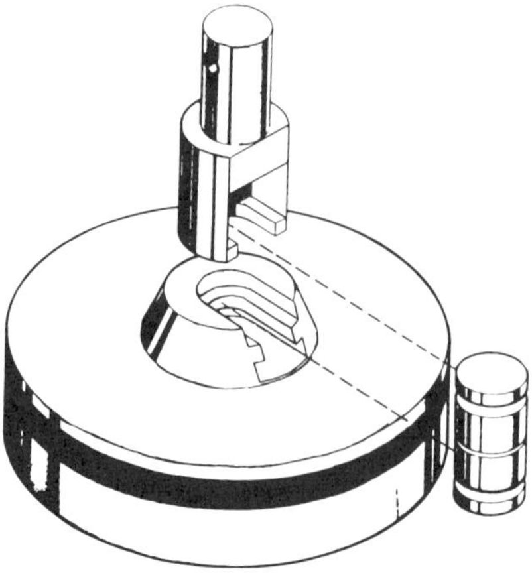

(b)

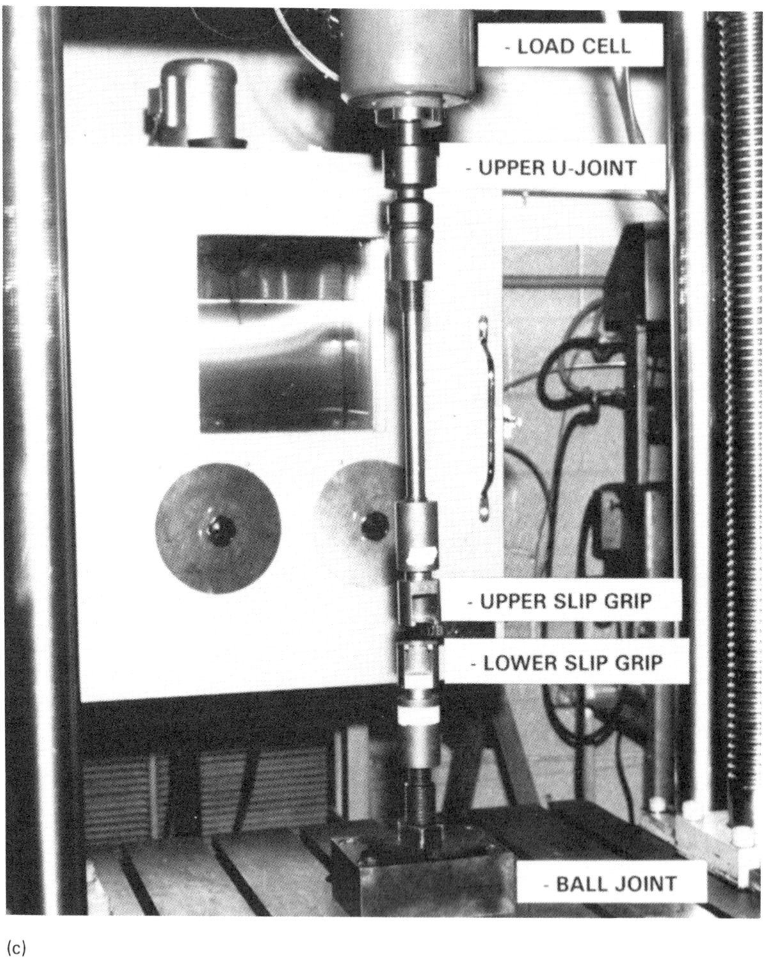
- LOAD CELL
- UPPER U-JOINT
- UPPER SLIP GRIP
- LOWER SLIP GRIP
- BALL JOINT

(c)

Table 3.6 Effect of Grip Type on Bond Strength and Variability

| Adhesive | Grip[a] | Adhesive thickness, mm (in.) | Bond stress capability, MPa (psi) | Coefficient of variability, % |
|---|---|---|---|---|
| EA934 | Standard | 1.65 (0.065) | 22.8 (3310) | 36 |
| EA934 | Standard | 0.15 (0.006) | 33.4 (4850) | 10 |
| EA934 | Modified | 1.65 (0.065) | 40.1 (5820) | 6 |
| EA934 | Modified | 0.15 (0.006) | 35.8 (5190) | 8 |
| EA913 | Standard | 0.15 (0.006) | 20.1 (2920) | 18 |
| EA913 | Modified | 0.15 (0.006) | 36.8 (5330) | 4 |

[a]See Figure 3.29.

cut to match the button slots. The top button slot is fitted into the standard grip, which in turn is attached through the U-joint to the load cell. As the sample is loaded, the rubber deforms under a very small load until the attached grip is aligned with the lower button slot. Then the remainder of the load is applied in a nearly axisymmetric manner. Shock absorbers were added to slow the rate of energy release after each bond is broken. Figure 3.30 shows the shock absorbers and the deformation of this rubber "link" at a load of approximately 19 kN (4300 lb). The average extension of the rubber at this load is about 300%. It is not our intent to suggest the use of this rubber link in standard tensile adhesion tests. It does, however, clearly demonstrate the role that moment loading plays in apparent adhesive strength reduction and data variability. Perhaps it will also motivate the adhesive community to design better testing techniques and joint designs.

The first four rows of data in Table 3.6 show the results obtained for a relatively brittle epoxy (EA934) with the standard linkages (Figure 3.29a) and with modified linkages (Figure 3.29b). Load rates were adjusted such that load time to failure was approximately the same for both grip types. The larger average failure stresses and much smaller coefficients of variation evident in the data indicate that the new grip did indeed serve its intended purpose. We feel that the lower coefficient of variation and relatively high debond load in the thin adhesive buttons tested with standard grips are due to the lower alignment index in these specimens. Similar results were obtained when the rubber-modified portion of the grip (Figure 3.29b) was simply replaced by a ball joint or a second U-joint (Figure 3.29c). The data in the last two rows of Table 3.6 were obtained using a higher-elongation adhesive (EA913).

Much data has been published [1,36], and our analyses as shown in Figure 3.26 predict an increase in tensile bond capability with decreasing adhesive thickness. Our limited data agree with this

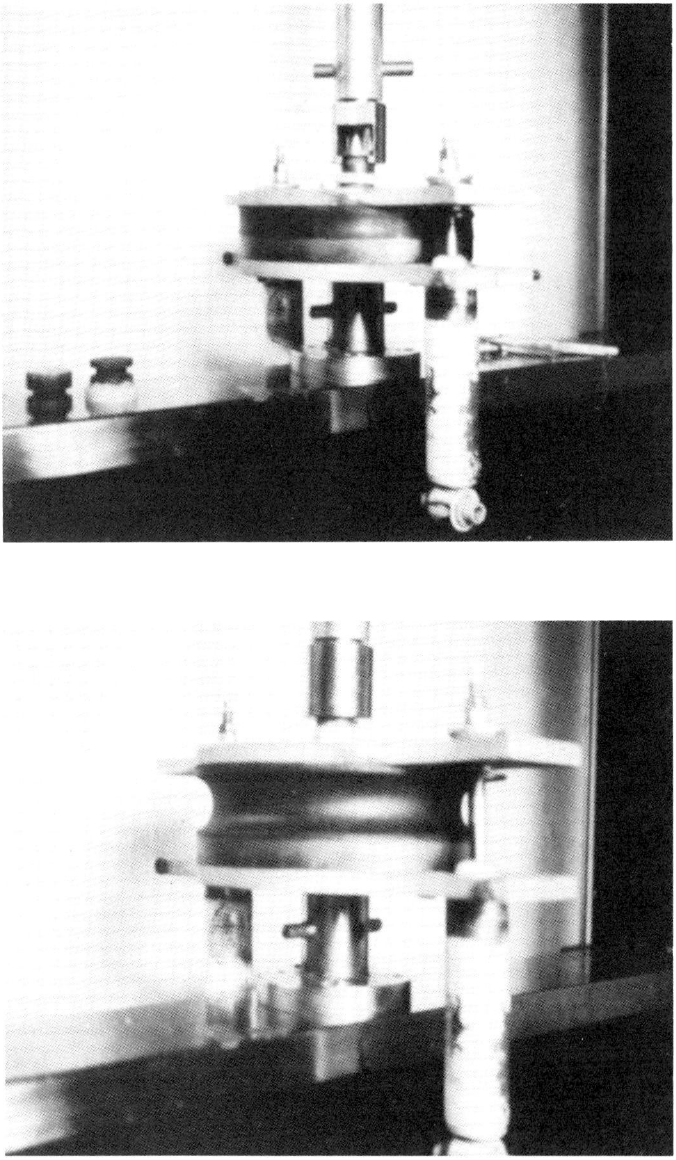

Figure 3.30 Modified tensile test load apparatus.

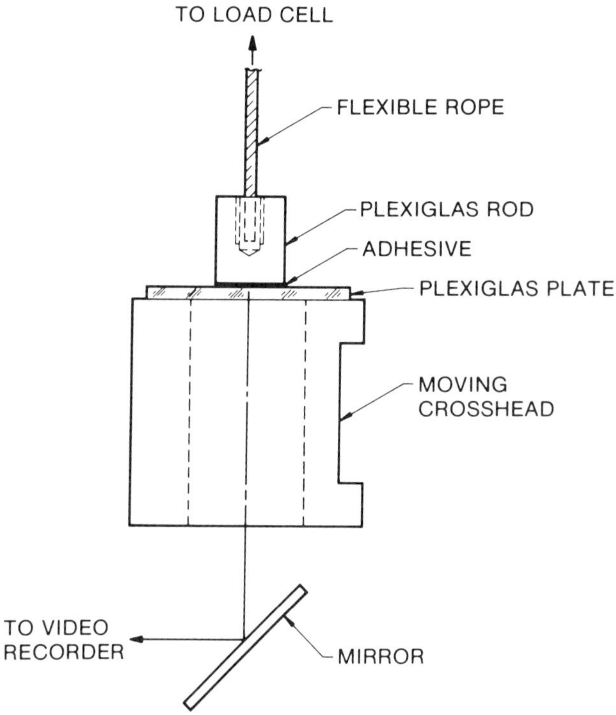


Figure 3.31 Transparent tensile bond test apparatus.

trend when standard grips are used (Table 3.6). However, when the modified grip is used, there is a slight trend in the reverse direction. The reason for the reverse trend is being investigated.

It was felt that it would be informative to explore experimentally the effect of adhesive thickness on the location of adhesive fracture initiation for comparison with the analytical predictions. For this task, "transparent" fixtures were constructed of clear acrylic (Plexiglas). A transparent polyurethane (Solithane 113, manufactured by Morton Thiokol, Inc.) was used as the adhesive to bond a 3.8 cm (1.5 in.) diameter acrylic rod to an acrylic plate. With the aid of a lathe, a center hole was accurately drilled and tapped partway into the rod. A "centric" fitting was machined and threaded to fit this hole and secured to a very flexible rope. The acrylic plate was attached to the moving crosshead of the Instron and the rope attached to the load cell, as shown in Figure 3.31. This arrangement applied a "pure" tensile loading to the rod. The rod was used as a light guide to direct light from photoflood lamps to the Solithane bond area. A mirror/television camera/video recorder system was used to observe the illuminated bond line between the Solithane and the acrylic plate.

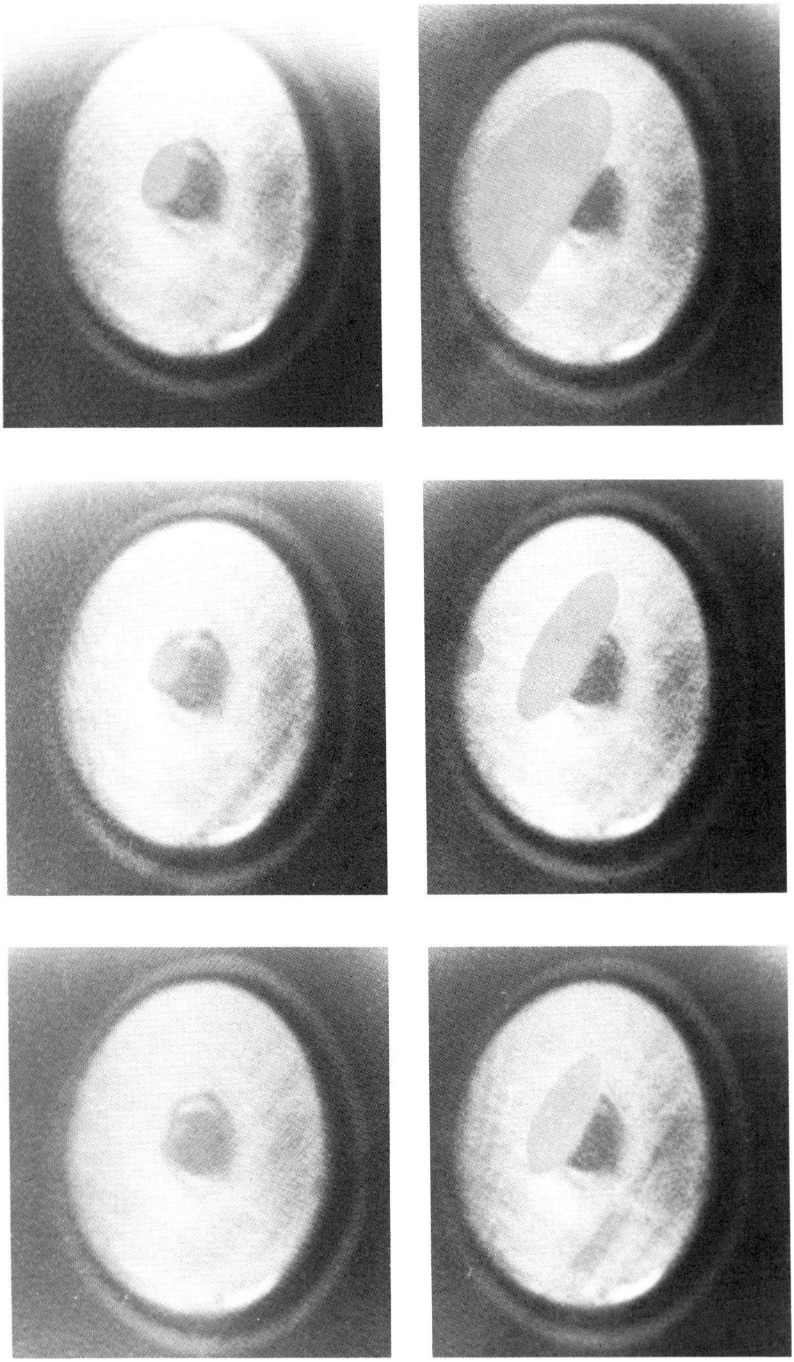

Figure 3.32 Debond initiation for relatively thin bond lines, incompressible adhesive.

The adhesive system was loaded at a rate of 1.25 mm/min (0.05 in./min). Although the actual failure took place very rapidly (time from initiation to complete debond was less than 1 second), the stop-action feature of the video recorder facilitated very accurate determination of the point of failure initiation. The growth of this initial debond region could also be followed. The Solithane was cured at 65°C (150°F). Adhesive thicknesses varying between 0.05 and 38 mm (0.002 to 1.5 in.) were tested.

The initial testing with the equipment shown in Figure 3.31 and reported in Ref. 37 produced failures in which the failure load increased with decreasing adhesive thickness up to a log D/h of 1.2 and then decreased. It was determined that the acrylic plates were bending as load was applied. This decreased the load-carrying capability of the bond greatly for thin bonds. The thickness of the acrylic plate was increased to 2.5 cm (1 in.) and the plate was reinforced with a 0.63 cm (0.25 in.) thick steel plate. The height of the acrylic rod was decreased from 51 mm (2 in.) to 32 mm (1.25 in.) to decrease load application bending loads. A complete discussion of testing is contained in Refs. 38 and 39. These changes greatly increased the failure load in thin bonds.

Figure 3.32 shows a series of photographs (taken from the television screen) of the growth in debond for an adhesive sample with a relatively thin bond line. For this and all samples with D/h greater than 17, the debonding always initiated at or near the bond line center. For samples with D/h less than 13, the adhesive failure inevitably began at the bond edge. The photographs shown in Figure 3.33 are for a sample with thick adhesive, illustrating typical edge-initiated debonds. For samples with D/h between 13 and 17, failure initiated at the bond center in some specimens and at the bond edge in other specimens. Figure 3.34 shows the experimentally observed values for the load $P_{cr}$ (circled data points) as a function of the adhesive thickness.

Three sets of five Solithane/PMMA buttons were tested using the test fixture of Figure 3.31. The following average failure loads, $P_{cr}$, were obtained:

| Adhesive thickness mm (in.) | Initial debond, mm (in.) | $P_{cr}$, N (lb) |
|---|---|---|
| 6.38 (0.251) | 3.30 (0.130) | 286 (64) |
| 6.17 (0.243) | 6.45 (0.254) | 138 (31) |
| 0.71 (0.028) | $a_0$ | 1430 (322) |

The average critical energy release rate from the first test set was 23 J/m$^2$ (0.13 in.-lb/in.$^2$). With this value of $G_c$, the failure load for the 6.45 mm (0.254-in.) flaw was predicted to be 156 N (35 lb). This value is close to the measured 138 N (31 lb) load.

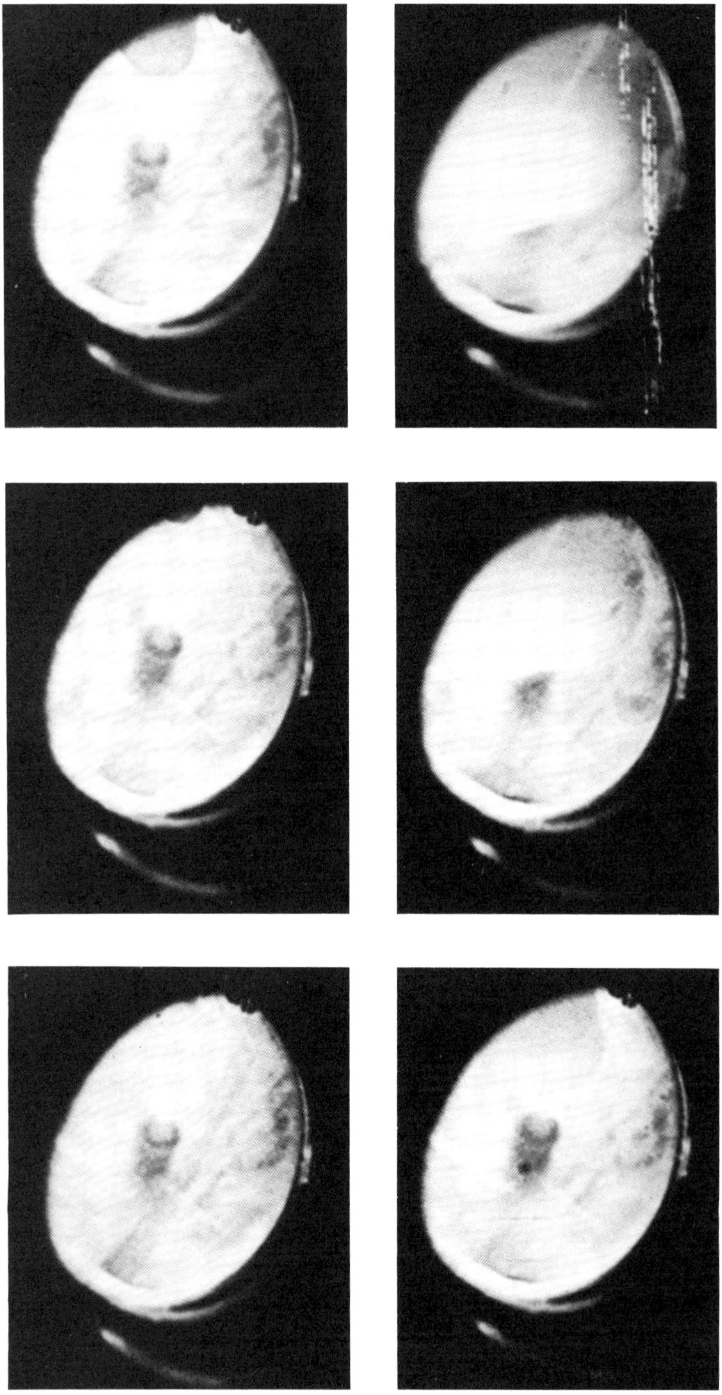

Figure 3.33 Debond initiation for relatively thick bond lines, incompressible adhesive.

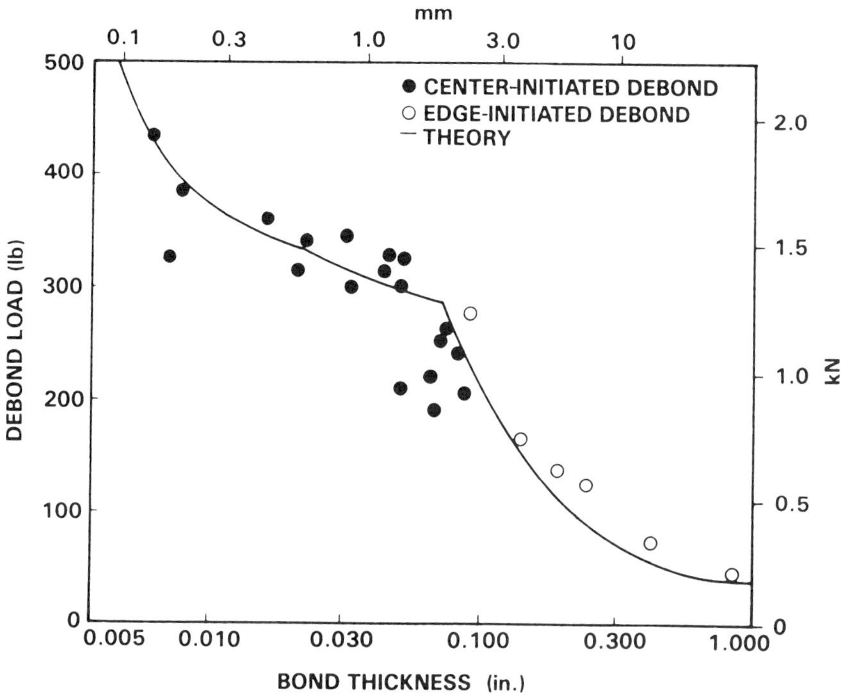


Figure 3.34 Effect of adhesive thickness on debond load—polyurethane.

The inherent flaw size $a_0$ is defined as the amount of unbond necessary to produce the proper critical energy release rate, 23 $J/m^2$ (0.13 in.-lb/in.$^2$), at the measured peak load, 1430 N (322 lb), in specimens with no initial unbond. For the Solithane-to-PMMA bond, the inherent flaw size was determined to be approximately 0.076 mm (0.003 in.).

Once $G_c$ and $a_0$ are known, the load capability of other bonded joints can be predicted. This is demonstrated by predicting the change in bond strength with adhesive thickness in butt joint tests (Figure 3.31). Adhesive thicknesses between 0.15 and 25 mm (0.006 and 1.0 in.) produced failure loads ranging from 1900 N (430 lb) for thin bonds to 180 N (40 lb) for thick bonds (Figure 3.34). Both the failure loads and debond initiation points were predicted using the curves of Figure 3.25 or 3.26 modified for a Poisson's ratio of 0.490. The predicted failure curve is superimposed on the experimental data in Figure 3.34.

The theory was extended to a nearly linear elastic brittle epoxy bonded to steel. A critical energy release rate of 32 $J/m^2$ (0.18 in.-lb/in.$^2$) was obtained for 29 mm (1.13 in.) diameter tensile buttons with an initial debond of 2.5 mm (0.1 in.) and a failure load of 7.1 kN (1600 lb). The critical load of 4.3 kN (967 lb) was then predicted for buttons with 5.1-mm (0.2-in.) initial flaws. This prediction was within 1% of the test results. An inherent flaw size $a_0$ of 0.025 mm (0.001 in.) was then determined using tensile buttons with no initial flaw. These specimens had an adhesive thickness of 1.7 mm (0.068 in.) and failed at an average of 21 kN (4760 lb). These data are summarized in the following table:

| Initial debond mm (in.) | Failure load kN (lb) | Result |
|---|---|---|
| 2.5 (0.1) | 7.1 (1600) | $G_c$ = 32 J/m$^2$ (0.18 in.-lb/in.$^2$ |
| 5.1 (0.2) | 4.3 (967) | 1% prediction error |
| $a_0$ | 21 (4760) | $a_0$ = 0.025 mm (0.001 in.) |

A series of tensile button tests was then completed where adhesive thickness ranged from 0.127 to 13.3 mm (0.005 to 0.525 in.). Average values (10 replications) from these tests are presented in Table 3.7. An attempt was made to adjust the load rate to obtain a constant load time to failure. However, the resulting load times were not constant. Thus additional testing was completed to allow an empirical correction of the first data set.

Three sets of 0.27-mm (0.060-in.) bond line tensile adhesion buttons were tested at different displacement rates to assess the effect of strain rate on tensile strength. The three displacement rates used were 0.022 mm/min (0.005 in./min), 0.22 mm/min (0.050 in./min), and 2.2 mm/min (0.500 in./min). Ten buttons were tested at each rate. The data from these tests were fit empirically by the following equation:

Table 3.7 Tensile Adhesion Tests with EA-934 Epoxy Adhesive

| Adhesive thickness, mm (in.) | Load rate, mm/min (in./min) | Failure load, kN (lb) | | |
|---|---|---|---|---|
| | | Measured | Adjusted | Predicted |
| 0.13 (0.005) | 1.27 (0.050) | 30.1 (6770) | 32.2 (7230) | 34.8 (7820) |
| 0.58 (0.023) | 1.27 (0.050) | 24.8 (5570) | 25.4 (5720) | 26.2 (5900) |
| 1.73 (0.068) | 1.27 (0.050) | 21.1 (4770) | 21.2 (4760) | 14.6 (3280) |
| 3.30 (0.130) | 1.91 (0.075) | 16.2 (3650) | 16.1 (3620) | 14.6 (3280) |
| 13.30 (0.525) | 2.54 (0.100) | 10.6 (2380) | 10.1 (2280) | 9.4 (2120) |

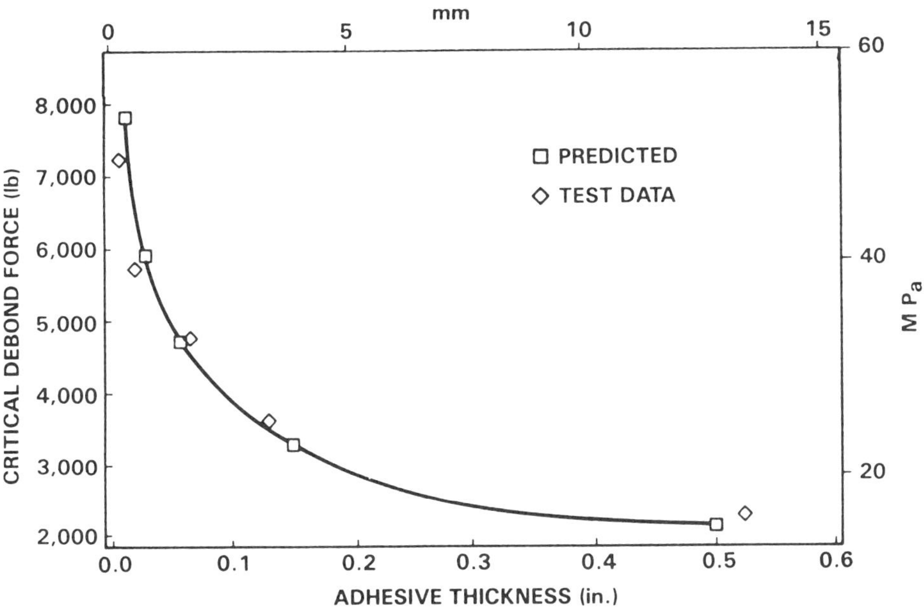


Figure 3.35 Effect of adhesive thickness on debond load—EA-934 epoxy.

$$P_{cr} = a\left(\frac{\dot{u}}{h}\right)^n \tag{32}$$

where $P_{cr}$ is the failure force and $\dot{u}$ is the displacement rate.

It was then assumed that failure force is dependent on average applied strain rate rather than displacement rate. The average strain rate is equal to the applied displacement rate divided by the adhesive thickness. Therefore, the ratio of critical loads $P_{cr1}/P_{cr2}$ for different rates and thicknesses may be expressed as

$$\frac{P_{cr1}}{P_{cr2}} = \left(\frac{\dot{u}_1 h_2}{\dot{u}_2 h_1}\right)^n \tag{33}$$

Using a reference thickness of 1.52 mm (0.060 in.) and a reference displacement rate of 1.27 mm (0.050 in./min), the test data were adjusted as shown in Table 3.7. These adjusted data are plotted as a function of bond thickness in Figure 3.35. Using the critical energy release rate of 32 J/m$^2$ (0.18 in.-lb/in.$^2$), the inherent flaw size of 0.025 mm (0.001 in.), and a curve of energy release rate as a function of bond thickness (Figure 3.36), the effect of adhesive thickness on bond strength was predicted. The predicted values presented in Table 3.7 and Figure 3.35 show very close agreement with the measured results.

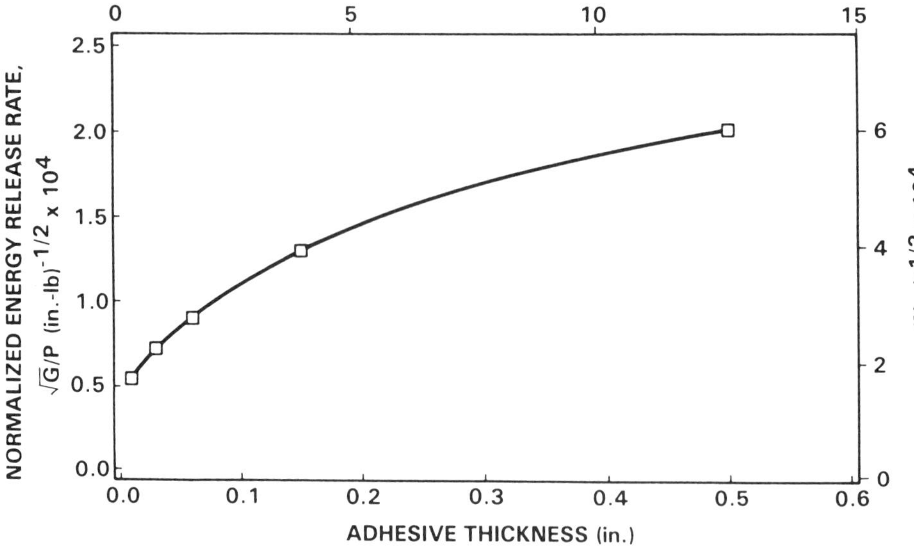


Figure 3.36 Energy release rate versus adhesive thickness; Poisson's ratio = 0.34, $a_0$ = 0.025 mm (0.001 in.).

For all tensile button tests with epoxy, edge-initiated failures are predicted because the energy release rate for an inherent flaw size of 0.025 mm (0.001 in.) is greater for an edge-initiated failure than for failures initiated internally. There was no direct experimental verification of the failure initiation point, as opaque adherends and adhesive were used. The observed failures were primarily cohesive within the adhesive layer. However, in most cases a small area of adhesive failure existed near the bond edge. Our hypothesis is that failure initiated at the adhesive-adherend interface adjacent to the bond edge. Failure then propagated into the adhesive to the bond axial center, then continued through the center of the adhesive layer.

## C. Discussion and Conclusions—Tensile Adhesion Tests

Much of the variability in the strength determined from tensile adhesion tests can be attributed to asymmetric loading of the sample and the resulting increase in local bond stresses. The asymmetric loading character comes from several sources, including variations in uniformity of sample thickness, misalignment of grips, tensile buttons, or testing machines, and slight machining tolerance differences, and is not consistent from sample to sample. Placing a more compliant (both tensile and rotationally compliant) link of soft rubber in the tensile string resulted in a 76% increase in the measured tensile strength of a 1.65 mm (0.065 in.) thick structural adhesive.

Associated with this increase in strength was a dramatic decrease in the data's coefficient of variation (from 36 to 6%). We attribute these differences to improvements in the uniformity of load application across the sample cross section.

The locus of fracture initiation in adhesives under axial tensile loading was monitored using transparent adherends and adhesives with the aid of a video camera. For a comparatively incompressible adhesive, fracture did not always initiate at the point of mathematically maximum stress, i.e., near the stress singularity at the outer boundary. Debonding occurred first and grew from the point where there would be the maximum energy release rate for growth of a small inherent flaw. For buttons with thick bond lines and incompressible adhesive, this point was at the outer boundary, but for specimens with a thin bond line, debond initiated close to the specimen center. This is, of course, consistent with the concepts of fracture mechanics.

It is recognized that at loads near failure, the response of many materials, at least locally, becomes markedly nonlinear. Nonlinear computer analyses are currently available but are still comparatively tedious and expensive to use. We were pleased with how well the simpler linear elastic fracture mechanics analyses agreed with experimental observations.

For the standard tensile buttons, serious problems are involved in controlling bond thickness, alignment, etc. These result in underestimates of the inherent bond strength and large data variability.

The combination of critical energy release rate and an inherent flaw was also successfully used to predict failure load in tensile buttons using a compressible epoxy bond system. With an epoxy adhesive (Poisson's ratio of 0.34), debonds are predicted to initiate at the bond edge except for unrealistically thin bonds.

## VI. PEEL TESTS

The general type of adhesive test to be discussed in this section is the peel test. As shown in Table 3.1, peel tests are perhaps the most popular method for testing bond strength.

Several of the standard peel tests are illustrated in Figure 3.37. All peel tests have the common characteristic that failure propagates from an initially debonded area. Thus linear elastic failure stresses are not defined for peel tests. Failure loads must be converted to critical energy release rate or an equivalent fracture mechanics parameter for use in engineering evaluations. In addition, specialized test procedures must be incorporated before the fracture mechanics parameters can be obtained. However, the peel specimen is in principle a very versatile geometry for obtaining adhesive fracture energy since various combinations of mode I and mode II loadings can be applied by varying the peel angle.

Because of the large deformations involved in peel tests, the analysis of such geometries is very difficult except under certain

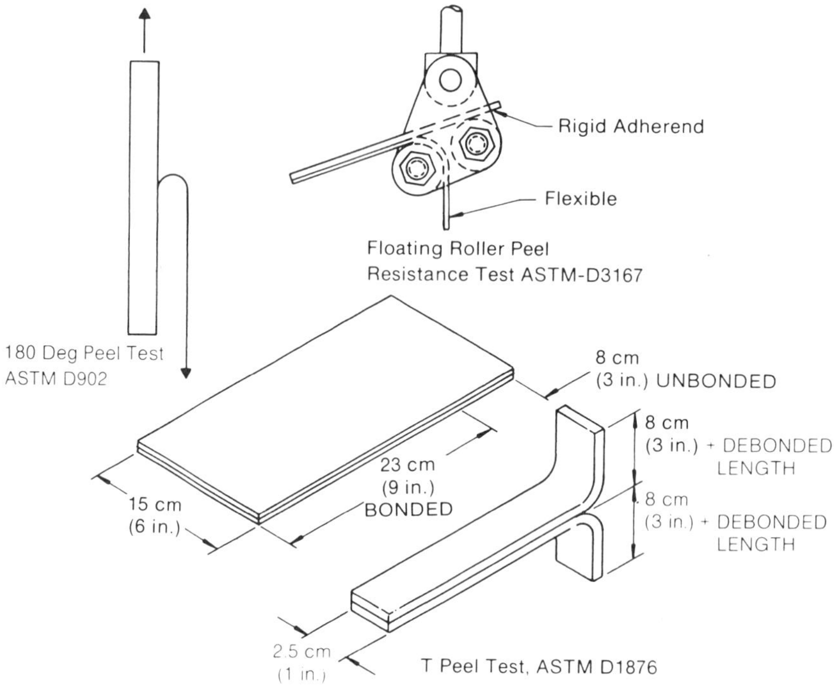


Figure 3.37 Peel test specimens.

conditions to be discussed later. Some very interesting and informative observations can be made on the basis of simplifying assumptions and approximations. Indeed, considerable useful work has been completed using peel tests. The informative work of Gardon [40] and Kaelble [41] is noteworthy. The polymer research group at Akron University under the direction of Professor A. Gent has been particularly adroit in applying peel techniques and the concepts of fracture mechanics to obtain critical information and insight into the behavior of adhesives (see Refs. 42 and 43). The quantity they call the "work of adhesion" is in essential details what we have called the "critical energy release rate." The stress analyses of Adams and Crocombe [44,45] provide added insight into the peeling mechanisms. They examined the stress distributions in peel specimens using elastic large-displacement finite-element analysis techniques.

## A. Structural Analysis—Peel Tests

Peel tests have been used for many years to compare relative strengths of adhesives and surface preparation effects [46]. These comparisons are normally made by applying a constant-displacement-rate loading to the specimen and recording applied force versus time. An average force (or force per unit bond width) after the specimen has reached a maximum load is generally reported as the peel strength.

One of the difficulties in using this specimen has been in providing accurate stress or energy values as required to convert critical load to adhesive fracture energy. Because of the large rotations involved in 90° and 180° peel tests, *normal* finite-element techniques have not been successfully applied to these specimens. This effort is presently being pursued by the authors using a finite-element program which allows large rotations and large deformations and incorporates an incremental loading procedure.

There are particular cases for which energy release rate analysis of peel specimens can be completed analytically. For instance, when a constant load is applied, the solution may be obtained in closed form with the following assumptions:

1. The material is linearly elastic.
2. The geometry, and therefore the energy stored within the region of large bending, does not change with peeled length.
3. Debond lengths are long enough that a portion of the peel specimen has no curvature.
4. The base plate is mechanically rigid (no energy is stored in the plate).

Assumption 2 is not valid for a constant-displacement load; however, if a constant force p (or constant force rate) is applied, this assumption is justified for specimens where one of the adherends is relatively flexible and debond lengths are relatively large. For example, consider two regions of the specimen which contain stored energy as load is applied. The first region contains the bonded portion and the part of the unbonded portion for which the curvature is nonzero (Figure 3.38). The second region contains the unbonded portion having zero curvature. When the initial debond length is sufficient to allow both regions to exist, the curved portion of the specimen will have the same curvature, and hence the same stored energy, independent of debond length (constant applied force). Note again that the first assumption is that the material is linearly elastic at all points; i.e., at no point has the yield stress been exceeded.

The energy stored per unit length in the zero curvature region is not affected by peeled length. Therefore, the total change in stored energy is that stored in the zero curvature portion after debond propagation of $\Delta A$ minus the energy in that section prior to debond propagation, or

$$\Delta U = \frac{p^2 \Delta a}{2AE} \tag{34}$$

If the force is applied at an angle $\theta$ to the plane of the specimen (Figure 3.38), the applied load will be displaced a distance $\Delta a - \Delta a \cos \theta$ in the direction of the applied load as the debond

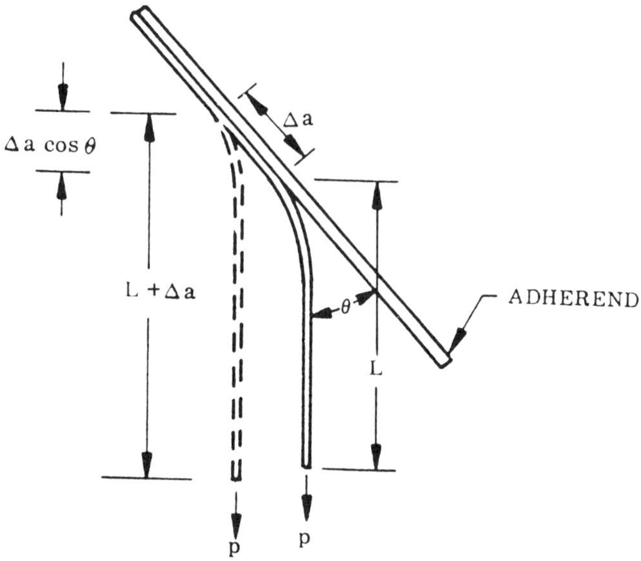


Figure 3.38 Peel test specimen.

propagates a distance $\Delta a$. In addition, the load will move a distance $\Delta a$ because of additional straining of the debonded segment. The increase in work input is therefore

$$\Delta W = p(\Delta a - \Delta a \cos\theta) + p(\varepsilon\,\Delta a) = p\,\Delta a(1 + \varepsilon - \cos\theta) \qquad (35)$$

where $\varepsilon = p/AE$ is the strain in the zero curvature portion of the peeled strip. The total energy lost in propagating a flaw through a distance $\Delta a$ with a specimen of width b is

$$\Delta\Gamma = \Delta a\, bG \qquad (36)$$

The total energy balance for the critical load is represented by the equation

$$\frac{\Delta U}{\Delta a} - \frac{\Delta W}{\Delta a} + \frac{\Delta\Gamma}{\Delta a} = 0 \qquad (37)$$

Substituting the above relations for $\Delta U$, $\Delta W$, and $\Delta\Gamma$ at the load required to propagate a debond ($p_c$) gives

$$\frac{p_c^2}{2AE} - p_c(1 + \varepsilon - \cos\theta) + bG_c = 0 \qquad (38)$$

or, solving for $G_c$,

$$G_c = \frac{p_c}{b}\left(1 + \varepsilon - \cos\theta - \frac{p_c}{2AE}\right) = \frac{p_c}{b}(1 + \varepsilon/2 - \cos\theta) \tag{39}$$

It is noted that for small extensions ($\varepsilon << 1$) in a 90° peel test, $\cos\theta = 0$ and Equation (39) reduces to

$$G_c \sim \frac{p_c}{b} \tag{40}$$

For a 0° peel test, $\cos\theta = 1$ and Equation (39) reduces to

$$G_c \sim \frac{\varepsilon p_c}{2b} = \frac{p_c^2}{2bAE} \tag{41}$$

## B. Testing of Peel Specimens

Since debond takes place slowly in many adhesives, it is often very difficult to determine the precise load at which debond starts to propagate. Detection of debond initiation in peel tests is extremely important since debonds can propagate at loads much lower than the maximum peel load. This point is demonstrated with test data determined using transparent adhesives and adherends so that the debond initiation load (time) could be monitored visually.

The results of two adhesive systems are presented. In the first, Trycite (Dow trade name for a polystyrene film) was bonded to an acrylic plate with an epoxy system. In the second system, a polyurethane (Solithane 113, as manufactured by Morton Thiokol, Inc.) was cast directly onto an acrylic plate.

In the Solithane/PMMA system, debond initiated long before the maximum load was reached. To study the phenomenon further, one specimen was cast with 13 cm (5 in.) of the specimen bonded initially. A constant displacement rate was applied 90° to the PMMA. As the displacement was applied, the load at debond propagation and the maximum load were noted. As soon as the maximum load was reached, the load was removed and the specimen repositioned for the next test. In this manner, 18 data points were obtained before the debond had propagated the length of the specimen.

The displacement rate was changed between tests so that a total of nine rates were obtained in a random order (one to three tests at each of nine displacement rates). The test machine crosshead rate (XHS) ranged from 0.064 to 51 cm/min (0.025 to 20 in./min). The data from these tests are plotted in Figure 3.39. Although the data exhibit some scatter, due in part to the small number of tests at each rate, their trend is clear, i.e., an increase in both the fracture force $P_{fracture}$ and the maximum force $P_{max}$ with load rate. The increase of mean peeling force with crosshead speed has been noted by many researchers (for example, see Figure 8-23 of Ref. 47). Several authors have reported a peak in the mean peel force curve with the loading rate, as shown later for the epoxy bond system.

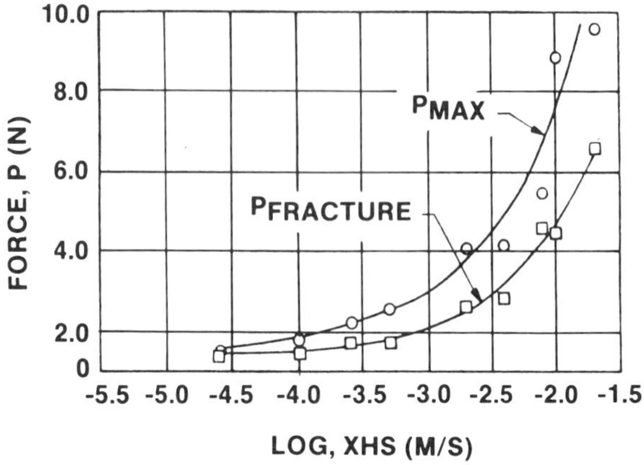


Figure 3.39 Rate dependence of Solithane-to-PMMA peel test (constant displacement rate).

The ratio of fracture force to maximum peel force was obtained from the data of Figure 3.39. This ratio is plotted as a function of crosshead speed in Figure 3.40. The ratio has little variation over the three decades of crosshead speed tested. These data indicate that, on the average, Solithane-to-PMMA debond is initiated at 54% of the maximum peel force for all crosshead rates tests.

A second study was performed for the epoxy system. The results are shown in Figure 3.41. The ratio of fracture force to maximum force for this system is again constant over the range of crosshead speeds tested (over two decades). However, the ratio of fracture force to maximum peel force is now 84% as compared

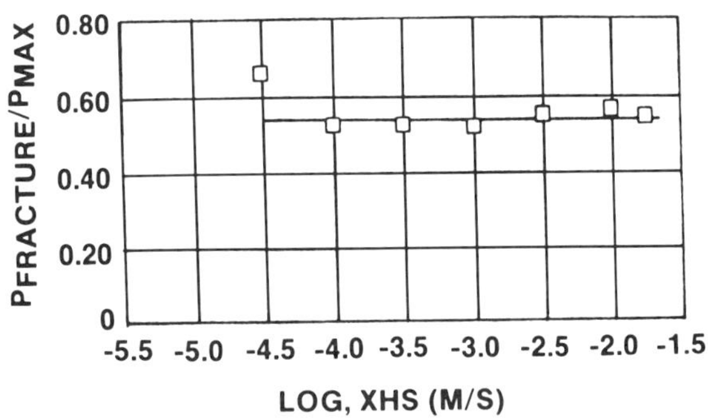


Figure 3.40 Ratio of fracture force to maximum force for Solithane-PMMA peel test.

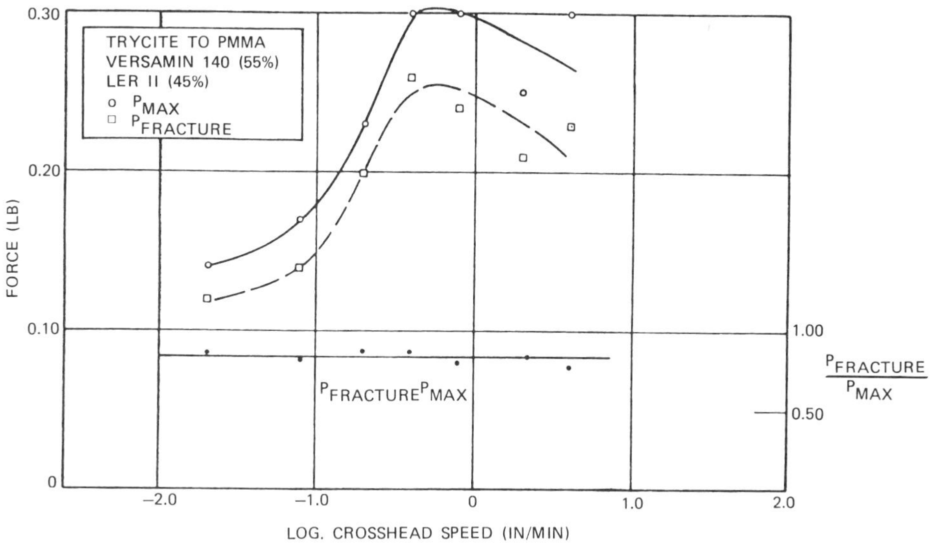


Figure 3.41 Ratio of fracture force to maximum force for an epoxy bond system.

with 54% for the Solithane. There was no apparent indication of the debond initiation force for either system on a force-time trace.

The adhesive fracture energy values for the Solithane system were then obtained at a constant applied force rate. These data were obtained for seven specimens, each cast from a different mix of Solithane at 1-week intervals. Equation (35) was used to obtain $G_c$ values from critical load. Between 10 and 13 values were obtained for each test specimen. The data scatter for values obtained within one specimen was relatively small, with an average standard deviation of 8% of the mean value. However, a slightly larger deviation (10%) is noted in the average specimen-to-specimen values. The total of 79 tests for the seven test specimens had an average $G_c$ value of 32.2 J/m$^2$ (0.184 in.-lb/in.$^2$) and a failure time of 0.60 min.

These data compare well with those reported in Ref. 43 for a blister test. When a mode I loading is applied to a Solithane/PMMA blister test system, a $G_c$ value of 30.0 J/m$^2$ (0.17 in.-lb/in.$^2$) is obtained at a load time of 0.65 min.

Thus, the $G_c$ values obtained for the two specimens are in close agreement. There are at least two possible causes of the slight difference. Elimination of either effect would bring the results into even closer agreement. The first effect is that caused by loading mode on the "apparent" $G_c$ values. The blister test data of Ref. 1 were extrapolated for mode I, while some mode II loading (amount not determined in the present study) is present in the 90°

peel tests. Since mode II loadings generally provide higher apparent $G_c$ values than mode I, a higher $G_c$ value may be expected in the peel specimens. The second effect is that of loading time. Since the blister specimens were loaded slightly more slowly than the peel specimens, a somewhat lower $G_c$ value is expected.

### C. Discussion and Conclusions—Peel Tests

The peel test is a versatile test that has been shown in at least one case to produce adhesive fracture energy values that are consistent with independently measured ones. We have demonstrated the possibility of now being able to interpret the masses of peel test data in terms of the fracture energy instability analysis. Additional effort will be required to demonstrate the capabilities and limitations of these methods for intermediate angles of pull and various material behaviors. In addition, methods of verifying the initial debond load for opaque systems are necessary to fully utilize the capabilities of this specimen. It is possible that acoustic emission, ultrasonic echo techniques, or other methods developed for nondestructive evaluation (NDE) may fruitfully be used for this purpose. Preliminary results [48] with acoustic emission have demonstrated the potential for using this type of equipment.

The authors again draw the reader's attention to the informative studies of Gent et al. (previously cited) in which the peel test has been used in conjunction with the concepts of fracture mechanics. Among other aspects, they have provided insight into time-temperature shift responses of adhesives [49] and the role of plasticity in adhesive bonds [50].

## VII. CONCLUSIONS

The three most common types of adhesive test geometries (i.e., lap shear, tensile adhesion, and peel) were discussed. Detailed numerical stress analyses of the lap shear and tensile adhesion tests were given. It was shown that the bond stresses are more complex than commonly assumed. Although test results are generally reported in terms of average stress at failure, the value of average stress may have little bearing on the actual failure of the adhesive. Failure generally initiates at points of high local stresses or, more properly, at points where large energy release rates are associated with crack or debond "growth."

In the case of the lap shear specimen, failure was shown to be more closely related to the induced mode I (tensile) stresses than to the shear stresses, although what is normally reported is the average bond shear stress for these tests.

In tensile adhesion tests the comparatively low measured adhesive strength and large data variability may often be attributed to slight sample nonuniformity, grip misalignment, etc. Inserting a compliant link in the load string results in a substantial increase in apparent bond strength and a significant decrease in the data's coefficient of variation.

For all three test geometries, application of the concepts of fracture mechanics is informative and provides insight into what might otherwise be puzzling observed phenomena. For example, in the case of thin adhesive bonds in tensile tests, crack initiation was found *not* to occur near the outer edge, where the calculated stresses were largest (i.e., elastically singular point). Rather, it occurred near the sample center, where the "energy release rate" from small inherent flaws was found to be larger even though the calculated stresses were smaller. Adhesive fracture mechanics also has the advantage that it should facilitate the use of results from standard tests in designing practical joints with geometries that may not exactly duplicate the test geometry.

## NOMENCLATURE

| | |
|---|---|
| a | Debond length |
| A | Area |
| b | Width of peel test specimen bonds |
| c | Subscript for critical load value (value for debond propagation) |
| D | Diameter |
| E | Equivalent Young's modulus |
| $E_d$ | Energy dissipated |
| G | Energy release rate |
| $G_c$ | Energy release rate required to propagate a fracture |
| $G_I$, $G_{II}$, $G_{III}$ | Mode I, II, and III energy release rate |
| h | Bond thickness |
| $I_0$, $I_1$ | Modified Bessel functions of the first kind |
| k | Bulk modulus |
| K' | Intercept value of stress intensity |
| $K_I$, $K_{II}$, $K_{III}$ | Mode I, II, and III stress intensity factors |
| n | Exponent describing geometry and material factors |
| p | Applied force |
| r, $\theta$ | Polar coordinates |
| u, v, w | Components of displacement |
| U | Strain energy |
| W | Work |

| | |
|---|---|
| z | Axial coordinate |
| $\varepsilon_{ij}$ | Strain components |
| $\nu$ | Poisson's ratio |
| $\bar{\sigma}$ | Average bond line tensile stress |
| $\sigma_{ij}$ | Normal stress components |
| $\tau_{ij}$ | Shear stress components |
| $\bar{\tau}$ | Average bond line shear stress |
| $\Gamma$ | Energy required to propagate a flaw |
| $\beta$ | Included material angle |
| $\lambda$ | Singularity exponent |

## ACKNOWLEDGMENTS

The authors appreciate the assistance of G. Sharon, K. C. Grammol, and S. H. Cardall, who helped obtain portions of the data included in this chapter. They also express appreciation to Professor M. L. Williams, who first introduced them to research in adhesive failure and has made many useful suggestions. The University of Utah portion of the research was supported by the National Science Foundation, polymer program DMR 79-25390.

## REFERENCES

1. G. P. Anderson, S. J. Bennett, and K. L. DeVries, "Analysis and Testing of Adhesive Bonds," Academic Press, New York, 1977.
2. I. N. Sneddon, The distribution of stress in the neighborhood of a crack in an elastic solid, *Proc. R. Soc. London Ser. A, 187*, pp. 229-260 (1946).
3. M. L. Williams, On the stress distribution at the base of a stationary crack, *Trans. ASME, J. Appl. Mech.*, Vol. 24, No. 1, pp. 109-114 (1957).
4. V. L. Hein and F. Erodogan, Stress singularities in a two-material wedge, *Int. J. Fract. 7*, 317 (1971).
5. M. L. Williams, *Int. J. Fract. Mech., 1*(4), 202 (1966).
6. E. H. Dill, A. L. Deak, and W. F. Schmidt, in "Handbook for the Structural Analysis of Solid Propellants," CPIA Publ. 214, Silver Spring, MD, 1971, Chap. 5.
7. A. A. Griffith, *Proc. lst Int. Congr. Appl. Mech.*, 55 (1924).
8. E. Orowan, *Rep. Prog. Phys., 12*, 185 (1949).
9. G. R. Irwin, *Trans. ASME, J. Appl. Mech.*, Vol. 24, No. 3, pp. 361-364 (1957).

10. E. F. Rybicki and M. F. Kanninen, A finite element calculation of stress intensity factors by a modified crack closure integral, *Eng. Fract. Mech. 9*, 921 (1974).

11. G. P. Anderson, K. L. DeVries, and M. L. Williams, The influence of loading direction upon the character of adhesive debonding, *J. Colloid Interface Sci.*, *47*(3), 600 (1974).

12. G. C. Trantina, Combined mode crack extension in adhesive joints, *J. Compos. Mater.*, *6*, 371 (1972).

13. W. S. Johnson and P. D. Manzalgire, Influence of the resin on interlaminar mixed mode fracture, NASA technical memorandum 87571 (1985).

14. O. Volkersen, Die Nietraftverteilung in Zugbeanspruchten Nietverblendugen mit Knastaten Laschenquerschntlen, *Luftfahrt forsch.*, *15*, 41 (1938).

15. M. Goland and E. Reissner, The stresses in cemented joints, *J. Appl. Mech.*, *11*, 17 (1944).

16. F. J. Plantema, De Schuifspanning in eme Lijmnaad, Rep. M1181, Nat. Luchtvaartlaboratorium, Amsterdam (1949).

17. G. R. Wooley and D. R. Carver, *J. Aircr.*, *8*(19), 817 (1971).

18. R. D. Adams and N. A. Peppiatt, Stress analysis of adhesive-bonded lap joints, *J. Strain Anal.*, *9*(3), 185 (1974).

19. P. S. Cooper and J. W. Sawyer, A Critical Examination of Stresses in an Elastic Single Lap Joint, NASA Tech. Rep. 1507, NASA Scientific and Technical Information Branch (1979).

20. S. Amijima, A. Yoshida, and T. Fujii, in *Proceedings of the Second International Conference on Composite Materials*, Toronto, Canada, Metallurgical Society of AIME, New York, 1978.

21. J. K. Sen and R. M. Jones, Stresses in double lap joints bonded with a viscoelastic adhesive: Part I. Theory and experimental corroboration, *AIAA J.*, *18*(10), 1237 (1980).

22. J. K. Sen and R. M. Jones, Stresses in double lap joints bonded with a viscoelastic adhesive: Part II. Parametric study and joint design, *AIAA J.*, *18*(11), 1376 (1980).

23. W. N. Sharpe and T. J. Muha, Comparison of theoretical and experimental shear stress in the adhesive layer of a lap joint model, in *Proceedings of the Army Symposium on Solid Mechanics*, Army Materials and Mechanics Research Center, Watertown, Mass., AMMRC-MS-75-8 (1974), p. 23.

24. W. J. Renton, The symmetric lap-shear test—what good is it? *Exp. Mech.*, *16*(11), 409 (1976).

25. T. R. Guess, R. E. Allred, and F. P. Gerstle, Comparison of lap shear test specimens, *J. Test. Eval.*, *5* (2), 84 (1977).

26. B. H. River, A method for measuring adhesive shear properties, *Adhes. Age*, *24*(12), 30 (1981).

27. A. N. Gent, Strength of adhesive bonds—plastic and viscoelastic effects, *Proc. 46th National Colloid Symposium*, Amherst, Mass., 1972.

28. W. D. Bascom and R. L. Cottington, Effect of temperature on the adhesive fracture behavior of an elastomer-epoxy resin, *J. Adhes., 1*, 333 (1976).

29. S. Mostovoy and E. J. Ripling, *J. Appl. Polym. Sci., 15*, 661 (1971).

30. K. L. DeVries, M. L. Williams, and M. D. Chang, A Fracture Mechanics Analysis of Adhesive Failure on a Single Lap Shear Joint, Preprint, Society for Experimental Stress Analysis, Westport, Conn.; Cleveland Meeting, 1972.

31. R. D. Adams and N. A. Peppiatt, Effect of Poisson's ratio strains in adherends on stresses of an idealized lap joint, *J. Strain Anal., 8*, 134 (1973).

32. A. D. Crocombe and R. D. Adams, Influence of the spew fillet and other parameters on the stress distribution in the single lap joint, *J. Adhes., 13*, 141 (1981).

33. S. S. Wang and J. F. Yau, Interface cracks in adhesively bonded lap-shear joints, *Int. J. Fract., 19*, 295 (1982).

34. G. H. Lindsay, Triaxial fracture studies, *J. Appl. Phys., 38*, 4843 (1967).

35. A. M. Messner, Stress Distribution in Poker Chip Tensile Specimens, Bull. of the Working Group on Mechanical Behavior, CPIA Publ. No. 27 (1963).

36. K. Ikegami and K. Kamiya, Effect of flaws in the adhering interface on the strength of adhesive-bonded butt joints, *J. Adhes., 24*, 1 (1982).

37. G. P. Anderson, K. L. DeVries, and G. Sharon, Evaluation of tensile tests for adhesive bonds, in "Delamination and Debonding of Materials," ASTM STP 876 (W. S. Johnson, ed.), American Society for Testing and Materials, Philadelphia, 1985, pp. 114-134.

38. K. C. Grammol, Debonding Characteristics of Incompressible Adhesives in Tension, M.S. thesis, Mechanical Engineering Department, University of Utah, Salt Lake City (1934).

39. K. L. DeVries, K. C. Grammol, and G. P. Anderson, Analysis of adhesive tensile test methods, *Conference Proceedings of the Society of Plastic Engineers*, April 1985, pp. 262-266.

40. J. L. Gardon, Peel adhesion, I. Some phenomenological aspects of the test, *J. Appl. Polym. Sci., 7*, 643 (1963).

41. D. H. Kaelble, Theory and analysis of peel adhesion: Mechanisms and mechanics, *Trans. Soc. Rheol., 3*, 161 (1959).

42. A. N. Gent and G. R. Hamed, Peel mechanics, *J. Adhes., 7,* 91 (1975).

43. A. N. Gent and G. R. Hamed, *J. Appl. Polym. Sci., 21,* 2817 (1977).

44. A. Crocombe and R. D. Adams, "Adhesion and Adhesives: Science, Technology and Applications," Plastics and Rubber Institute, London, 1980, p. 10.1.

45. R. D. Adams and A. Crocombe, *J. Adhes., 12,* 127 (1981).

46. J. J. Engel and R. N. Fitzwater, in "Adhesion and Cohesion," (T. Weiss, ed.), Elsevier, Amsterdam, 1962, p. 89.

47. J. L. Gardon, in "Treatise on Adhesion and Adhesives," (R. Patrick, ed.), Marcel Dekker, New York, 1967, p. 267.

48. K. L. DeVries and G. P. Anderson, Analysis and design of adhesive-bonded joints, Lecture Series No. 102—Bonded Joints and Preparation for Bonding, AGARD-LS-102, NATO (1979).

49. A. N. Gent and R. P. Petrich, Adhesion of viscoelastic materials to rigid substrates, *Proc. R. Soc. London Ser. A, 310,* 433 (1979).

50. A. N. Gent, Energy dissipation in stretching filled rubbers, *J. Appl. Polym. Sci., 18,* 1397 (1974).

# Four

# The Fracture of Epoxy and Elastomer-Modified Epoxy Polymers

W. D. BASCOM *Hercules Inc. Bacchus Works, Magna, Utah**

D. L. HUNSTON *National Bureau of Standards, Gaithersburg, Maryland*

## I. INTRODUCTION

The fracture behavior of epoxy-based polymers in bulk, as adhesives, and as matrix resins in fiber-reinforced composites has assumed considerable importance as these materials are increasingly used for structural components in aircraft, automobiles, ships, and housing. In many instances the component is load bearing so that its strength, stiffness, and toughness are critical to the function and reliability of the structure.

About 25 years ago it was recognized that in order to design adhesively bonded structures there was a need to develop reliability failure criteria for adhesive joints. The peel and lap-shear tests normally used to characterize adhesive strength are too dependent on specimen geometry to be useful for design purposes. Instead, it was judged that fracture toughness is the appropriate criterion for adhesive failure for essentially two reasons. First, the adhesive resins are generally brittle materials that fail by crack initiation and propagation rather than by yielding and flow. Second, even if the adhesive resin does exhibit yielding, it does so within the confines of a thin bond line which is usually small compared to the overall joint dimensions, so that the joint as a whole fails in a linear elastic, "brittle," fashion. Consequently, a program was initiated in the early 1960s by the U.S. Navy to develop the methodology for adhesive fracture testing and to build a data base on the adhesive fracture resistance of epoxy resins and commercial structural adhesives. The results of this program have been recorded in the publications of Mostovoy, Ripling, Corten, Patrick, and Bersch [1-11].

About 15 years ago, a research program was started at the Naval

*Current affiliation: Department of Materials Science and Engineering, University of Utah.

Research Laboratory (NRL) to use the techniques developed by Mostovoy et al. to investigate the micromechanics of the adhesive fracture for unmodified epoxies, rubber-modified epoxy resins which model structural adhesives, and commercial structural adhesives. The initial scope of the NRL program was later broadened to include the fracture of epoxy-based polymers in bulk and as the matrix resin in fiber-reinforced composites [12-23]. During the past 5 years, this program has continued with work at the National Bureau of Standards [24-32]. The results of these studies form the basis of much of this chapter. In addition to this work, a great many other programs have studied adhesive bond fracture during the past 25 years. These efforts, which include major programs by the Air Force, the Office of Naval Research, and NASA, have made major contributions to the field. For the most part, however, this chapter does not cover these studies since they focus on aspects of bond failure other than those that form the main theme of this review.

The base resins from which essentially all structural adhesives are formulated are thermosetting polymers which, when cured, have the minimal long-term creep and the high tensile strength that are required to bond load-bearing structures. Moreover, they begin as liquids or semisolids that can easily be applied to adherend surfaces. Unfortunately, cured thermosets are relatively brittle materials, having fracture toughness values two or three orders of magnitude less than those of the metals and composites they are called on to bond. Consequently, all structural adhesives are highly formulated resins, generally complex mixtures of the base resin with elastomers, plasticizers, and inorganic particulate fillers. Indeed, structural adhesive formulation is very much an art that is rarely revealed even in the patent literature.

Despite the complexity (and secrecy) of adhesive formulations, they generally have a common morphological feature: an elastomeric phase dispersed in and bonded to a matrix of the thermoset resin [33]. In this morphology the elastomer imparts a high toughness to the material without serious losses in its tensile strength, modulus, heat distortion temperature, and similar properties relative to the matrix resin itself. Morphologically, these adhesives are related to the high-impact polystyrenes and other resins having a low-modulus phase dispersed in a high-modulus matrix. In fact, there are several similarities in the fracture behavior for all of these multiphase polymer systems, as has been demonstrated by McGarry and co-workers [34,35] in their studies of rubber-modified polyester and epoxy polymers.

## II. FRACTURE ENERGY AND FRACTURE DESIGN CRITERIA

A few comments are in order about the concept of fracture energy and fracture energy measurements. The depth of the discussion is necessarily limited, and the reader is referred to a voluminous literature on these subjects for more detailed treatments [36,37]. Modern fracture mechanics began with the Griffith criteria for fracture based on equating the straining energy content with the energy required to create new surface area at the instant of crack propagation. Thus, for a tensile-loaded, edge-notched bar,

$$\sigma_c = \left(\frac{2E\gamma}{\pi a}\right)^{1/2} \tag{1}$$

where $\sigma_c$ is the stress at initiation of crack growth, a is the notch depth, $\gamma$ is the surface fracture energy, and E is the tensile modulus. The subscript c is used here and elsewhere to indicate the critical value corresponding to the onset of crack growth.

Except for exceedingly brittle materials, the experimentally determined values of $\gamma$ are orders of magnitude greater than can be accounted for by just the rupture of chemical bonds at a crack tip. Actually, crack initiation involves a number of energy-consuming processes within the immediate vicinity of the crack tip, including viscoelastic and plastic deformation. Irwin [38] demonstrated that the Griffith equation is formally correct if the surface energy term, redefined by Irwin as the critical strain energy release rate $G_c$, is taken to include all these energy dissipative processes, so that

$$\sigma_c = \left(\frac{EG_c}{\pi a}\right)^{1/2} \tag{2}$$

where $G_c$ is an index of fracture toughness and in the ideal, brittle limit, $G_c = 2\gamma$. The deformation processes are limited to a small volume at the crack tip. The dimensions of the region, e.g., the radius of the deformation zone ($r_c$) or the crack opening displacement at failure ($\delta_c$), can also be taken as critical conditions for the onset of fracture (Figure 4.1). Analytical relationships exist between $G_c$ and these dimensions, but their explicit forms depend on the model assumed for failure and yielding in the deformation zone. For example, an elastic-plastic model gives for $G_c$

$$G_c \cong 6\pi\sigma_y\varepsilon_y r_c \tag{3}$$

in plane strain conditions, where $\sigma_y$ and $\varepsilon_y$ are the yield stress and strain, respectively.

None of the failure parameters ($\delta_c$, $G_c$, or $r_c$) is a true material constant for polymers. In general, they are time and temperature dependent. In addition, they vary with material thickness since the fracture energy under plane strain conditions (when triaxial stresses are generated at the crack tip by large sample thicknesses) is different from that measured under plane stress conditions in thin sections. The fracture energy of isotropic solids is also a function of the stress state, so it is necessary to define strain energy release rates for each of the three principal modes of failure (Figure 4.2)—opening mode, $G_{Ic}$; in-plane shear, $G_{IIc}$; and torsional shear, $G_{IIIc}$—and a relationship to predict failure for mixed-mode loading. Once all of these effects are considered, however, the result is a completely general failure criterion for very brittle materials.

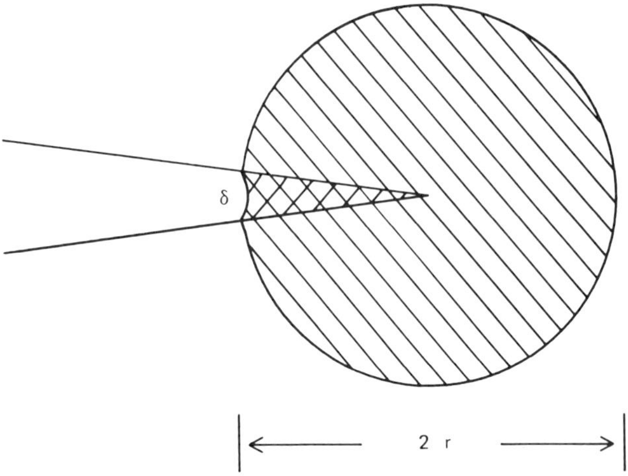


Figure 4.1 Schematic of the crack tip deformation zone for an elastic-plastic material. The radius of the zone, r, and the crack opening displacement, δ, are shown.

Another parameter used in fracture mechanics is the stress intensity factor, K, a measure of stress intensity in the vicinity of a crack tip. For an isotropic material the general expression for K is

$$K = YPa^{1/2} \tag{4}$$

where P is the applied load, a is a flaw size, and Y is a function determined by the stress conditions and the geometry of the structure. The exact form of Y must be determined from a stress analysis. For a linear elastic material the strain energy release rate is related to K by

$$\mathcal{G}_I = \frac{K_I^2(1 - \nu^2)}{E} \tag{5}$$

for the case of mode I loading and plane strain conditions in an isotropic material ($\nu$ is Poisson's ratio). Thus for a critical value of $\mathcal{G}_{Ic}$ there is a corresponding value of $K_{Ic}$.

In order to predict failure load, $P_c$, the explicit form of Equation (4) must be determined, $K_{Ic}$ or $\mathcal{G}_{Ic}$ must be measured experimentally, and the size of the most severe flaw (internal crack, surface scratch, etc.) must be known. The field of fracture mechanics is devoted to developing stress analyses and the test methodology for fracture toughness determinations and flaw detection. The application of fracture mechanics to adhesive joint design is complicated by the fact that the explicit form of Equation (4) is generally not known.

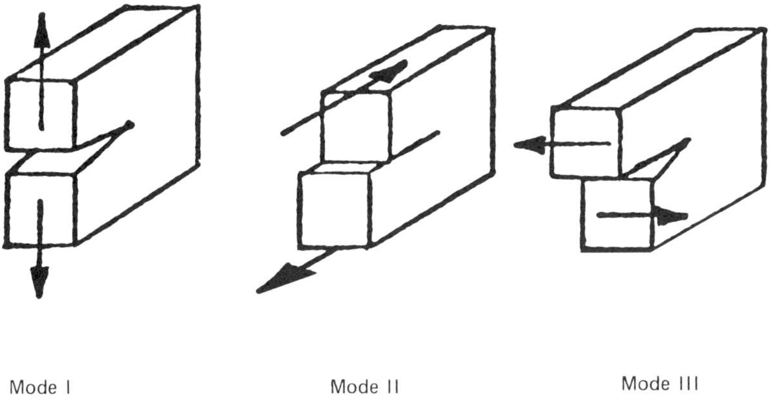


Figure 4.2 The three basic modes of loading.

The problem is that solutions require analyses for the stress distribution at the crack tip, and until recently such analyses have not been available for a flaw in an adhesive layer. Fortunately, this has now changed. To illustrate with just a few examples, Arin and Erdogan [39] have addressed the problem, and the application of the results to a simple butt joint in tensile loading has been described by Hilton and Gupta [40]. More recently, Wang et al. [41] developed a stress analysis for a crack in an adhesive layer and found a marked deviation from classical linear elastic theory. They note that this analysis obviates the strict use of linear elastic fracture mechanics for joint design, as had been surmised earlier [1].

The problem can be approached in an alternative, albeit less precise, way if the design analysis is based on the strain energy release rate $G$ rather than on the stress intensity factor. The general expression for $G$ is given by

$$G = \frac{P^2}{2b}\frac{dC}{da} \tag{6}$$

where b is the material thickness in the vicinity of the crack, and dC/da is the change in the compliance, C, of the structure with crack length a. Since dC/da is a "global" property of the specimen, Equation (6) can be determined without knowing the crack tip stress distribution. In order to develop Equation (6) in this way, however, dC/da must be determined experimentally [42], and this is tedious and often inaccurate. Alternatively, there are analytical solutions to Equation (6) for simple geometries, and in recent years dC/da has been obtained by finite-element analyses for many of the more complex specimens. Once the proper expression for $G$ has been obtained, the load, $P_c$, for fracture in the adhesive bond line can be computed from an experimentally determined $G_c$ and the flaw size, a. Unfortunately, the critical strain energy release rate, $G_c$, of a resin when measured as an adhesive layer may be different from that obtained when it is measured as the bulk resin. The situation is even more complex when measuring the $G_c$ of state-of-the-art structural adhesives. As described here and in Refs. 13-15, 22, 23, and 30, the adhesive fracture behavior of structural adhesives includes a very pronounced bond thickness dependence and bond angle dependence, both of which are affected by temperature and loading rate. Consequently, before adhesive bonds can be designed using a fracture mechanics approach, it is essential that basic research studies develop a scientific understanding of these effects and effective ways to model them.

## III. FRACTURE ENERGY MEASUREMENT

There is a variety of fracture test specimen geometries [43,44], many of which were developed for specific applications. For general use, a widely employed configuration is the compact tension (CT) specimen [43,44]. Shown in Figure 4.3, it can be rectangular or circular in shape and is essentially a short double cantilever beam with a saw cut about one-third of the way through the midsection. A precrack is made at the end of the saw cut. Precrack insertion is often the most crucial aspect of the fracture experiment. When a specimen with a sharp crack is loaded, the crack tip will blunt before reaching a critical radius of curvature at which failure occurs. The failure can proceed by stable or unstable crack growth. In stable growth the crack is driven by the crosshead motion on the test machine and stops when the crosshead is stopped. With unstable growth the crack rapidly propagates ahead and the load drops abruptly. Under some conditions the crack will arrest before the specimen is completely broken so that continued motion of the crosshead will reload the specimen, thus giving a stick-slip motion of the crack.

The critical crack tip radius at failure depends not only on the material (larger values for tougher polymers) but also on parameters such as temperature and loading rate. When generating the precrack, it is essential that the crack tip radius is less than this critical value and that the material surrounding the crack tip not be damaged. Only under these conditions will the measurement be independent of the precracking procedure. This also explains why precracking must be done more carefully in a brittle material than a tough material.

In the work described here, two precracking methods were employed. For certain adhesive bond tests, the precracks were made by inserting either a very thin polytetrafluoroethylene (PTFE) shim or a PTFE-coated razor blade into the specimen prior to curing the adhesive. In these cases, however, the crack exhibited either slow growth or stick-slip behavior, and thus all but the first point from each specimen corresponded to a naturally developed crack tip.

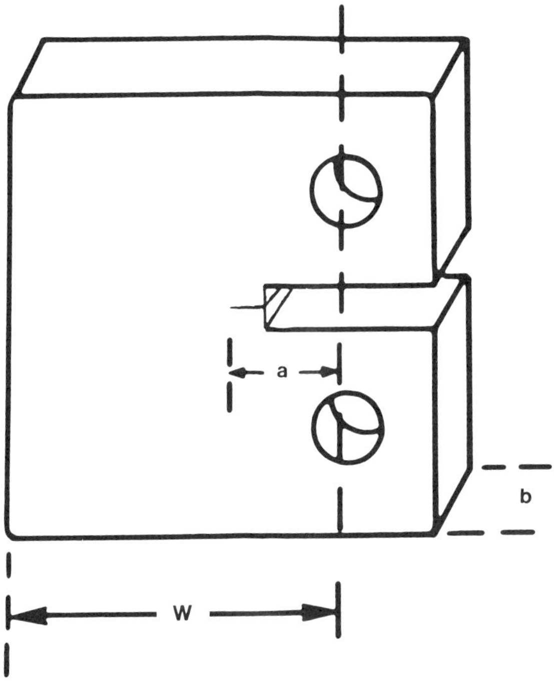


Figure 4.3 Compact tension specimen (rectangular).

For the remaining specimens the precracks were made by tapping a knife edge against the specimen to generate a growing crack, which then arrested. To assist in crack arrest, compressive forces ahead of the saw cut can be used with brittle materials, and in tough materials the region at the end of the saw cut can be cooled in liquid nitrogen to make it more brittle.

For the CT specimen the equation for $\mathcal{G}_{Ic}$ is

$$\mathcal{G}_{Ic} = \frac{Y^2 P_c^2 a}{E_b W^2 b^2 (1 - \nu^2)} \quad (7)$$

where the dimensions W and b are given in Figure 4.3. $E_b$ is the bending modulus, and the geometry factor Y is

$$Y = 29.6 - 186 \frac{a}{W} + 656\left(\frac{a}{W}\right)^2 - 1017\left(\frac{a}{W}\right)^3 + 639\left(\frac{a}{W}\right)^4 \quad (8)$$

for rectangular specimens and

$$Y = 30.0 - 162\frac{a}{W} + 493\left(\frac{a}{W}\right)^2 - 664\left(\frac{a}{W}\right)^3 + 405\left(\frac{a}{W}\right)^4 \quad (9)$$

for circular specimens. The inclusion of the term $(1 - \nu^2)$ presumes a plane strain condition, which was the case for most of the testing reported here.

The compact tension specimen has two shortcomings. First, fracture usually occurs catastrophically for tough materials because, although the load falls as the crack extends, the sample dimensions are usually too small to permit crack arrest. The load-displacement diagram looks like Figure 4.4A. Use of a very stiff test machine can help promote crack arrest so that more than one value can be obtained for each specimen, but generally this can be achieved only

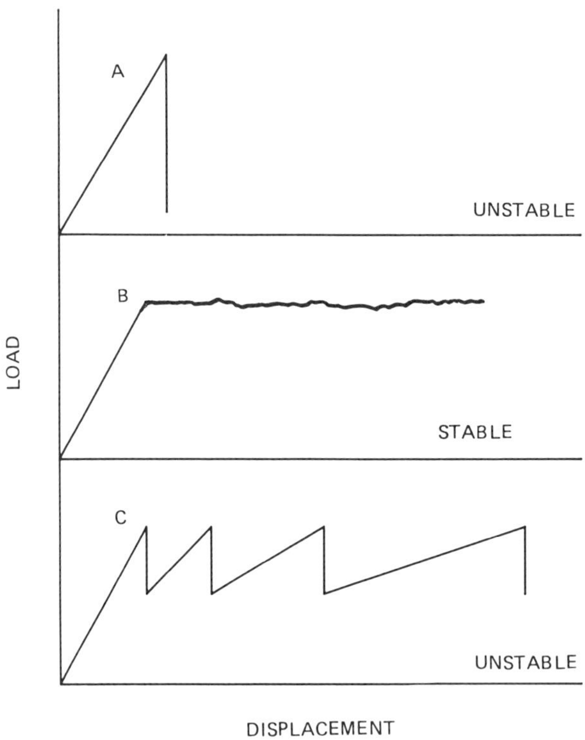


Figure 4.4 Typical load-displacement curves for fracture tests. (C) Stick-slip crack growth.

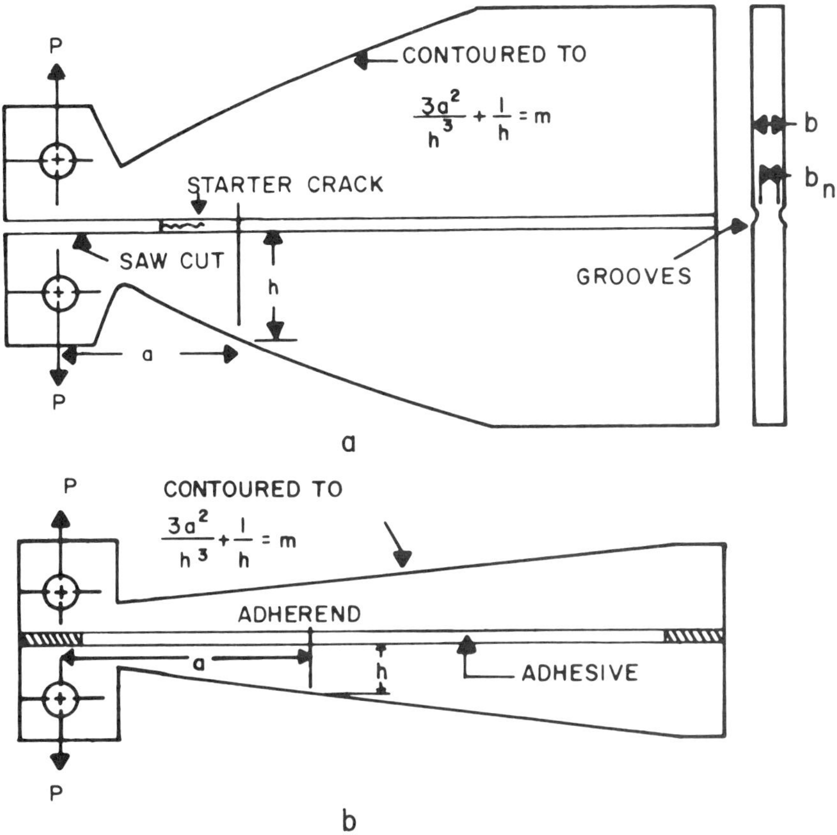


Figure 4.5 Tapered double cantilever beam specimen for bulk (top) and adhesive (bottom) samples; plots (B) and (C) are for specimen geometries where the fracture load is independent of crack length.

with brittle materials. Second, the calculation of $G_{Ic}$ using Equation (7) requires that the crack length be determined, and this can be difficult to do unambiguously in some cases.

These problems can be partially avoided by using a double cantilever beam (DCB) specimen that has been tapered for a constant change in compliance with crack length. Mostovoy and co-workers [2-9] used this specimen geometry in all of their work on mode I adhesive fracture. The explicit form of Equation (6) for a double cantilever beam is

$$G_{Ic} = \frac{4P_c^2}{b^2E_b}\left(\frac{2a^2}{h^3} + \frac{1}{h}\right)^{1/2} \tag{10}$$

where h is the beam height measured normal to the crack plane at the crack tip. If the beams are tapered as in Figure 4.5 so that the term in parentheses is a constant (m), then $G_{Ic}$ is independent of crack length, and only the load at crack initiation need be measured. Two constant compliance-gradient beams are illustrated in Figure 4.5: m = 3 $cm^{-1}$ and m = 90 $cm^{-1}$. It is important in all such tests to avoid inelastic bending of the beam arms, and for this reason the bulk polymer specimens (Figure 4.5A) are made quite stiff, m = 3 $cm^{-1}$. Owing to the much higher modulus and yield stress of the metal adherends compared to the polymer, the adhesive specimen (Figure 4.4B) need not be as steeply tapered as the bulk specimen, m = 90 $cm^{-1}$. For neat specimens it is sometimes convenient to use side grooves to help control the crack path [6]. For this geometry Equation (10) is modified by replacing $b^2$ with $bb_n$ (see Figure 4.5).

Equation (11) has the interesting property that for the tapered DCB specimen the load-displacement curve should show crack growth at constant load as in Figure 4.4B. This will be the case if the fracture energy is independent of crack velocity. However, many polymers are sensitive to strain rate; that is, $G_{Ic}$ decreases with increasing crack growth rate. In this case, once the crack starts to grow it will propagate spontaneously, and the strain energy in the specimen must drop substantially before it is insufficient to force continued crack propagation. At that point the crack arrests.

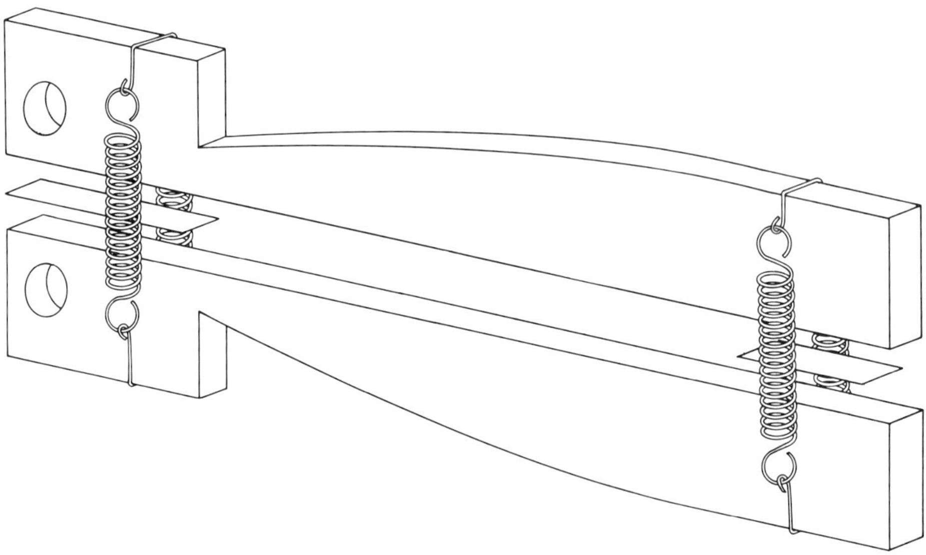

Figure 4.6 Assembly for making tapered double cantilever beam specimen for adhesive testing. Spacers for control of bond thickness and springs to hold specimen together during cure are shown.

The result is a peaked load-displacement curve as in Figure 4.4C, with the peaks corresponding to an initiation energy and the valleys to an arrest energy.

The tapered DCB specimens for bulk polymer fracture are prepared by casting and curing the resin in silicone rubber molds or coated metal molds. Grooves are then machined along each side of the specimen to guide the crack through the center plane (Figure 4.5). A saw cut is made at the loading end, and a starting crack is formed at the end of the saw cut. The adhesive fracture specimens are assembled as shown in Figure 4.6. Two metal halves are clamped together with polytetrafluoroethylene spacers at each end to establish the desired bond thickness. One side of the bond is closed with pressure-sensitive tape, and the liquid polymer is poured into the resulting cavity. After heat curing, the tape is removed and any excess resin is machined or abraded off. When testing commercial structural adhesives, which are generally supplied as films of partially cured resin, the specimen is prepared by placing strips of the film between the tapered beams. These films usually contain a support or skrim cloth, which establishes the bond thickness without any need for spacers. The bonding pressure recommended by film adhesive suppliers may not be obtainable using spring-loaded clips as shown in Figure 4.6, so vacuum bagging or mechanical jigging as described by Mostovoy and Ripling [11] can be used. Details concerning the tapered DCB specimen and its use, as well as information on adhesive testing, can be found in Refs. 2-15. In all of the work discussed here the adherends were 5086 or 2024 aluminum alloys and were given an acid-chromate etching pretreatment prior to bonding. The choice of alloy had no effect on the results.

It is important to test adhesive fracture under stress conditions other than mode I loading since most adhesive joints are designed to minimize peeling (mode I) loads in favor of shear loads. Two specimens designed for applying both shear and tensile forces are illustrated in Figure 4.7. In the first of these, the scarf joint (Figure 4.7A), the applied load is resolved in the bond line into mode I and mode II components and their ratio changes with the bond angle, $\phi$. In the second specimen (Figure 4.7B) the shear and tensile loads are applied independently, and this geometry is referred to as the independently loaded mix mode (ILMM) specimen. Mostovoy and co-workers [2] have described this specimen, and the equations for the two components of the fracture energy are

$$\mathcal{G}_{\mathrm{I}} = \frac{4P_{\mathrm{I}}^2}{b^2h^3E}\,[3(a + 0.6h)^2 + h^2] \tag{11}$$

$$\mathcal{G}_{\mathrm{II}} = \frac{P_{\mathrm{II}}}{b^2hE} \tag{12}$$

These specimens are prepared and assembled in much the same way as the tapered double cantilever beam specimen. The two side arms

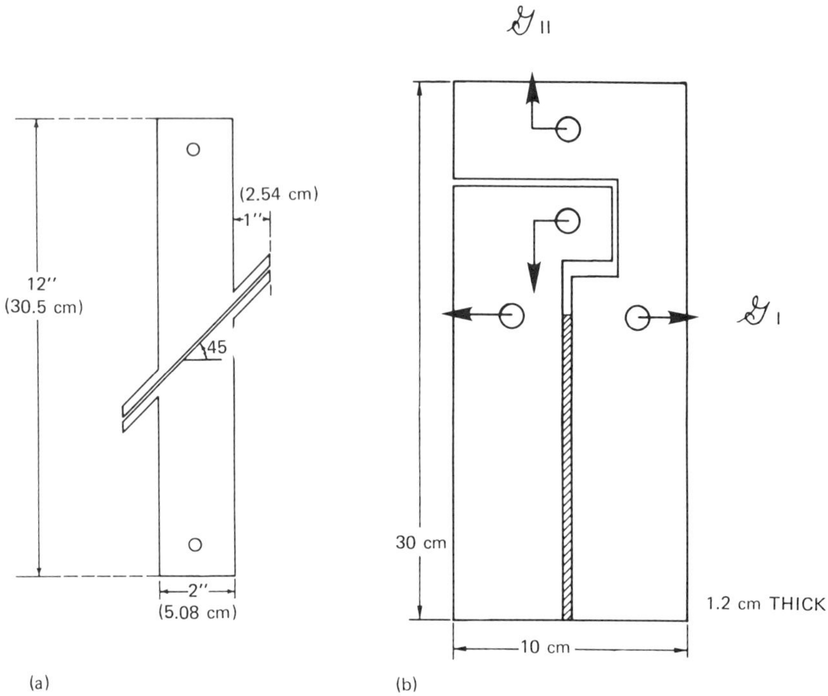


Figure 4.7 Mixed-mode test specimens. (Left) scarf joint (bond angle at 45° to load); (right) ILMM specimen.

on the scarf-joint specimen are clamped with spring clips to hold the two halves together during sample preparation.

Trantina [45-47] developed a finite-element stress analysis for the scarf-joint from which the adhesive fracture energy can be computed. He presented his results in terms of the dimensionless parameter f,

$$f = \frac{KbW^{1/2}}{P} \tag{13}$$

where K is the stress intensity factor, P is the tensile load, and b and W are the specimen thickness and width, respectively. The parameter is tabulated for various crack length-to-width ratios (a/W) and bond angles in Ref. 46.

For the tests performed in this work, the fracture energy (total critical strain energy release rate) is calculated using Equations (5) and (13). It should be noted that Trantina's analysis was for a monolithic specimen with an edge crack at various angles. Although Equation (13) contains the stress intensity factor, the values of K obtained for a monolithic specimen are probably not correct for the adhesive bond specimen since K describes the local crack-tip stress field. The fracture energy, however, depends on a calculation of the energy stored elastically in the specimen, and thus if the bond line is thin and contributes little to the energy storage, $G_c$ values calculated for adhesive bonds using Trantina's analysis should be good approximations. The parameter that is involved in the calculation of $G_c$ is the derivative of compliance with respect to crack length (see Equation 6). Experiments performed on a bonded and monolithic specimen with a sawed notch at a 45° angle to the loading directions gave values for the compliance gradient that were identical within experimental error. This supports the use of Trantina's analysis for calculating fracture energies. There are, however, several other assumptions involved in the use of this specimen, and their validity may not be as well established. This will be discussed later when the results are presented.

With organic-matrix, fiber-reinforced composites an important problem is the growth of cracks in the matrix between the layers of fibers, i.e., delamination. To study one aspect of this problem, an interlaminar fracture energy was determined using a double cantilever test specimen in which the width, rather than the height, was tapered to eliminate crack length from the fracture energy equation. The width-tapered beam (WTB) specimen is shown schematically in Figure 4.8, and a photograph of a specimen in the test machine is shown in Figure 4.9. The design and analysis of the WTB specimen for adhesive fracture testing have been described by

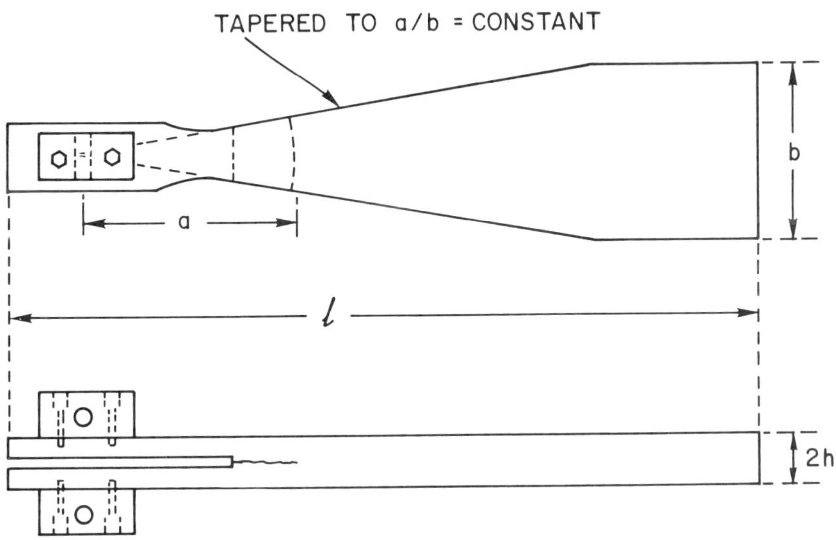


Figure 4.8 Width-tapered double cantilever beam specimen for evaluation of composite.

Figure 4.9 Width-tapered double cantilever beam specimen in tensile test machine.

Mostovoy and co-workers [48]. From Equation (6) and simple beam analysis, the interlaminar fracture energy can be shown to be

$$G_c = \frac{12P^2}{E_b h^3} \left(\frac{a}{b}\right)^2 \tag{14}$$

and if the specimen is width-tapered for a constant ratio of a/b, the $G_c$ can be determined from $P_c$ and $E_b$ alone without measuring the crack length, a. Some composite experiments were also performed using the straight-sided cantilever beam specimen, and for these tests it was necessary to measure the crack length. In addition, the specimens were periodically unloaded during the experiment to determine the compliance as a function of crack length. As a result, it was not necessary to perform a separate experiment to measure the bending modulus. The fracture energies were determined [27] from these tests using Equation (6). In addition, the areas under the load-displacement curves were used to determine fracture energies, and in all cases the two methods gave similar results [27].

Although these test specimens are designed to test mode I, it is problematic in the testing of composites whether the crack front experiences pure opening mode stresses [49]. The utility of this test, however, is in providing relative values for the resistance to a certain type of interlaminar crack growth for various composite matrix resins.

## IV. MATERIALS*

The general morphological requirements for the two-phase, toughened material are well documented, i.e., elastomeric particles dispersed in a rigid matrix. The effects of the details of the morphology, however, are not clearly understood. It has been shown that the fracture energy increases with increasing concentration of the elastomer up to about 18% elastomer. Beyond that level it is difficult to obtain the desired particle-matrix morphology. Below that level a wide range of morphological features can be produced. It is generally found that high fracture energy systems have a distribution of particle sizes in which many particles are in the 1 to 5 μm range. The particles contain both epoxy and elastomer molecules, but the role (if any) of the particle composition in determining fracture energy is still a topic of discussion. The interfacial bond

---

*Certain commercial materials are identified in this chapter in order to specify adequately the experimental procedure. In no case does such identification imply recommendation or endorsement by the National Bureau of Standards, nor does it imply that the materials are necessarily the best available for the purpose.

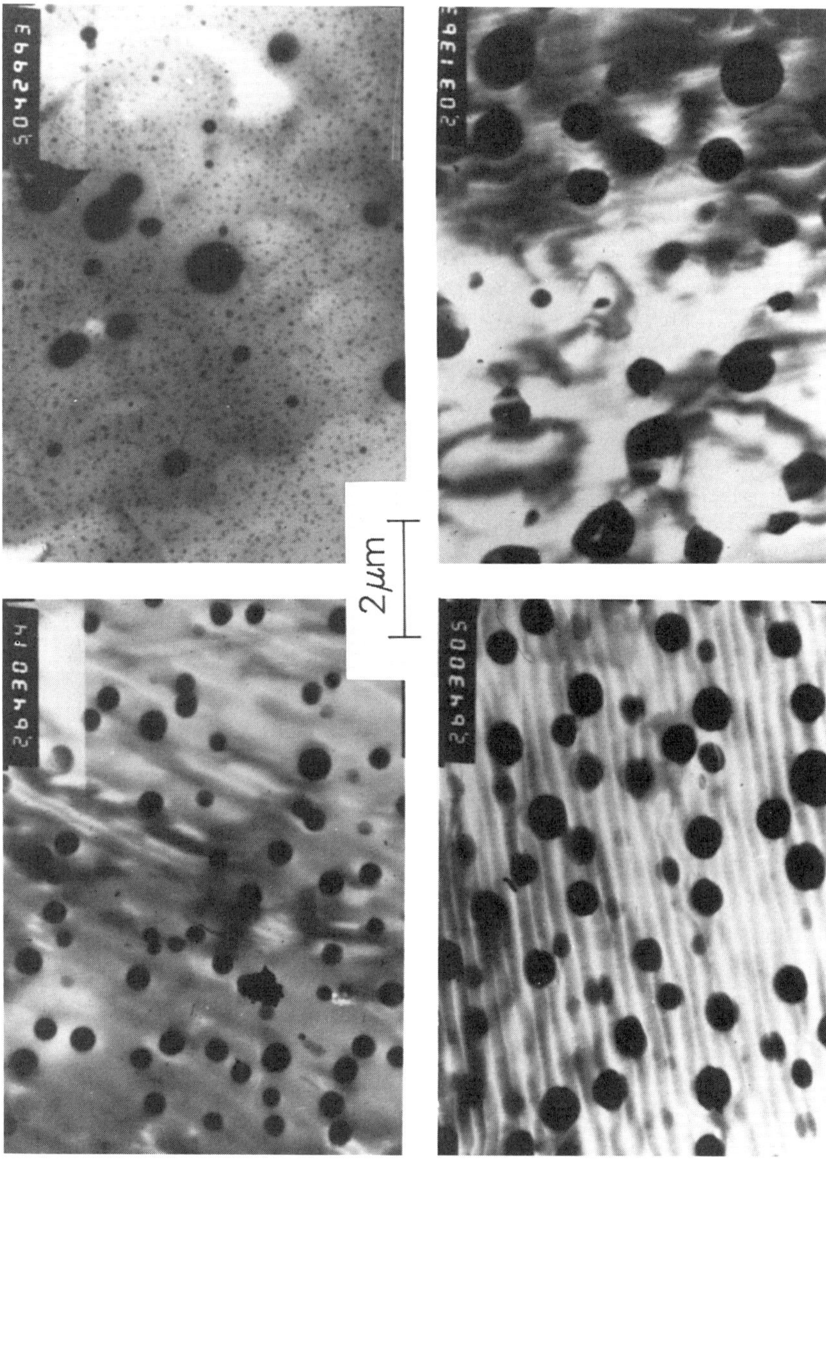


Figure 4.10 Electron micrographs for four different formulations of rubber-modified epoxies. Picture at upper right shows bimodal distribution of particle sizes; other pictures show conventional distributions. For details on materials see Ref. 31.

strength between the matrix and the particles is another factor whose role has not been clearly established. Finally, the matrix often contains some elastomer that is not phase separated or is phase separated in particles that are too small to see with the electron microscope. The role of the matrix characteristics in fracture behavior is also under study by several groups, and Yee and Pearson [50] have already shown that the cross-link density in the matrix can be very important.

Another factor that may be influential is the distribution of particle sizes. In this regard one type of morphology is of particular interest. When the cure chemistry is adjusted in the proper way, both large (>0.5 μm) and small (>0.2 μm) elastomeric particles are obtained. As a result, these formulations are often said to have a bimodal distribution of particle sizes. This is illustrated in Figure 4.10, which shows transmission electron microscope pictures for thin slices of a variety of formulations. Osmium tetroxide was used to stain the elastomer particles [31]. In reality, the actual distribution may be quite complex. Nevertheless, this chapter follows the general practice of using the term "bimodal" to identify materials in which the particle sizes are predominantly distributed around a central size value. This is done with the acknowledgment, however, that these terms may not represent a completely accurate description of the distribution.

Although the particle size distribution provides the most obvious distinguishing feature of the bimodal systems, it is important to note that other differences are also present. To understand this, one must consider how these materials are prepared. In most cases the elastomer is added to the epoxy before cure, and they are compatible. During cure the epoxy polymerizes with itself and with the elastomer. As the molecular weights increase, the compatibility decreases and eventually phase separation occurs. To change the particle size distribution, one must change the chemistry, and this can of course, alter other features such as the degree of phase separation and the nature of the particle and matrix phases. Consequently, it is very dangerous to attribute differences in behavior obtained for these systems simply to particle morphology. A number of studies [51-53] have tried to address this problem by using preformed particles, but so far the results are limited. Pursuit of this approach, however, may provide a better basis for establishing structure-property relationships in the future.

What makes the bimodal systems of particular interest is that, for a given concentration of elastomer, they often show a significantly higher fracture energy than systems with the conventional morphology. As will be shown in this chapter, however, a comparison of fracture energies obtained using a single set of experimental conditions is not sufficient to characterize the failure behavior of toughened systems and can, in fact, be very misleading.

The work discussed here involved a variety of modified and unmodified epoxies. The major difference in the unmodified systems is the curing agent used. For the elastomer-modified epoxies, results are reported for a variety of formulations in which both the elastomer concentration and morphology are varied. Information and references on the composition, preparation, and morphology of these systems will be given in the text as the materials are discussed.

## V. FRACTURE OF UNMODIFIED EPOXIES

### A. Bulk and Adhesive Fracture

Some perspective on the magnitude of polymer fracture energies relative to other materials is offered in Figure 4.11. Values of $\mathcal{G}_{Ic}$ for various polymers in bulk and as adhesives are scaled between the very brittle inorganic glasses and the relatively tough metals. Note that the epoxies and other cross-linked thermosetting polymers are very brittle, whereas the thermoplastics (e.g., polycarbonate and polysulfone) are much tougher. Nevertheless, Figure 4.11 shows that the fracture energy of an epoxy can be greatly increased if it contains a dispersed phase of elastomer particles.

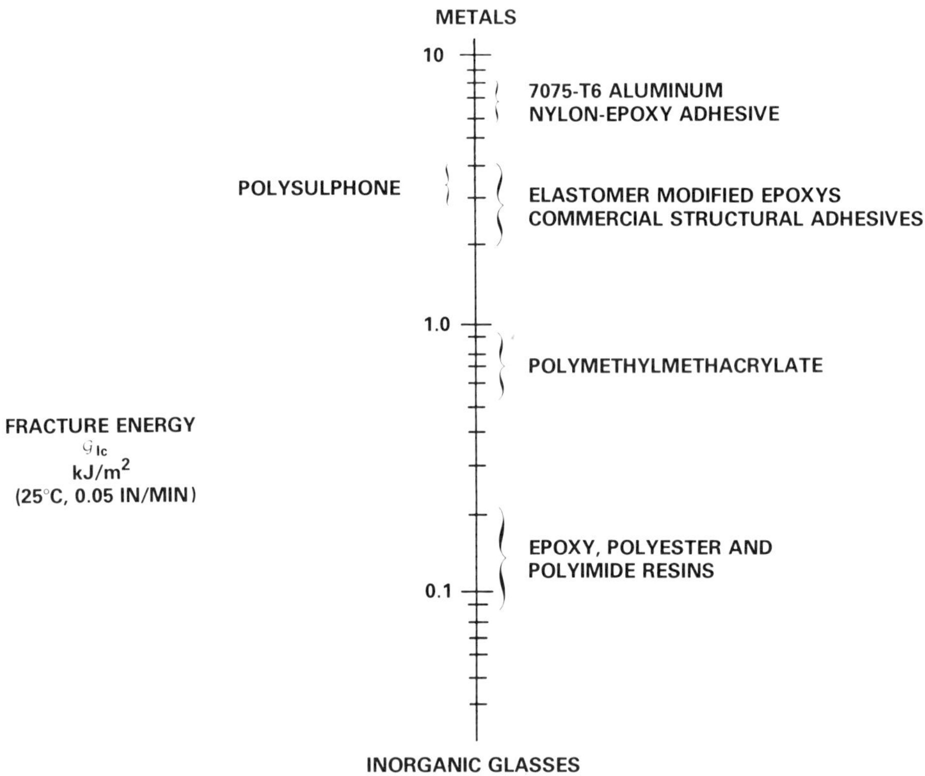


Figure 4.11 Diagram indicating relative fracture energies for different materials.

Table 4.1 Bulk Fracture Energies[a] of Unmodified Epoxy Resins[b]

| Curing agent | $G_{1c}$ (J/m$^2$), bulk |
|---|---|
| Hexahydrophthalic anhydride[c] | 136 |
| Hexahydrophthalic anhydride[d] | 153 |
| "Nadic" methyl anhydride[e] | 86 |
| Piperidine[f] | 154 |
| Tetraethylpentamine[d] | 88 |
| *m*-Phenylenediamine[e] | 148 |

[a]25°C, 45% relative humidity (RH), crosshead speed of 0.13 cm/min.
[b]Diglycidyl ether of bisphenol A (DGEBA).
[c]Reference 12.
[d]Reference 6.
[e]Unpublished data.
[f]Reference 13.

A comparison is made in Table 4.1 of the bulk fracture energies for a few epoxy resins based on a simple epoxy monomer, diglycidyl ether of bisphenol A (DGEBA), and cured with various curing agents. Values range between 80 and 160 J/m$^2$ with no obvious systematic dependence on the type of curing agent. The magnitudes of the fracture energies clearly indicate that inelastic dissipative processes are involved in the initiation of crack growth. The rupture of chemical bonds can account for no more than 10 J/m$^2$, less than 10% of the measured energies [54]. Even in these very brittle polymers most of the fracture energy involves viscoelastic and plastic deformation of the resin structure at the crack tip. Postfailure scanning electron microscopy (SEM) examination of the fracture surfaces reveals clear evidence of plastic deformation in the region of crack growth initiation [13,55]. Intuitively, plastic yielding and flow of these highly cross-linked resins are difficult to accept in view of their apparently brittle behavior in tensile tests. Nonetheless, plastic flow does occur prior to failure in the small, highly stressed region of the crack tip. The stress conditions in this region are triaxial, and therefore one suspects that the process involves void formation followed by shear yielding.

The adhesive fracture energy of an unmodified epoxy is generally equal to the corresponding bulk fracture energy as long as the fracture is a cohesive, center-of-bond failure. In some cases, however, there have been reports of significant differences between adhesive and bulk fracture energies. Mostovoy and Ripling [3] found bulk

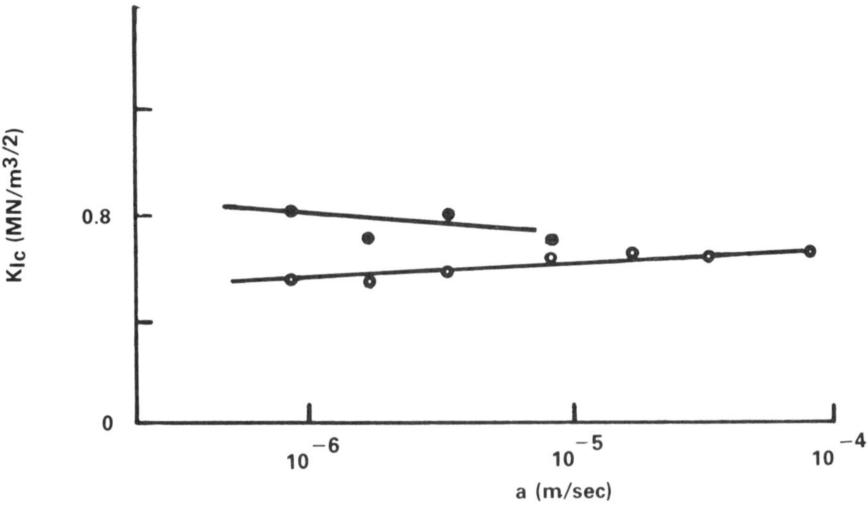


Figure 4.12 Critical stress intensity factors for initiation (filled circles) and arrest (open circles) of crack growth in an epoxy (see Ref. 56) as a function of crosshead speed, å.

$G_{Ic}$ to be greater than the adhesive $G_{Ic}$ for a tetraethylpentamine (TEPA) cured DGEBA over a wide range of curing agent concentrations. At a given TEPA concentration the adhesive toughness became greater than the bulk value at a postcure temperature of 230°F. Gledhill et al. [56] reported differences between bulk and adhesive fracture energies (determined at different laboratories) for TEPA-cured DGEBA polymers. On the other hand, tests performed in this study on a piperidine-cured DGEBA gave essentially the same value for adhesive and bulk $G_{Ic}$. Differences in the two toughness values may reflect the inherent uncertainties generally encountered in fracture measurements. Experimental scatter of ±10% or more is not uncommon. Also, even if the cure conditions are ostensibly the same for both bulk and adhesive specimens, they are not necessarily identical. For example, the rate of heating or cooling of the resin or the loss of a volatile component can be different for a resin in a thin bond line than for a bulk casting.

The epoxies exhibit either stable or unstable crack propagation (see Figure 4.4) depending on temperature, strain rate, and composition. The specimen geometry has an understandable influence on crack growth stability, but until recently the effects of material properties and test conditions have not been understood. Now, however, the question has been partially answered in papers by Phillips et al. [57], Gledhill et al. [56], and Kinloch and Williams [58]. The data given in Figure 4.12 exemplify the observed variations in behavior, in this case as a function of crosshead speed.

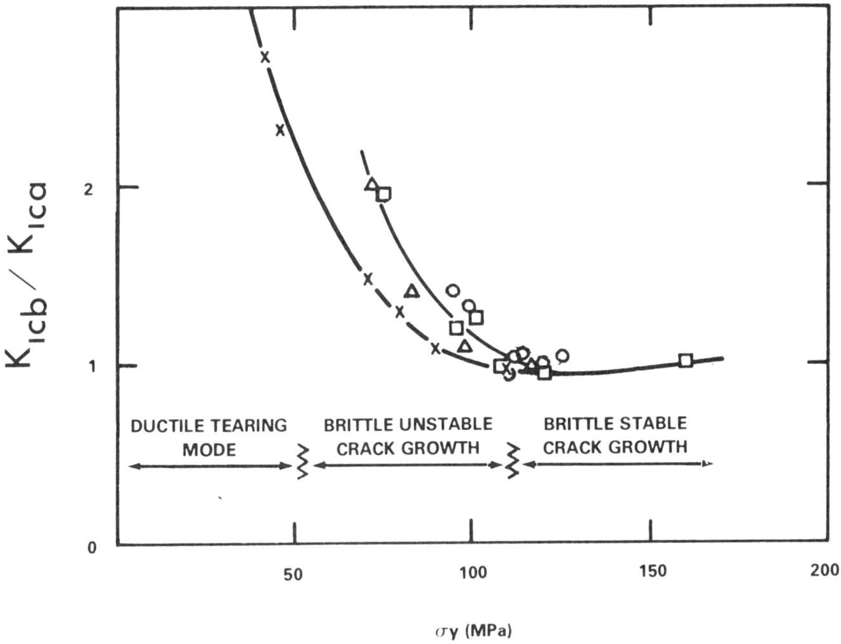


Figure 4.13 Normalized critical stress intensity factors for various epoxies (open symbols) and modified epoxies (X) measured at different temperatures and plotted against their yield stress values (see Ref. 58).

At crosshead speeds up to about $10^{-5}$ m/sec fracture is unstable, and there is a stress intensity factor for both initiation and arrest, i.e., $K_{Ici}$ and $K_{Ica}$ (i.e., the peaks and the troughs in the load-displacement diagram). Above a crosshead speed of $10^{-5}$ m/sec, cracking is stable and there is only one $K_{Ic}$ for each loading rate. Gledhill et al. [56] related this behavior to the critical crack opening displacement, $\delta_c$, in Figure 4.1, which is given for an elastic-plastic material in terms of the fracture energy by

$$\delta_c = \frac{G_c}{\sigma_y} \tag{14}$$

where $\sigma_y$ is the yield stress. They were able to demonstrate that for the bulk and adhesive fracture of TEPA-cured DGEBA polymers, cracking was stable and at a constant $G_{Ic}$ for $\delta < 1.0$ µm but became unstable with a systematic increase in $G_{Ic}$ when $\delta > 1.0$ µm. Stated phenomenologically, at low yield stresses (and/or given sufficient time) the crack tip blunts and the material exhibits a high toughness.

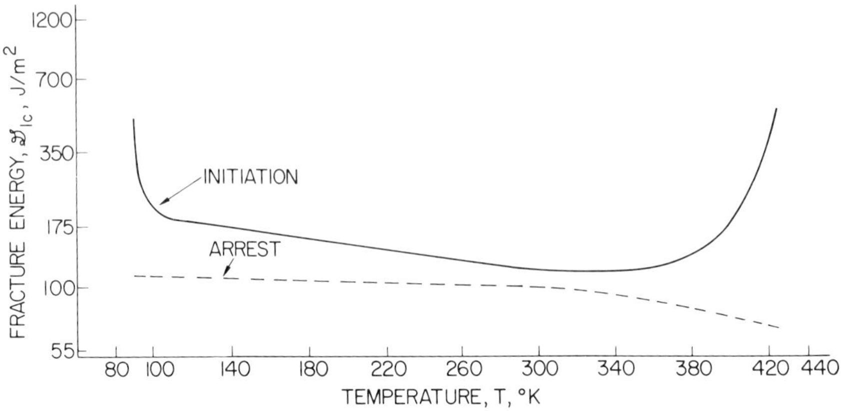


Figure 4.14 Initiation and arrest fracture energy for a typical epoxy as a function of temperature.

They then argued that once a blunt crack begins to grow rapidly it sharpens because the short time scale minimizes yielding. Consequently, a lower stress intensity factor is needed to maintain crack growth, and thus unstable growth is produced.

Kinloch and Williams [58] developed a theoretical analysis of crack tip blunting of epoxy polymers. They assumed that in the absence of crack tip blunting the critical stress intensity factor for initiation is equal to the arrest value, $K_{Ica}$. If the crack initiation stress intensity factor measured in the experiments is designated $K_{Icb}$, then the ratio $K_{Icb}/K_{Ica}$ is a measure of crack blunting. This ratio is equal to one for stable propagation. In Figure 4.13, adapted from Ref. 58, $K_{Icb}/K_{Ica}$ is plotted against $\sigma_y$ for various unmodified and elastomer-modified epoxies tested in bulk and as adhesives. The influence of yield stress on blunting is clearly illustrated. Kinloch and Williams chose to draw a single curve through all the data points. A better fit of the data is obtained, however, if separate curves are drawn for the unmodified epoxies and the elastomer-modified epoxies. This distinction points up the fact that the details of the crack tip yielding are different for the two types of polymers, as will be discussed in the next sections.

The interdependence of fracture toughness and yield stress strongly suggests a time-temperature dependence of $\mathcal{G}_{Ic}$. With the unmodified epoxies, the variation of $\mathcal{G}_{Ic}$ with the strain rate and temperature has been found to be small except near the glass-to-rubber transition, i.e., high temperatures/low strain rates. In Figure 4.14, adapted from work by Mostovoy and Ripling [59], $\mathcal{G}_{Ic}$ (initiation) for a hexahydrophthalic anhydride-cured DEGBA is essentially flat up to about 380 K, beyond which the toughness increases sharply as the test temperature approaches the $T_g$ of this resin. The increase in $\mathcal{G}_{Ic}$ at very low temperatures may be

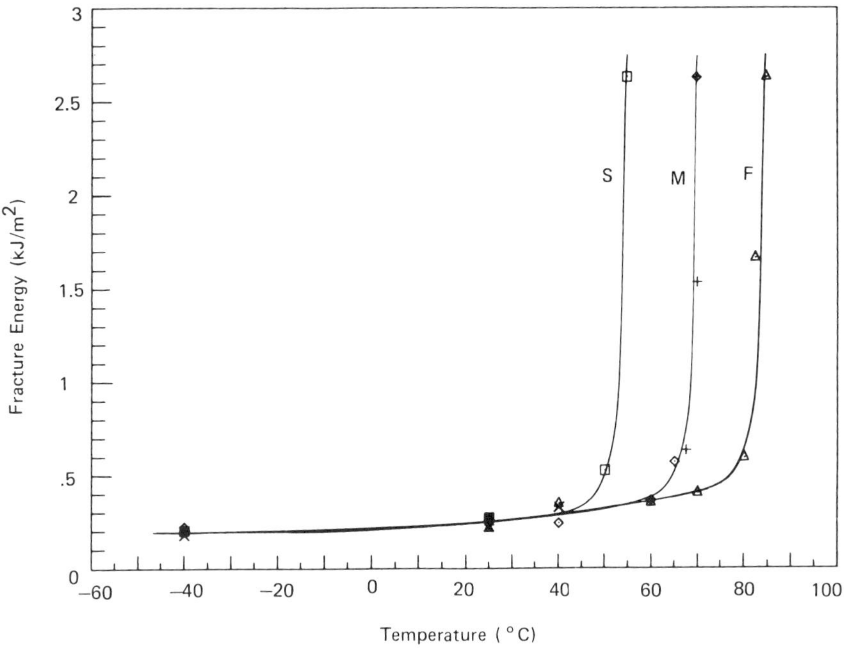


Figure 4.15 Fracture energy of the piperidine-cured DGEBA as a function of temperature and crosshead speed; S = 0.05 mm/min, M = 1.25 mm/min, and F = 50 mm/min.

related to a limiting crack velocity or may be the result of a plasticizing effect of the liquid nitrogen coolant, as discussed in Ref. 15. Note in Figure 4.14 that the arrest energy is relatively constant even at the higher temperatures. This is consistent with the argument put forth by Kinloch and Williams that the strain rate dependence of fracture is related to the $\sigma_y$ dependence of the initiation toughness, i.e., $K_{Icb}$ in their notation.

In a study of the effect of temperature and strain rate on $\mathcal{G}_{Ic}$ of a piperidine-cured DGEBA, Hunston and Kinloch [60] observed a very steep and sudden increase in toughness as the glass-to-rubber transition is approached either by increasing the temperature or decreasing the loading rate, Figure 4.15. One might think that the enhanced toughness was due to increased molecular mobility near the transition; however, a change in molecular mobility is only part of the story. Examination of the high-$\mathcal{G}_{Ic}$ specimens shows evidence of large-scale yielding at the crack tip. The pattern of the flow reflects a ductile failure. Consequently, the effect of the increase in molecular mobility is magnified because it produces a change in the failure mode, i.e., a brittle-to-ductile transition.

### B. Composite and Mixed-Mode Adhesive Fracture

Unmodified epoxies have also been studied extensively as matrix resins in fiber-reinforced composites and as adhesives in bonds subjected to mixed-mode loading. The resulting fracture energies vary from slightly higher than to much higher than the corresponding mode I values for the neat resin. Such behavior is quite reasonable, but a detailed discussion of these results will be delayed until the section on elastomer-modified materials, since comparisons of values for modified and unmodified materials are of the most interest.

## VI. ELASTOMER-MODIFIED EPOXIES

### A. Bulk Fracture

Addition of a liquid carboxy-terminated polybutadiene acrylonitrile (CTBN) elastomer to DGEBA epoxy resins can increase the fracture energy by factors of 30× or 40× over the unmodified resin. In Table 4.2, $\mathcal{G}_{Ic}$ values are listed for piperidine-cured DGEBA resins with increasing amounts of a liquid CTBN. The CTBN is initially soluble in the epoxy but phase separates during cure. In the concentration range up to about 18 wt %, the elastomer precipitates as spherical particles, which, for the piperidine-catalyzed epoxy-CTBN system cured at 120°C, have diameters of 1 to 3 μm [13,15,61-63]. These particles can be seen in the fracture surface by using SEM. At concentrations above approximately 18% the elastomer does not form a fine dispersion but forms an inhomogeneous blend, and there is a corresponding decline in the fracture energy.

One of the merits of elastomer-epoxy dispersions is that the large gain in toughness is accomplished with relatively little loss in high tensile strength, modulus, thermal mechanical properties, etc. relative to those of the epoxy matrix resin. This is illustrated in Figure 4.16 for the piperidine-cured DGEBA system. Above 18% CTBN,

Table 4.2 Bulk Fracture Energy and Deformation Zone Size for CTBN-Modified Epoxy Resins[a]

| CTBN concentration (wt %) | Fracture energy, $\mathcal{G}_{Ic}$ (J/m$^2$) | Deformation zone size, $2r_c$ (cm) |
|---|---|---|
| 0 | 121 | 0.0008 |
| 4.5 | 1050 | 0.0124 |
| 10 | 2720 | 0.021 |
| 15 | 3430 | 0.033 |
| 20 | 3590 | 0.070 |

[a]From Ref. 13.

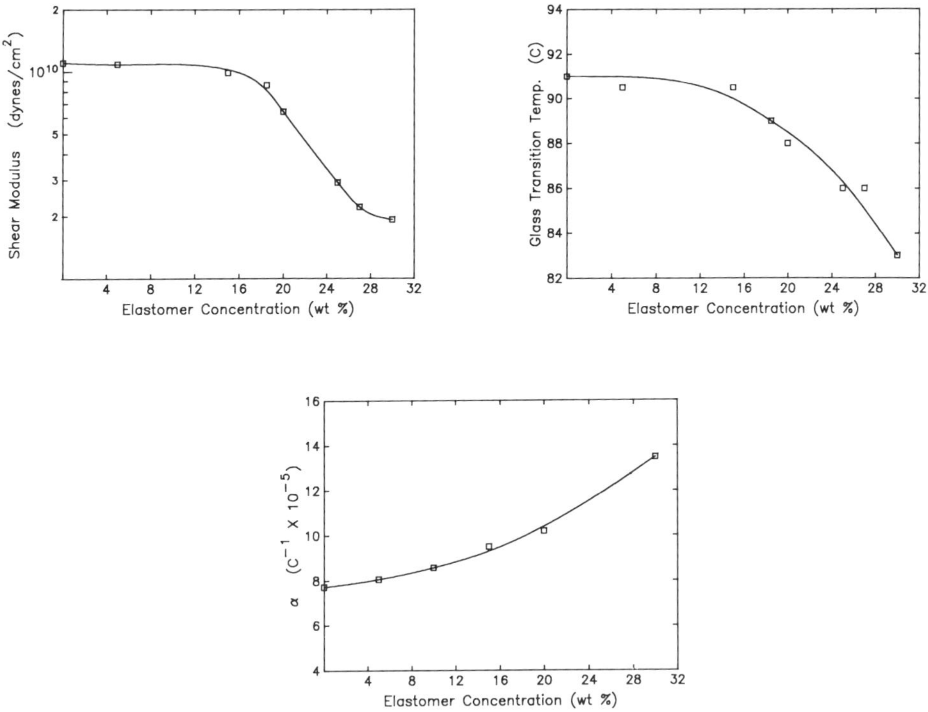


Figure 4.16 Plotted against elastomer concentration in a piperidine-cured DGEBA epoxy are dynamic shear storage modulus (22°C and 0.6 Hz), glass transition temperature, and linear thermal expansion coefficient, $\alpha$.

where the resin composition passes from a dispersion to a blend, there is an abrupt decline in the properties. It is this retention of properties at the lower concentrations that causes the dispersed elastomer morphology to be sought in formulating structural adhesives and composite matrix resins.

The mechanism of toughening in elastomer-modified epoxies has been developed through a series of papers beginning in 1970 and recently refined and summarized in Ref. 26. These papers and the references cited in them give detailed descriptions of the proposed mechanism and the evidence for it. The following is a brief summary of this mechanism. As with the unmodified epoxy resins, the CTBN-epoxy polymers exhibit deformation markings in the region of crack initiation. However, the extent of deformation is many times larger for the modified materials. In fact, a distinct stress whitening can be seen developing ahead of the tip of the precrack as the load on the test specimen is increased. An SEM examination

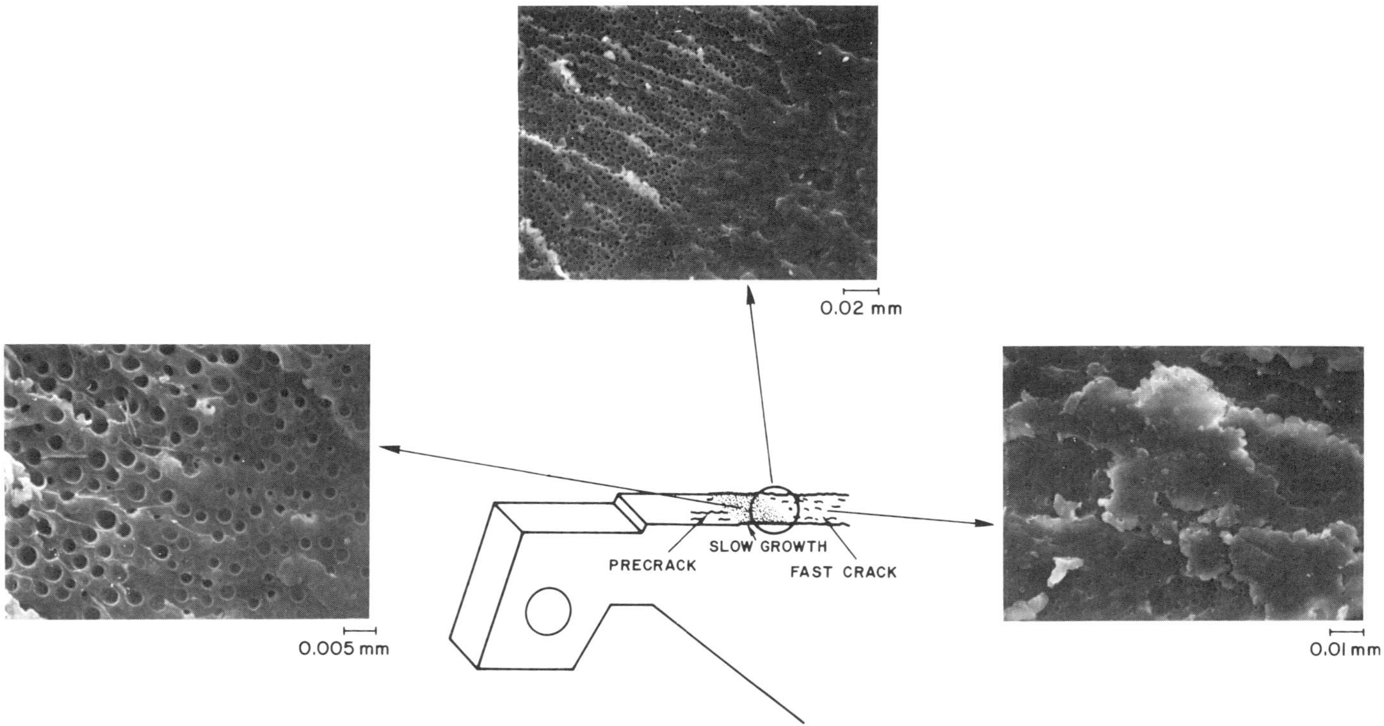


Figure 4.17 SEM picture of fracture surface showing region of initiation of rapid crack growth.

of the fracture surface in the region of stress whitening reveals closely spaced holes, as shown in Figure 4.17. These holes correspond to rubber particles that have cavitated. In the region of fast crack propagation beyond the stress whitening, the elastomer particles appear as filled or partially filled holes, and the matrix exhibits much less deformation than in the stress-whitened region. The rapidly propagating crack appears to slice through the material, allowing little time for deformation and yielding.

The fractographic features of the stress-whitened region indicate a number of deformation processes that contribute to toughening. The hole formation suggests that there has been a dilational deformation, since the holes are larger in diameter than the undeformed particles seen in the region of rapid crack growth. This is consistent with the triaxial stress field that is known to be present at the crack tip. These stresses lead to rupture of the elastomer particles, which then relax back into the enlarged cavities, thus giving the appearance of holes. Results from several different experiments support this idea, but the most convincing are those of Kinloch. He exposed the stress-whitened region of the fracture surface to an organic fluid to swell the rubber. The SEM then revealed that the holes had become hills.

The local stress concentrations associated with the particles and holes nucleate local shear yielding of the epoxy matrix, and this is a significant part of the crack tip deformation process. The holes may assist the yielding by relieving the triaxiality of the stresses. The tear markings in the stress-whitened zone (see Figure 4.17) are evidence of shear yielding. Strong supporting evidence for this mechanism was provided by the photographs of Yee and Pearson [50], which clearly show shear bands between cavitated rubber particles in modified epoxy materials.

The final stages of the deformation process involve the elongation and rupture of any elastomer particles that have not cavitated. Figure 4.18 is an SEM view into a crack showing "pillars" of elastomer stretched across the open crack tip. The insert in Figure 4.18 shows the specimen configuration; a thin strip of polymer was held in a U-shape and a notch cut at the top of the U. The microscope was then focused into the notch.

Kunz-Douglas et al. [64] were the first to report elongation, bridging, and rupture of the elastomer particles in the fracture of CTBN-epoxy polymers. They concluded that this process was the major contributor to the fracture energy. The modified epoxies that were the original basis for their work, however, exhibited only modest improvements in toughness and no stress whitening. The extension of their model to the highly toughened systems of interest here, on the other hand, involves a number of uncertainties and difficulties, which have been outlined in Ref. 26. Consequently, at the present time, it appears that although the rubber particle bridging mechanism may be a contributing factor, it does not represent the major energy dissipation process for the very tough materials.

The CTBN-epoxy system described in the previous paragraphs is a relatively simple composition compared to the commercial

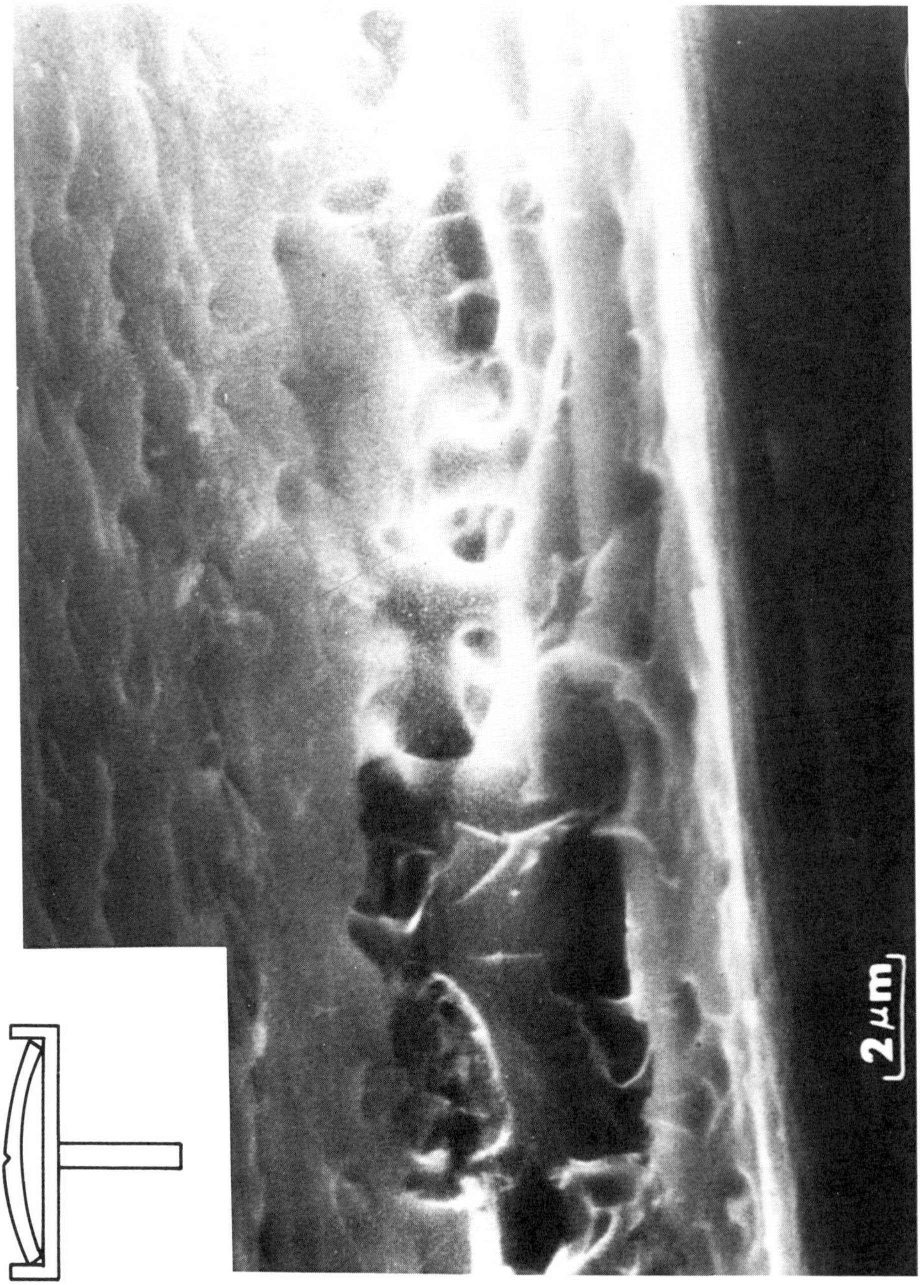
2 μm

formulations used as adhesives, coatings, encapsulants, and matrix resins. Many of these commercial resins are developed to have the bimodal distribution of particle sizes mentioned earlier. This can be obtained by "precooking" the epoxy (DGEBA) with bisphenol A (BPA) to form an adduct. When this adduct is diluted with CTBN dissolved in DGEBA and cured with piperidine, the dispersed elastomeric phase includes small particles <0.5 μm in diameter and large particles 1 to 5 μm in diameter.

As mentioned earlier, the bimodal distribution systems often have a higher toughness for a given rubber content when measured under the usual laboratory test conditions. Also, in the bimodal systems a larger proportion of the rubber is often precipitated rather than remaining dissolved in the epoxy matrix. Consequently, because there is a lower rubber content and less dissolved rubber to decrease the modulus and lower the $T_g$, such systems are often considered superior to those with the conventional particle size distribution. Some of the results from the work of Riew et al. [63] on bimodal CTBN-epoxy resins are given in Table 4.3. The chemistry of these materials is outlined in Ref. 63. Note that at the concentration of BPA giving the highest toughness, the $T_g$ is higher than for the material with a conventional particle size distribution obtained when the BPA concentration is zero.

Detailed investigations of the failure mechanisms for CTBN-epoxy polymers with bimodal size distributions have been undertaken recently [23,31,65]. One of these studies [65] involved an SEM examination of the fracture surfaces for specimens of a piperidine-cured, CTBN-DGEBA-BPA epoxy. The photomicrographs in Figure 4.19 give views of the stress-whitened region. Figure 4.19A shows the boundary between the precrack (dark area) and the stress whitening. The massive yielding in the stress-whitened region is shown in closer detail in Figures 4.19B and C. As in the case of fracture surfaces for CTBN-epoxy materials with a conventional particle size distribution, there are holes in the stress-whitened region which presumably result from dilation and rupture of the elastomer particles. Most of the larger holes, however, contain round inclusions (Figure 4.19D), which are believed to be hard (high epoxy content) regions formed in the particles. Both large and small particles were visible in the region of fast crack growth (beyond the stress whitening), but they have not been as severely deformed as in Figure 4.19 C).

Figure 4.18 SEM view into a crack in a 15% CTBN-modified DGEBA cured with piperidine. Insert shows specimen.

The massive shear yielding of the matrix resin evident in Figure 4.19 is difficult to explain if the epoxy has not been plasticized by the rubber. However, the fact that the $T_g$ and the modulus are essentially the same as those for the unmodified epoxy is not indicative of plasticization. Kinloch et al. [26] argue that the increased yielding is a result of local stress concentrations associated with the particles acting to initiate multiple localized shear bands over a large volume at the crack tip.

Another possible explanation or contributing factor can be hypothesized on the basis of the work of Sayre et al. [66] and Kunz et al. [67]. They used scanning transmission electron microscopy (STEM), nuclear magnetic resonance (NMR), and other techniques to determine the volume fraction of CTBN in both the particulate phase and the epoxy matrix for a diethanolamine-cured DGEBA-CTBN polymer.

Table 4.3 Some Properties of Epoxy-Bisphenol A-CTBN Polymers[a]

| Concentration of bisphenol A (phr) | $G_{Ic}$ (kJ/m$^2$) | $T_{hd}$[b] (°C) |
|---|---|---|
| 0 | 2.1 | 78 |
| 6 | 2.1 | 83 |
| 12 | 3.3 | 78 |
| 24 | 8.6 | 82 |
| 36 | 3.7 | 74 |
| 45 | 0.7 | 71 |

[a]Base material contains DGEBA (100 parts), CTBN (5 parts), and piperidine (5 parts) in addition to bisphenol A and is cured at 120°C for 16 hr.

[b]Heat distortion temperature.

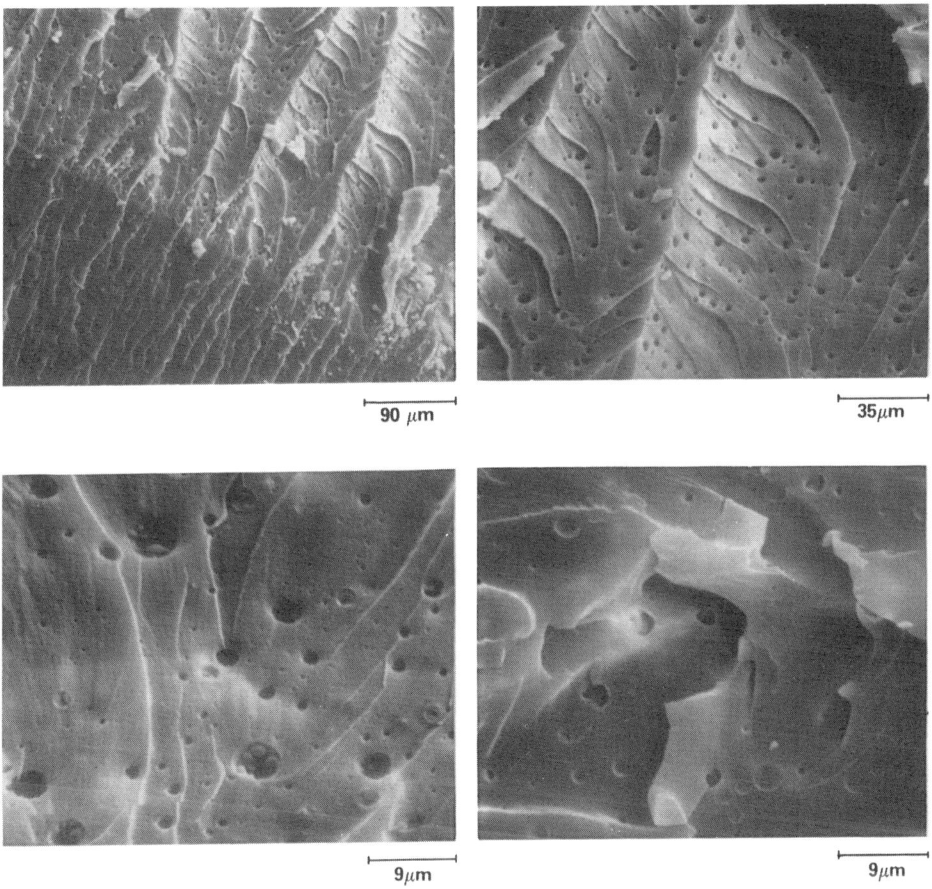


Figure 4.19 SEM pictures of fracture surface for material with a bimodal distribution of particle sizes.

This determination is necessary because the phase-separated particles contain significant amounts of epoxy. Consequently, even though the volume fraction of phase-separated particles is much larger than the volume of CTBN in the sample, this does not mean that all of the elastomer is in the particles. For the systems they studied, Sayre et al. concluded that only 50% of the CTBN could be accounted for in the particles. For a number of reasons, outlined in their paper, they suggested that the matrix contains a fine dispersion of the elastomer inclusions with dimensions below the resolution of STEM, i.e., a few hundred angstroms. It would be extremely interesting to have their analysis extended to other systems to determine how common such a morphology may be. At present, however, any role that such a morphology might have in promoting the extensive yielding that is observed in the fracture surface is only speculative.

The extensive yielding evident in the fractography of CTBN-epoxy materials with both conventional and bimodal particle size distributions suggests that their fracture behavior should be time and temperature dependent. This was shown to be the case by Hunston et al. [22,23,31], who determines $G_{Ic}$ using CT specimens in experiments at constant crosshead speeds between 0.002 in./min (0.05 mm/min) and 2.0 in./min (50 mm/min) and temperatures between -60 and +60°C. They propose that the failure process is dominated by viscoelastic deformations, and thus there should be an inverse relationship between test temperature and loading rate. In fact, the experiments do show that increasing the temperature produces effects similar to those obtained by decreasing the loading rate. If the behavior is viscoelastic and the relationship between these two parameters is simple, the data can be analyzed using time-temperature superposition. Hunston et al. selected time to failure, $t_r$, as the parameter to characterize the loading rate; $t_f$ is the time from initial application of the load until the failure point in the experiment at constant crosshead speed. Some results of this study are given in Figure 4.20 for two piperidine-cured DGEBA-CTBN compositions and a piperidine-cured DGEBA-BTA-CTBN composition. Indeed, as shown in this figure, the data can be successfully shifted using the principle of time-temperature superposition so that $G_{Ic}$ is presented as a function of a reduced time to failure, $t_f/a_T$, where $a_T$ is a function of temperature. Low values of $t_f/a_T$ correspond to low temperatures and high strain rates, while high values correspond to high temperatures and low strain rates.

There is some interesting information in Figure 4.20. The unmodified epoxy exhibits very little rate or temperature dependence over the range of test conditions presented here (data shown as line at bottom of graph). The abrupt increase in $G_{Ic}$ expected of an unmodified epoxy in Figure 4.15 occurs outside the test conditions covered in Figure 4.20. At the higher temperature and lower strain rates, $G_{Ic}$ of the 5% elastomer-modified epoxy with the bimodal distribution of particle sizes greatly exceeds that of the 5% CTBN-epoxy with the conventional particle size distribution and is almost as high as that of the 15% composition. This observation has been made in

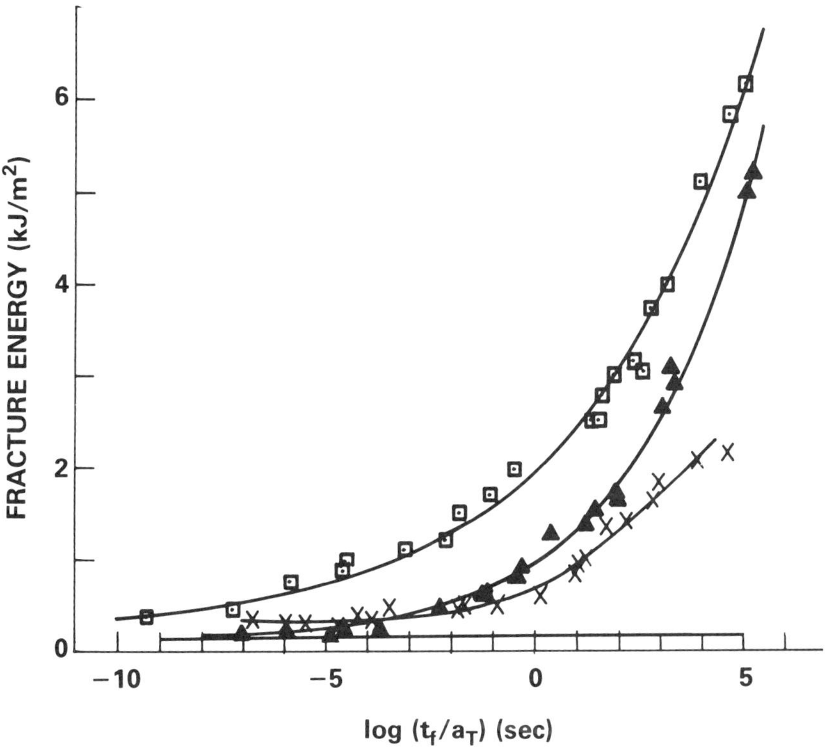


Figure 4.20 Master curves of fracture energy versus reduced time to failure for three different formulations of elastomer-modified epoxy. With no BPA, ⊡ = 15% CTBN and X = 5% CTBN; with BPA, △ = 5% CTBN. Solid line at bottom of graph represents unmodified epoxy.

the past and cited as an advantage of the bimodal systems. A reduction in the reduced time to failure of only three decades, however, brings the bimodal system to near equivalence with the conventional 5% CTBN composition. The curves come together at strain rates that correspond to low-temperature impact testing. In this region there appears to be no advantage of the bimodal system. Moreover, if designs are based on measurements made in the range where the bimodal system is superior, this higher rate sensitivity can lead to significant overestimates of toughness for the bimodal system. Further reductions in temperature (or increases in loading rate) bring the toughness of both modified-epoxy systems down to a value only slightly greater than that of the unmodified epoxy.

The most interesting point raised by the curves in Figure 4.20 concerns the mechanism of failure. Studies of the fracture surfaces suggest that quite similar deformation processes are involved for both the bimodal and conventional systems. Moreover, Hunston and Bullman [31] found the temperature dependences, i.e., the expressions for $a_T$, to be identical within experimental scatter for all of the systems shown in Figure 4.20. The interesting question, then, is why there are differences in the rate dependences. At the moment there is no clear answer to this intriguing question.

In analyzing the data in Figure 4.18, Hunston et al. [22,29,30,31] suggest that the results can be represented empirically by

$$\mathcal{G}_{Ic} = A(t_f)e^{\Delta H/RT} - \mathcal{G}_{IcB} \qquad (15)$$

This equation contains four constants, and although they are strictly empirical, they do have meaning in terms of the measured behavior. Examination of the equation shows that A provides a measure of the magnitude of the toughening, m assesses the rate sensitivity, $\Delta H$ measures the temperature dependence, and $\mathcal{G}_{IcB}$ is the limiting toughness at low temperatures and/or high loading rate.

The data presently available are not sufficient to clearly establish relationships between these parameters and the morphology of the materials. Nevertheless, the following observations can be offered for the four materials systems that have been examined. The thermal parameter, $\Delta H$, is quite similar for all four materials and is identical to that obtained by applying a similar time-temperature superposition process to yield data obtained for two of the materials [68]. Consequently, there is a close connection between yielding and fracture, as might be expected in light of the mechanism discussed above. The parameter m for the bimodal system is significantly different from that for the three conventional distribution systems tested, the latter having identical values for m. The parameter A increases as the rubber concentration increases but is also a function of the morphology.

The results, at present, are quite limited. If, however, the analysis procedure described above continues to be applicable as additional data are obtained, reliable correlations can be developed with Equation (15) or some other relationship. Once such a relationship has been established, only a few test conditions will be

Table 4.4 Composition of the Base Epoxy Resin of a Commercial Modified Epoxy

| Component[a] | Approximate weight % |
|---|---|
| Expoxides<br>(diglycidyl ether of bisphenol A)<br>(epoxidized novolac, epox, equivalent weight 165) | 73 |
| Diphenols<br>(bisphenol A)<br>(tetrabromobisphenol A) | 20 |
| Catalysts<br>(dicyandiamide)<br>(substituted urea) | 7 |

[a]Hexcel Corp.

needed to evaluate the parameters for new materials. A data base can then be established that will be useful in designing new and improved polymer systems.

The CTBN-epoxy resins discussed thus far contained only a liquid CTBN additive. Many commercial adhesive or matrix resin formulations use both a liquid CTBN and a solid rubber in an attempt to obtain a multiple particle size distribution. Bascom et al. [17,18,21] studied a rubber-modified epoxy of this type and determined the effect of the amount of each type of elastomer additive on fracture energy as a function of strain rate. The composition of the base epoxy resin is given in Table 4.4. To this base resin different amounts of a liquid CTBN and a solid CTBN rubber were added. The solid rubber was precooked with the constituents of Table 4.4 before the liquid CTBN was added. The composition and numerical designation of the epoxy materials are listed in Table 4.5.

Fracture energies of the polymers listed in Table 4.5 are plotted as a function of relative strain rate in Figure 4.21. Because the strain rate at the crack tip cannot be determined, the inverse of the time to failure (time between the initial application of the load and the fracture point) was used as a relative measure of strain rate. The data include results obtained from impact tests using an instrumented Izod impact tester (data points at $\sim 7.5 \times 10^2$ sec$^{-1}$). The impact energies were converted to $G_{Ic}$ values using the analysis of Plati and Williams [69]. The results in Figure 4.21 are in many ways similar to those discussed previously for the simpler rubber-epoxy compositions: (1) the unmodified epoxy (composition HX-205) has a low toughness, and $G_{Ic}$ exhibits very little strain rate dependence; (2) all of the rubber-modified compositions are many times tougher than the base epoxy, and $G_{Ic}$ is strain rate dependent;

Table 4.5 Elastomer Content of Epoxy Resins

| Sample[a] | Liquid CTBN[b] (wt %) | Solid elastomer[c] (wt %) |
|---|---|---|
| HX-205[d] | 0 | 0 |
| HX-206 | 8.1 | 0 |
| HX-207 | 8.1 | 1.0 |
| HX-210 | 0 | 8.1 |
| F-185 | 8.1 | 5.4 |

[a]Hexcel Corp.

[b]B. F. Goodrich Hycar 1300 × 13, molecular weight ≈ 3500.

[c]B. F. Goodrich Hycar 1472, molecular weight ≈ 260,000.

[d]Base epoxy system given in Table 4.4.

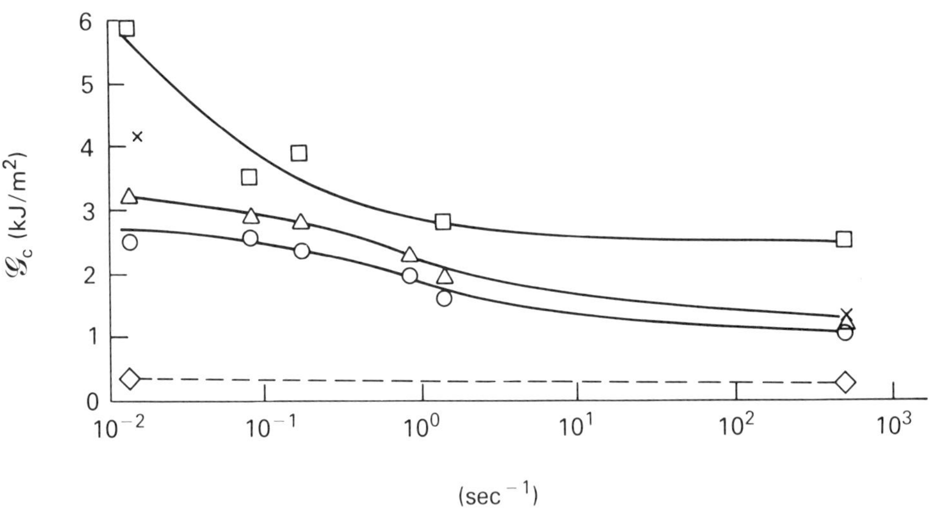


Figure 4.21 Fracture energy versus reciprocal of time to failure for ◇ = HX-205, ○ = HX-206, X = HX-207, △ = HX-210, and □ = F-185 (see Tables 4.4 and 4.5 and Ref. 17).

(3) composition F-185, which contains both liquid and solid rubber, exhibits relatively high toughness in low-strain-rate testing but has a higher strain rate sensitivity than the other compositions. These are similar to the results for the CTBN-DEGA-BPA system discussed above. In both cases the morphology involves both large and small particles.

## B. Adhesive Fracture—Mode I

There are some significant differences between the adhesive fracture of elastomer-modified epoxies and their bulk fracture behavior. All of these differences are related to the adhesive joint geometry—bond thickness, bond angle, etc. The simplest example, and the first to be recognized, was the effect of bond thickness on adhesive $\mathcal{G}_{Ic}$. In Figure 4.22 the fracture energies are given as a function of bond thickness for a 15% CTBN, piperidine-cured, DGEBA epoxy tested as an adhesive at temperatures between 233 and 323 K [15]. At each temperature $\mathcal{G}_{Ic}$ goes through a maximum (or approaches a maximum in the 233 K experiments). Failure occurred by stable crack growth at bond thicknesses less than the maximum, but cracking was unstable at bond thicknesses greater than the maximum. The maximum in the data shifted to smaller bond thicknesses with decreasing test temperature.

The appearance of the stress-whitened zone as the bond line thickness was reduced provides another important feature of the failure. In Figure 4.23, the stress-whitened zone is shown schematically. In specimens with bond thickness above the maximum point, stress whitening is contained within the bond line (Figure 4.23A), and the failure is center-of-bond. Below the maximum, the stress whitening fills the bond line, and the locus of failure is near the interface, with the crack jumping from one side of the bond to the other (Figure 4.23B) as it propagates. It was established by SEM fractography that, while failure occurred near the interface, the crack growth was decidedly in the resin [13].

These observations led to the following proposal for the bond thickness effect. As Figure 4.23B suggests, the zone of stress whitening is increasingly constrained as the bond thickness is decreased, and this effectively reduces $r_c$, the damage zone radius, and thus, according to Equation (3), decreases $\mathcal{G}_{Ic}$. It was argued [13] that the maximum in adhesive $\mathcal{G}_{Ic}$ occurs when the bond thickness, h, is essentially equal to $2 \times r_c$, the critical diameter of the damage zone measured in bulk specimens. This hypothesis was given some support by using Equation (3) to calculate $r_c$ at the different test temperatures from measured values of bulk $\mathcal{G}_{Ic}$, $\sigma_y$, and $\varepsilon_y$. In Table 4.6 values of $h_{max}$ and $2r_c$ are compared, and the agreement is reasonably good considering the assumptions involved. It should be noted, however, that this explanation of the bond thickness effect does not account for the decrease in adhesive $\mathcal{G}_{Ic}$ at bond thicknesses greater than $h_{max}$. This point will be addressed in more detail later in this chapter.

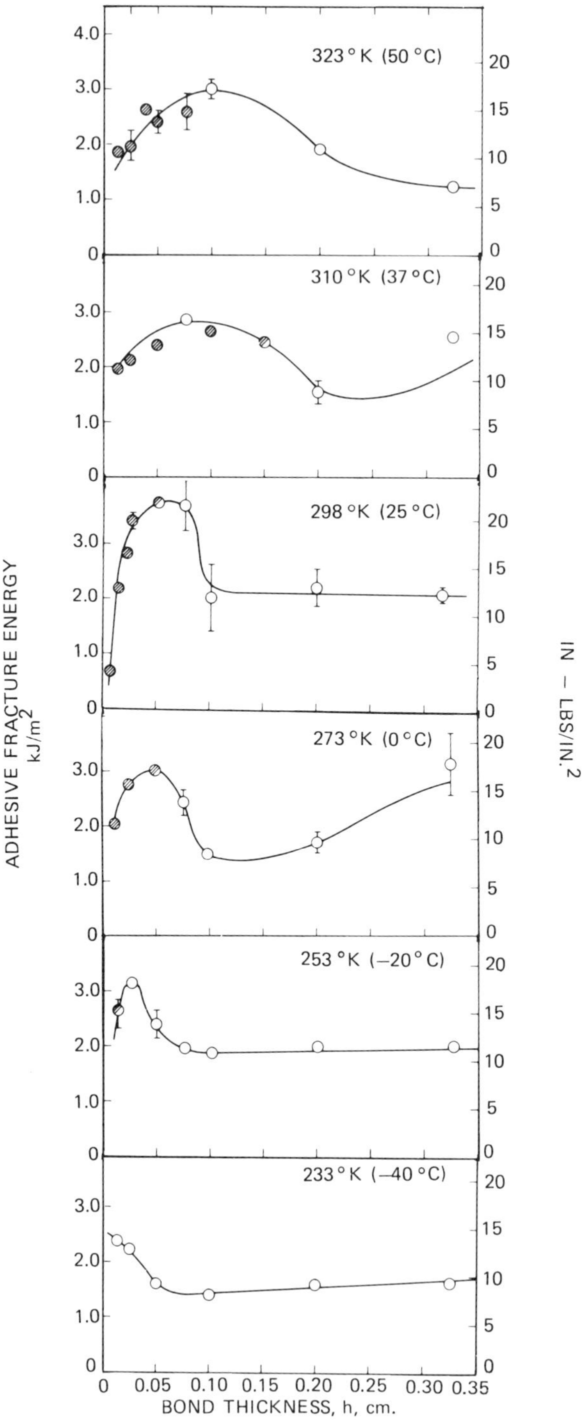
323 °K (50 °C)
310 °K (37 °C)
298 °K (25 °C)
273 °K (0 °C)
253 °K (–20 °C)
233 °K (–40 °C)
ADHESIVE FRACTURE ENERGY kJ/m²
IN – LBS/IN.²
BOND THICKNESS, h, cm.
0
0.05
0.10
0.15
0.20
0.25
0.30
0.35
1.0
2.0
3.0
4.0
5
10
15
20

Figure 4.22 Fracture energy as a function of bond thickness at a series of temperatures (15% CTBN-modified DGEBA cured with piperidine).

In more recent work, Hunston et al. [22,23] studied the bond thickness dependence of the effect of loading rate as well as temperature. Based on the results for the fracture of bulk samples as a function of rate and temperature, they suggested that the bond thickness dependence should be affected in a predictable way by changes in loading rate; i.e., increasing the loading rate should give results similar to decreasing the temperature. Using data at selected temperatures and crosshead speeds, they were able to show that their results were consistent with this idea.

This hypothesis was later confirmed by Kinloch and Shaw [70], who obtained complete bond thickness curves similar to Figure 4.22 but at a fixed temperature and a series of different crosshead speeds. The adhesive used was identical to that employed by Bascom and Cottington [15] but with a slightly lower elastomer concentration. Kinloch and Shaw found that the maximum in $G_{Ic}$ shifted to lower bond thickness values with increasing loading rate, and this is consistent with the previously observed shift in the maximum to a lower bond thickness with decreasing temperature (Figure 4.22). To make a quantitative comparison, Hunston et al. [30] proposed using results corresponding to the same time to failure as a better means of comparing bulk resin and adhesive tests. Using this procedure, they found good agreement between $2r_c$ and the bond thickness at which the adhesive $G_{Ic}$ is a maximum when $r_c$ is calculated assuming plane stress conditions [30].

At first glance, it is difficult to understand how plane stress conditions could prevail in a highly constrained bond line. In fact, the actual situation is more complicated than either plane stress or plane strain since the material contains two phases with different moduli. As a result, the details of the stress field in the material

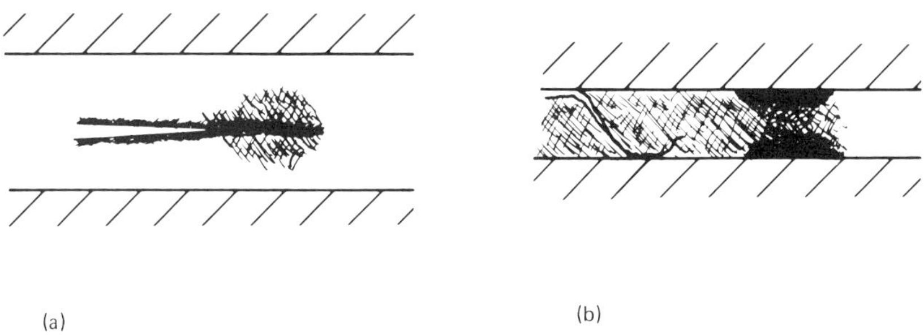


Figure 4.23 Schematic of deformation zone in adhesive bonds of different thicknesses.

Table 4.6 Comparison of Bond Thickness, $h_{max}$, at Maximum Adhesive Fracture Energy with Crack Tip Deformation Zone Diameter, $2r_c$, for Bulk Resin Fracture[a]

| Temperature (K) | $h_{max}$ (cm) | $2r_c$ (cm) |
|---|---|---|
| 323 | 0.10 | 0.12 |
| 310 | 0.075 | 0.078 |
| 298 | 0.060 | 0.033 |
| 273 | 0.050 | 0.013 |
| 253 | 0.025 | 0.008 |
| 233 | <0.01 | 0.004 |

[a]From Ref. 15.

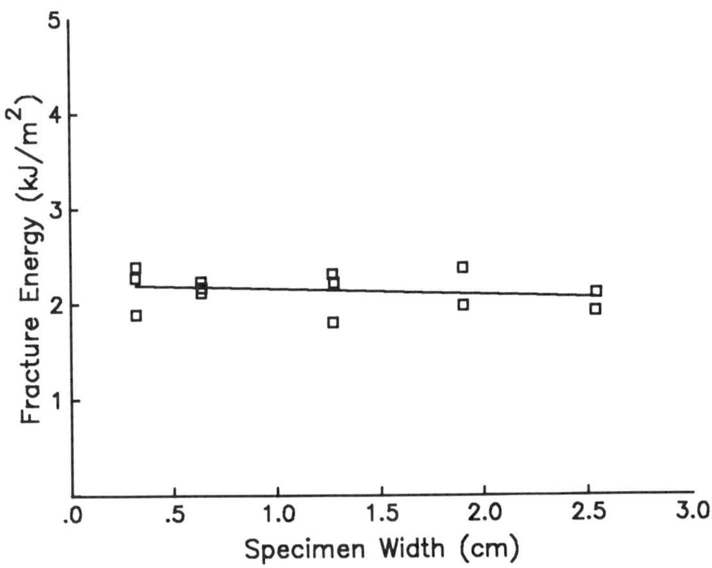


Figure 4.24 Fracture energy versus specimen width for bulk samples of 15% CTBN-modified DGEBA cured with piperidine.

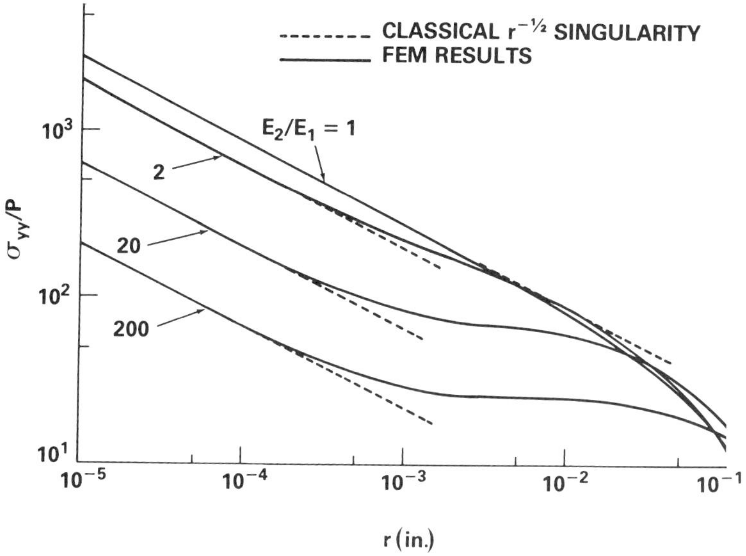


Figure 4.25 Local tensile stress normalized with applied load as a function of distance, r, ahead of crack tip. Results from Ref. 41 for different ratios of moduli for adherend and adhesive ($E_2/E_1$).

are not the same as those for a homogeneous solid. For example, if bulk samples of rubber-modified epoxies are examined for a dependence of fracture energy on specimen width, the expected transition from plane stress to plane strain is not observed; in fact, the fracture energy shows very little dependence on width [70,71] over the range that has been measured. This is illustrated in Figure 4.24, which shows fracture energy measurements on bulk samples of piperidine-cured DGEBA with 15% CTBN. Based on a simple elastic-plastic model for homogeneous materials [43,44], the transition would be expected at widths below 1 cm. A second complicating factor is the particle cavitation that is thought to occur. If this relieves, to some degree, the triaxiality of the crack tip stress field, it is no longer clear that a plane strain model would be appropriate. Finally, the simple model described above does not explain the falloff in adhesive fracture energy at thicknesses above that corresponding to the maximum. Consequently, it is clear that although a simple plane stress model has had some success in predicting the optimum bond thickness, more realistic models are needed before the thickness effect can be understood quantitatively.

In recent years considerable progress has been made in analyzing the mode I failure of the tapered double cantilever beam adhesive fracture specimen. This work provides a qualitative explanation for the maximum in $\mathcal{G}_{Ic}$ versus bond thickness. It begins with the early studies of Wang et al. [41] on the stress distribution in the

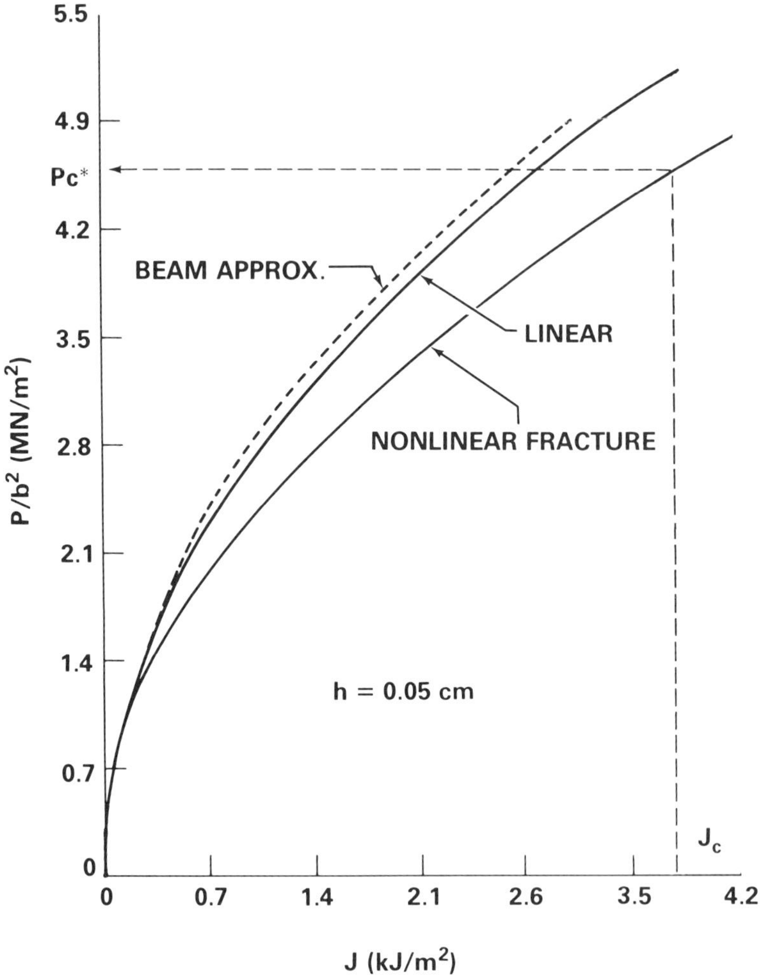


Figure 4.26 Comparison of results for load normalized with specimen width plotted against applied J [72]. Bond thickness, h, is 0.05 cm.

bond line of a TDCB adhesive specimen. They showed with a linear elastic finite-element analysis (FEA) that the crack tip stress distribution does not decrease as $r^{-1/2}$ ahead of the crack, but instead decays more slowly than expected along the bond line as shown in Figure 4.25. This effect increases as the bond thickness decreases and is due to the difference in moduli between the metal adherend

($E_2$) and the adhesive ($E_1$). For aluminum and epoxy, $E_2/E_1 = 20$. Subsequently, Wang [72] conducted an FEA analysis using a nonlinear constitutive equation for the adhesive based on data for commercial and model materials. The results are presented in terms of a J-integral, which can be viewed in a simplistic way as a generalization for nonlinear systems. In Figure 4.26, Wang [72] has shown a comparison of results obtained with simple beam theory, linear FEA, and nonlinear FEA for a TDCB adhesive specimen. As illustrated in the figure, beam theory and linear FEA give essentially the same result. The curve for a more realistic constitutive equation, however, suggests that both may underestimate the toughness of the bond when nonlinear (high fracture energy) materials are involved. Nevertheless, the Wang analysis found that the trends such as dependence of adhesive $J_{Ic}$ on bond thickness were unaltered.

Most recently, Wang, in cooperation with Hunston and Kinloch [30,72], reexamined the effects of bond thickness on adhesive fracture. The FEA of Wang suggests that two opposing effects are active in the bond line as the thickness is reduced. First, there is a progressive reduction in the volume in which the damage zone can develop due to the constraint in the thickness direction. This is the same as the effect proposed earlier by Bascom et al. and studies by numerous authors, as discussed above. In addition to this effect, Wang's analysis finds that the zone extends farther down the bond line ahead of the crack tip as the thickness is decreased. This is caused by the high stresses farther than expected ahead of the crack tip as shown in Figure 4.25. The analysis shows that both effects are present and interact in the following way. At large bond thicknesses ($\gg 2r_c$), the adhesive and bulk samples show similar behavior. As the bond thickness is decreased, the extension of the deformation zone down the bond line causes the zone size to increase. This produces an increase in the $J_{Ic}$ of the adhesive, although the exact relation between $J_{Ic}$ and zone size is complex since the details of the stress-strain field in the zone change as the bond thickness is decreased. When the bond thickness is further decreased the constraint imposed by the interaction between the adherends and the zone becomes important. This eventually produces a decrease in the zone size (i.e., a reduction in height) and a related decrease in $J_{Ic}$. The net result is that $J_{Ic}$ first increases and then decreases as the bond thickness is reduced.

At present it is not possible to quantitatively predict the $J_{Ic}$ curve with Wang's analysis using any simple failure criterion. On the other hand, if $J_I$ is regarded as the driving force for failure, the analysis does predict a minimum in this driving force at any given load, and this minimum occurs at the bond thickness where $J_{Ic}$ is a maximum.

In an effort to test the hypothesis proposed by Wang's analysis, Hunston and Rushford [73] took high-speed movies of the crack tip region during loading and failure. They used the size of the stress-whitened region as a relative measure of the deformation zone size. The results qualitatively support Wang's analysis. In thick bonds ($h \gg 2r_c$), the stress-whitened region in the adhesive at the failure

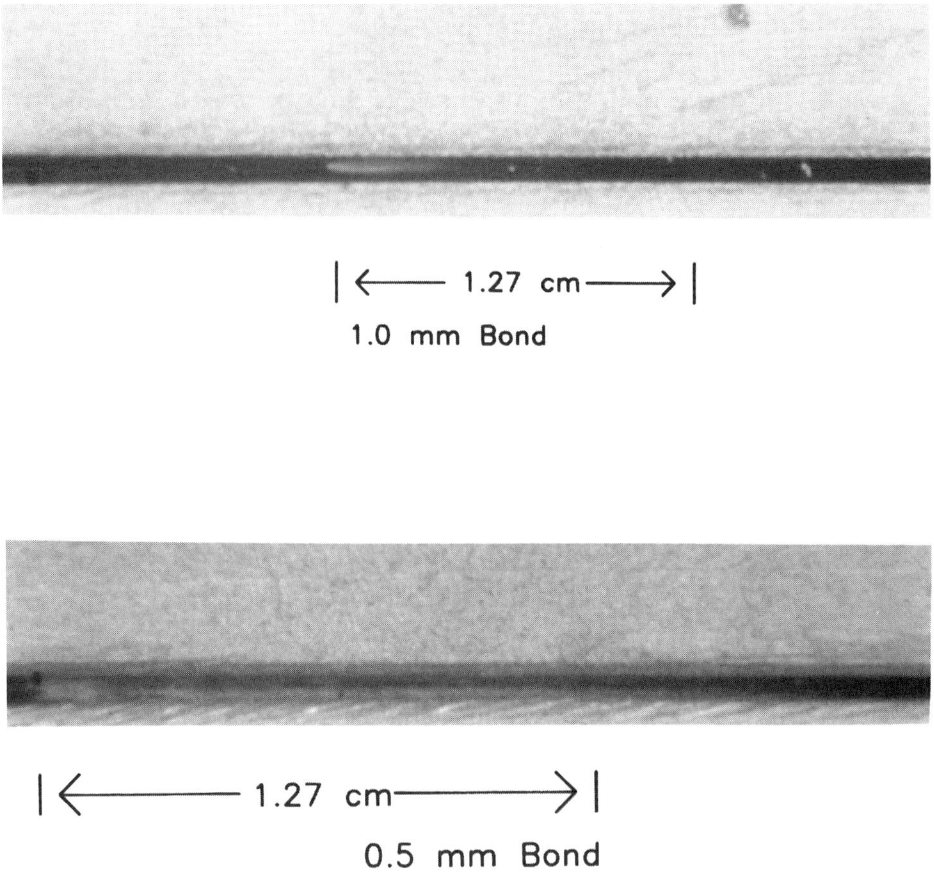


Figure 4.27 Photographs of stress-whitened zone in adhesive (15% CTBN-modified DGEBA cured with piperidine) just before failure. (Top) $h = h_m$; (bottom) $h = 2h_m$.

point is contained within the bond and looks quite similar to that in bulk samples. When the bond thickness is near the optimum ($h = 2r_c$), however, the movie shows that the stress-whitened zone at failure fills the bond line and extends ahead of the crack tip for a much longer distance than would be expected from observations on the bulk resin. Two still pictures from the movie as shown in Figure 4.27.

The temperature, loading rate, and bond thickness effects discussed above can shed new light on an adhesive property that has been known for many years but has not been clearly understood.

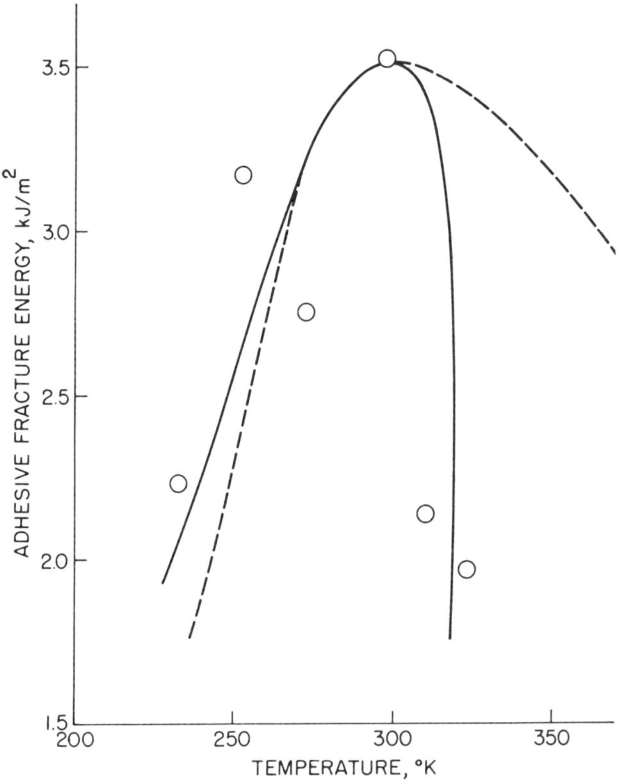


Figure 4.28 Adhesive fracture as a function of temperature for commercial material (dashed line) and model system from Figure 4.22 (points and solid line).

It is customary to designate a temperature range in which an adhesive exhibits its maximum fracture behavior. For example, Mostovoy and Ripling [74] determined the adhesive $G_{Ic}$ as a function of temperature for a number of commercial structural adhesives, and in most cases there was a maximum in toughness. The dashed line in Figure 4.28 represents one of the commercial adhesives tested by Mostovoy and Ripling. This is typical of the type of data often provided for adhesives in that only a single bond thickness is used, i.e., the thickness recommended by the manufacturer for applications. To understand this behavior, however, the influence of bond thickness (and loading rate) as well as temperature must be taken into account. For example, the data points and solid curve in Figure 4.28 are taken from Figure 4.22 at a bond thickness of 0.25 mm. Although a completely rigorous comparison would include

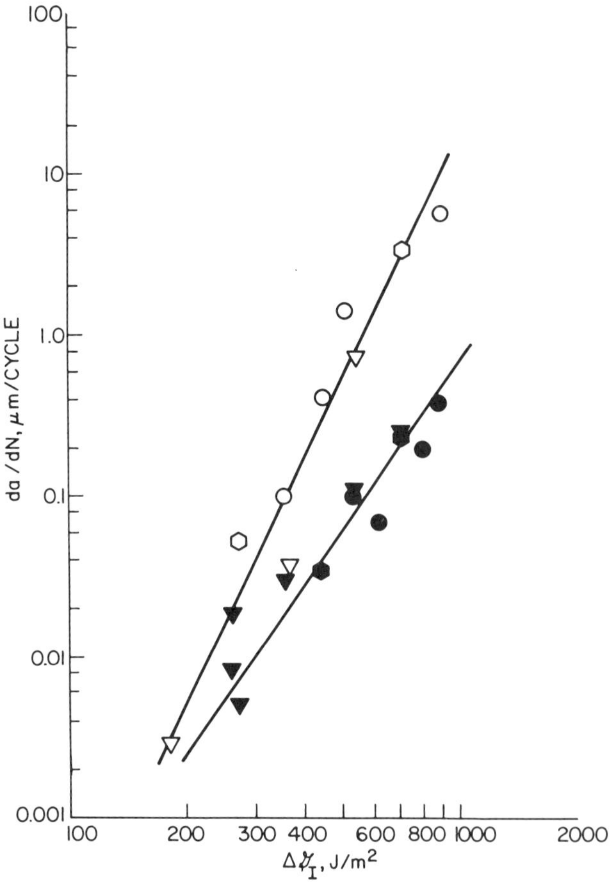


Figure 4.29 Fatigue growth plots for 15% CTBN-modified DGEBA cured with epoxy and measured at two load ratios: R = 0.6 for open symbols and R = 0.1 for filled points [78].

the influence of loading rate as well, the curve for the model adhesive is quite similar to that for the commercial material.

Another area of great importance is fatigue behavior. There have been innumerable studies of fatigue failure for structural adhesives, including their fatigue crack growth characteristics. Research sponsored by the Air Force, particularly work at General Dynamics [75], examined most of the commercial materials available at the time. More recently, NASA [76,77] has conducted a wide-ranging program on fatigue with a focus on failure criteria. Although fatigue has received a good deal of attention, relatively few of these studies

Table 4.7 Comparison of Power Law Constants[a]

| Adhesive chemical type/support[a] | A | n |
|---|---|---|
| CTBN-DGEBA/none | $4.85 \times 10^{-17}$ | 3.56 |
| DGEBA-MPDA[c]/none | $8.87 \times 10^{-13}$ | 2.76 |
| Nitrile-spoxy/Dacron | $6.28 \times 10^{-14}$ | 3.78 |
| Nylon-epoxy/none | $4.56 \times 10^{-14}$ | 2.99 |
| Nitrile-phenolic/none | $2.42 \times 10^{-22}$ | 5.80 |
| Modified epoxy/glass | $1.20 \times 10^{-16}$ | 3.90 |
| Modified epoxy/none | $2.05 \times 10^{-16}$ | 3.58 |

[a] $da/dN = A(\Delta G_{Ic})^n$ for R = 0.6, units are μm/cycle and $J/m^2$.

[b] Scrim cloth.

[c] *m*-Phenylenediamine, 14.5 parts per hundred parts of resin by weight.

have focused on the micromechanics of failure in the adhesive. One such study was conducted by Bascom and Mostovoy [78] and determined the fatigue life of a 15 wt % CTBN-DEGBA piperidine-cured epoxy using the TDCB adhesive specimen (bond thickness of 0.025 cm). A power law plot of crack growth/cycle (da/dN) against the applied $\Delta G_I$ is given in Figure 4.29 for two levels of load differential ($R = P_{min}/P_{max}$). The power law constants for this adhesive are compared in Table 4.7 with results obtained by Mostovoy and Ripling [74] for an unmodified epoxy (DGEBA-*m*-phenylenediamine) and various commercial adhesives. The results for the CTBN-DEGBA material are comparable to the constants for the commercial adhesives, especially the modified epoxies. Scanning electron microscopy examination of the fatigue fracture surface strongly suggests that the failure occurs by microcracks ahead of the main crack front. At the lower $\Delta G_I$ separation occurs by coalescence of the microcracks, while at the higher load differential separation involves a well-defined crack front advancing through the microcracks. In all cases, the locus of failure is in the adhesive but near one of the interfaces.

## C. Adhesive Fracture—Combined Mode

Rarely, if ever, is a structural adhesive joint designed to experience pure mode I loading. In general, the bond line and the flaws within the bond line will see some combination of mode I, II, and possibly III loads (Figure 4.1). In the case of an isotropic specimen, e.g., bulk fracture, it has been determined [79,80] that

Table 4.8 Effect of Fracture Mode on Adhesive Energy of Epoxy Resins[a]

| Resin | Fracture energy (J/m$^2$) | | |
|---|---|---|---|
| | $G_{Ic}$ | $G_{IIc}$ | $G_{(I,II)c}$ 45° |
| TEPA[b]-DGEBA | 70 | 1450 | 130[d] |
| HHPA[c]-DGEBA | 116 | — | 140[d] |
| Piperidine-DGEBA | 121 | — | 255[d] |

[a]Data from Refs.

[b]TEPA, Tetraethylpentamine.

[c]HHPA, Hexahydrophthalic anhydride.

[d]Data from scarf joint with bond line at 45° angle to direction of applied load.

mode I is the lowest-energy fracture mode, and thus the crack always propagates along a path normal to the direction of maximum resolved tension. In other words, a crack in an isotropic plate will propagate in mode-I-like fracture regardless of how the initial flaw is oriented with respect to the applied load. This is not necessarily the case in adhesive fracture. Here the flow propagation will be constrained to the resin layer if tough adherends are used, and this can force the crack to propagate in the bond direction at or near the interface regardless of the orientation of the bond line. Consequently, attention must be given to adhesive fracture in combined stress conditions since measurements on bulk samples may be of little use in predicting adhesive behavior.

Ripling et al. [2] determined the mode II adhesive fracture of a simple unmodified epoxy resin (DGEBA-TEBA) using the specimen illustrated in Figure 4.7B. Trantina [45-47] determined the mixed-mode adhesive fracture behavior of the epoxy resin using the scarf-joint specimen (Figure 4.7A), and Bascom et al. [12] used the same specimen to study the mixed-mode adhesive fracture behavior of two other unmodified epoxy resins (DGEBA-HHPA and DGEBA-piperidine).

The results of these tests are listed in Table 4.8 with the notation $G_{IIc}$ and $G_{(I,II)c}$ for mode II and mixed-mode fracture energies, respectively. The mode I values are also presented and are in all cases less than the $G_{(I,II)c}$ values. The rationale for the higher $G_{(I,II)c}$ values comes from observations of crack growth. It was established by Trantina [47] and Bascom et al. [12,19] that the precrack (AB in Figure 4.30) advances across the bond line (BC in Figure 4.30) during loading, i.e., a mode-I-like fracture, but arrests when it reaches the adhesive/adherend interface. The

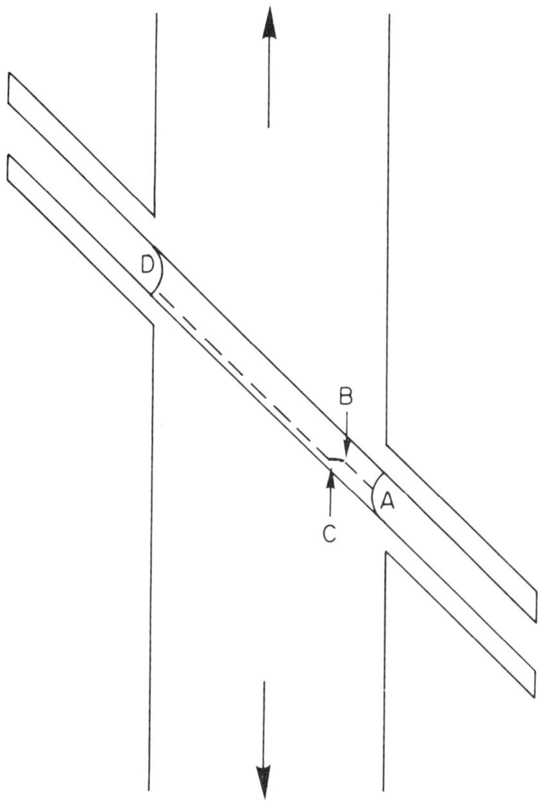


Figure 4.30 Diagram of crack growth path in scarf joint. Loading direction is indicated by arrows at top and bottom.

specimen must then be loaded to a higher level to force the crack to grow in an unfavorable direction (CD in Figure 4.30).

The pure shear fracture values, $G_{IIc}$, were found to be more than a factor of 10 greater than either the mode I or mixed-mode values. Observations by Ripling et al. [2] and at NRL [81], however, suggest that failure in the shear specimen may not occur by simple mode II fracture. Rather, it seems to proceed via the formation throughout the length of the bond of microcracks across the bond in the mode-I-like direction, and these microvoids then join by failure of the material in between. As a result, many scientists have concluded that if adhesive $G_{IIc}$ results have any meaning, values such as those in Table 4.8 probably represent lower limits.

In other work [12,19] using the "scarf-joint" specimen (Figure 4.7A), it was shown that the roughness of the adherend surface can strongly influence the value of $G_{(I,II)c}$. This roughness dependence can be understood in terms of the locus of the crack growth in the scarf joint. Since the crack propagates first over to the interface, the surface roughness can affect the toughness by influencing the crack tip deformation process or the local direction of crack growth. Generally—but not always—$G_{(I,II)c}$ is greater for a rough surface than for a highly polished adherend [12]. The proposed explanation lies in the assertion that the crack tip can be trapped in the surface roughness, and the extra energy required to divert it out depends on the type and degree of the roughness. In certain cases, however, the asperities of the roughness may provide local stress concentrations that add to the crack tip stress field and thus reduce the measured toughness [12,19].

In the scarf-joint test, the locus of failure is apparently interfacial, but by using radioactive tracers in the adhesive it was established [19] that the fracture actually occurred in the resin, albeit deceptively close to the interface. As outlined above, the nature of the mixed-mode stress field directs the crack toward the interface, and this results in a locus of failure near but not at the interface. Failure to recognize this common trait of mixed-mode tests has led some people to conclude erroneously that there was some weakness of the adherend/adhesive bond.

Although the mixed-mode adhesive fracture behavior of brittle, unmodified epoxies is understood qualitatively, the behavior of the CTBN-elastomer epoxy compositions and commercial structural adhesives is a very different situation. Compared to the mode I fracture energies, some of the literature results show mixed-mode fracture energies that are higher [74,82], others show lower values for mixed-mode fracture [12,14,19,20,24], and still others show little difference at least until the mode II component is very high [76,77]. A complete scientific explanation of this complex area will require much more detailed research. Nevertheless, it is useful to examine the results that are available.

The first question that might be asked is whether the tough materials behave in some unusual and unexpected manner in mixed-mode loading. To investigate this behavior in a qualitative way, plates with centrally located precracks were made from both modified and unmodified epoxies and loaded in tension. The precracks were inserted at different angles to the applied load to obtain various types of mixed-mode loading. The failed specimens were then examined to determine the direction of crack initiation. The results [20,29] showed that in all cases the initiation directions in the modified and unmodified materials were in good agreement with predictions based on initiation in a direction perpendicular to the maximum resolved tensile stress at the crack tip. In conclusion, therefore, at least in these tests, the modified epoxy exhibited no unusual behavior.

Why then do the bonded specimens give conflicting results? Unfortunately, there is no completely satisfactory explanation at present, but some interesting observations can be made. The tests that gave high mixed-mode fracture energies [74,82] generally involved a scrim cloth (a thin layer of cloth used in some adhesives to control bond thickness), and the failure was center-of-bond along the scrim cloth. Moreover, failure often occurred when the load component perpendicular to the bond line reached the mode I value.

On the other hand, the tests that gave low mixed-mode fracture energies [12,14,19,20,24] involved adhesives without a scrim cloth. This includes experiments conducted as part of this work using both the scarf joint and the ILMM. These results were reported in detail elsewhere [19,24] and will only be summarized here. There was good agreement between the data for the two geometries, and experiments designed to check the validity of the tests—for example, studies to ensure that the fracture energies were independent of precrack length—gave positive results. In these tests the crack was driven toward the adhesive-adherend interface and then propagated along the interface in the adhesive. This is, of course, what would be expected for the mixed-mode stress field.

There is, however, an important uncertainty in these tests and, in fact, in most mixed-mode tests. Although the overall path of crack growth is known, the growth is unstable, and thus the fracture energy depends only on the direction for the first increment of rapid crack growth. For the unmodified material the crack position can be followed during loading, so there is at least some evidence for choosing this direction. With the modified materials, on the other hand, the large extent of deformation in the crack tip region obscures the details of the initial crack growth. Since the equations for calculating the fracture energy must make assumptions about the initial crack growth direction, this adds an important uncertainty to the results.

The most recent data [76,77] were obtained with the cracked lap-shear specimen (see Refs. 76 and 77 for specimen description), and these results suggest that for a given material failure occurs when the total fracture energy reaches a particular value. For both mode I and mixed-node loading the authors found that this value provided a unique failure criterion as long as the fraction of mode II loading was not too large. In these tests the crack path was also near the interface but in the adhesive.

Clearly, the literature shows a wide range of results for mixed-mode fracture, and therefore generalization is difficult at present. Nevertheless, based on these results some interesting speculations can be offered. For the first set of data mentioned above (mixed-mode fracture energy higher than mode I), the scrim cloth may have provided either a change in the stress field or a preferred path (i.e., weaker) for the crack. This keeps the crack in the center of the bond despite the fact that the normal mixed-mode stress field would drive the crack toward the interface. Under these conditions, it might be predicted that the initiation of rapid crack growth would occur when the load component normal to the scrim cloth plane reached the mode I failure load. On the other hand, for tests where a scrim cloth was not present, the crack was driven toward the interface, as was seen in the unmodified epoxy experiments. If either of the latter two sets of results is correct and the fracture energies are lower than expected (equal to or less than the mode I values), this may be related to some type of interaction between the interface and the crack tip deformation zone that restricts the toughness. Clearly, there is a need for a more detailed examination of mode I and mixed-mode failure behaviors both with and without a scrim cloth to establish a more complete understanding of these important effects.

## D. Interlaminar Fracture of Composites

Another area of interest for toughened epoxies is composite matrix resins. State-of-the-art composites are more susceptible to the initiation and growth of interlaminar cracks than is desirable. One possible approach to improving this situation is the use of tougher matrix resins. Intuitively, it is reasonable to suspect that the resistance to initiation and/or growth of interlaminar cracks would be improved if tougher resins were used as the matrix. On the other hand, the results discussed above for adhesive bonds might suggest that little is to be gained by toughening the resin since the very small fiber-fiber spacing could greatly constrain the deformation zone size and hence the composite toughness.

Evidence in the early literature concerning this question is rather scarce and at times ambiguous [83-85]. One problem is that factors other than matrix toughness can influence the interlaminar fracture energy, and it is generally not possible to control all of these factors. Consequently, unless the range of matrix toughness covered is very large, it is difficult to separate resin effects from the influence of other factors. Another complication is that some of the previous studies using test specimens, such as a short beam shear test, confused the issue because they assumed that failure occurred by interlaminar fracture. There is evidence that this assumption was often ill-founded [86].

Eventually, a conclusive comparison [18] was made as part of the work being reviewed here. It employed two model systems where the bulk resins exhibited a large difference in toughness (a factor of 20 in the fracture energy). These experiments employed epoxy matrix composites with and without elastomeric modifiers. The base epoxy used is described in Table 4.4 and designated as composition HX-205 in Table 4.5. Composites were fabricated from glass and graphite woven cloth using the HX-205 and the F-185 resin, which is the base epoxy with a liquid CTBN and a solid elastomer (Table 4.5). Interlaminar $G_c$ values were determined using the width-tapered specimen (Figure 4.8), and the results are given in Table 4.9. The improvement in the interlaminar fracture energy with increased matrix toughness is clear: a factor of 4 for the glass-reinforced composite and of 8 for the graphite-reinforced material.

Fractography of the failed surfaces of the composite [18] using SEM revealed that when the matrix was the unmodified epoxy, the failure process was largely restricted to the single interlayer of resin between the fibers on either side of the crack plane.

Table 4.9 Effect of Elastomeric Modifiers on Resin and Composite Toughness

| Material | Fiber volume (%) | Modulus (GPa) | Fracture energy $G_{Ic}$ (kJ/m$^2$) |
|---|---|---|---|
| | | Tensile | Neat resin |
| 205 resin | 0 | 2.8 | 0.27 |
| F-185 | 0 | 2.2 | 5.1 |
| | | Flexural | Interlaminar |
| 205/glass cloth | 59.9 | 31.6 | 1.0 |
| F-185/glass cloth | 59.7 | 24.0 | 4.4 |
| 205/graphite cloth | 61.0 | 54.1 | 0.60 |
| F-185/graphite cloth | 57.9 | 42.1 | 4.6 |

However, when the matrix resin contained the elastomer modifier, there was considerable deformation and/or damage in a number of layers above and below the plane of the failure. This is an important observation because it demonstrated that the deformation zone was not restricted by the fibers on either side of the crack to just one layer of resin. This is consistent with the fact that an improvement in interlaminar fracture energy could be obtained with high-toughness resins. On the other hand, a factor of 20 improvement in the resin produced only a factor of 4 to 8 improvement in the composite. Thus, there is a less than complete transfer of toughness from the resin to the composite. It has been suggested that this can be explained by invoking some degree of constraint by the fibers on the deformation zone size or shape. As of now, however, no quantitative analysis of this has been presented.

At high magnifications the fracture surface for these model systems shows clear evidence of plastic deformation of the matrix, particularly around the fibers (Figure 4.31). The elastomer-modified resin shows much more extensive yielding of the matrix than the unmodified resin and shows evidence of dilatation and cavitation on the elastomer particles (as in Figures 4.31 to 4.33). In composites fabricated using woven reinforcement, there are resin-rich regions at the crossovers of the warp and fill stands. In the absence of an elastomer additive, these regions show relatively little evidence of plastic deformation. However, when the matrix contains the liquid and solid rubber additives, the fracture surfaces at the crossover points show extensive deformation similar to that exhibited by the same resin on fracture surfaces of bulk specimens (Figure 4.32).

Initially, it was suspected that the matrix-rich regions were very important to the improvement in interlaminar fracture energy. This

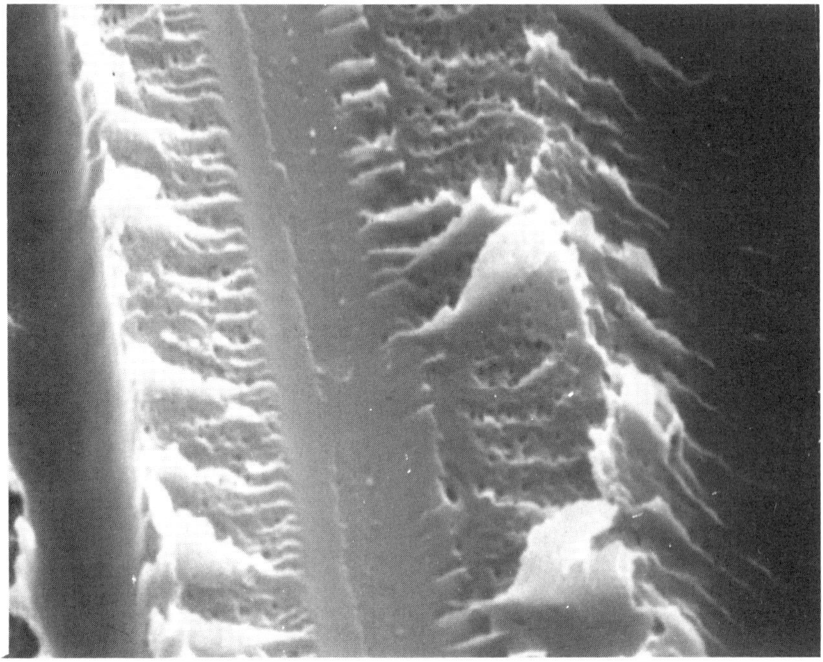

Figure 4.31 Fracture surface of F-185 composite showing deformation near region where fiber had been pulled out.

suggests that unidirectional lay-ups would not show the same improvement with increases in matrix toughness. Recent work in a number of laboratories [28 and references therein] using the same resins, however, has shown that unidirectional composites exhibit improvements that are as large or almost as large as those for woven composites. Extensive deformation has been observed in the resin for many plies above and below the crack plane [87]. Thus the resin-rich areas at the crossover points in the weave are not essential for toughening in the composite.

Since the general concept of using a very tough resin to improve the resistance of the composite to interlaminar fracture has been established, it is of interest to examine the detailed relationship between the fracture behaviors of resins and their composites. As a first step, a study [28,32,88] was conducted to compare the mode I bulk resin fracture energies for a series of different resins with the interlaminar fracture energies as measured with the DCB specimen for composites made with these resins. The results have been reported in detail [28,32,88] and will only be summarized here.

Figure 4.34 shows the results of this study with the resins separated into four general types (details of the materials can be found in Refs. 28 and 88. Although there is significant scatter in the

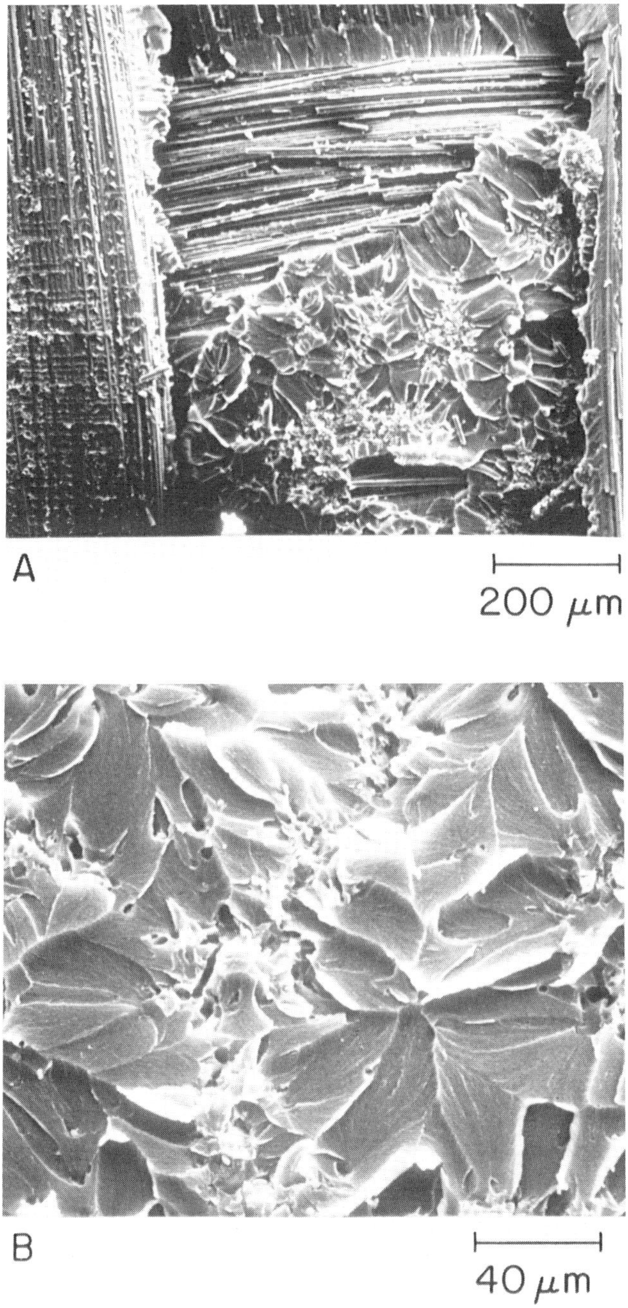


Figure 4.32 Fracture surface showing deformation of resin in crossover region of weave in F-185 composite.

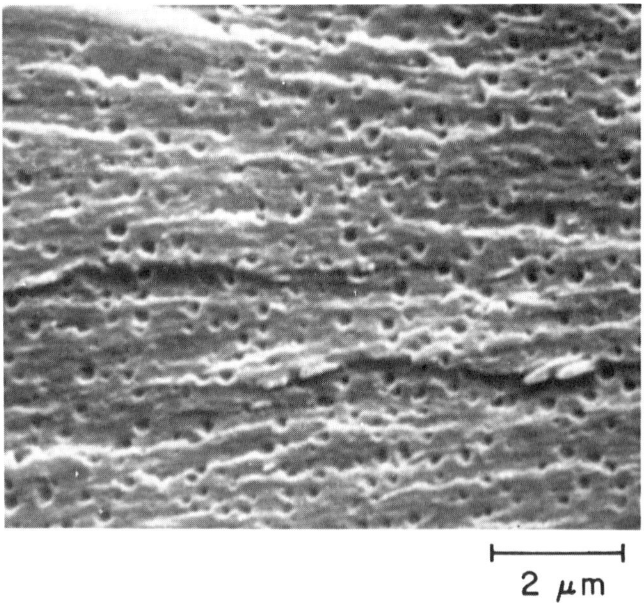


Figure 4.33 Fracture surface showing cavitation of rubber particles in F-185 composite.

data, general trends can clearly be seen. These trends are indicated by the two straight lines in the figure, but the lines are not meant to imply that the trends should necessarily be linear. For the brittle systems the composites have fracture energies which are, on average, about twice those of the bulk resins. The rationale for this invokes a number of mechanisms [28]. For example, the fiber layers in the composite are not perfectly planar, and thus the crack must go around fibers and break fibers to propagate. As a result, it is more difficult to drive the crack in the composite than in the bulk resin.

For the tougher matrices, there is a less advantageous trade-off. Increasing the resin fracture energy by 3 $J/m^2$ produces only about a 1 $J/m^2$ improvement in the composite. As with the results discussed above, this relationship is qualitatively attributed to a constraint imposed by the fibers on the deformation zone. Clearly, more work needs to be done so that the exact nature of this constraint can be understood.

The transition between the two types of behavior outlined above occurs somewhere in the vicinity of the point where the two lines in Figure 4.34 intersect. This is the same general range where the sizes of the deformation zone and the spacing between plies in the composite are similar. This is consistent with the idea that the

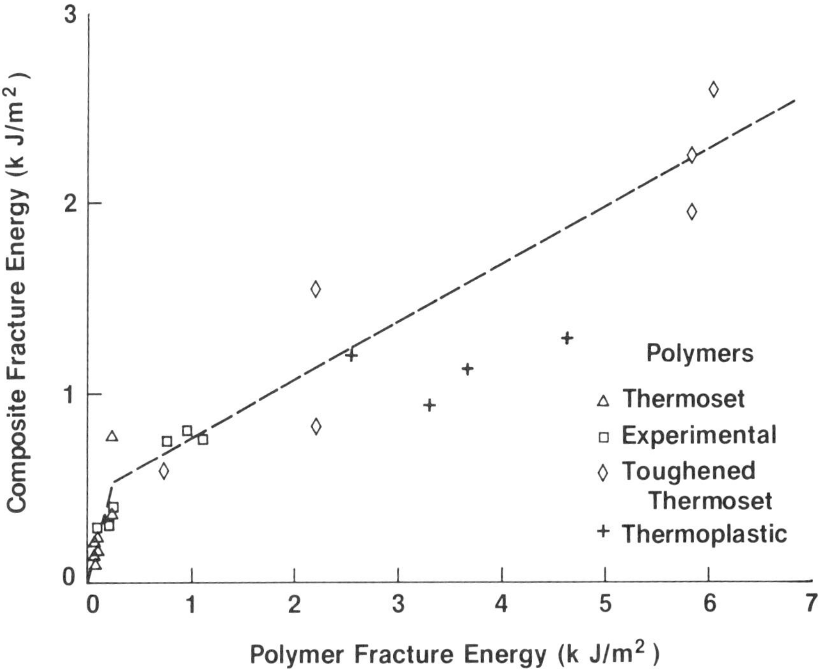


Figure 4.34 Comparison of resin fracture energies and composite interlaminar fracture energies for many different resin systems. Lines indicate general trends but are not meant to indicate that the trends are necessarily linear.

trade-off between matrix and composite toughnesses changes at the point where the deformation zone size first begins to interact with the fibers.

Although a general correlation between the resin and composite fracture behaviors can be seen in Figure 4.34, there is also considerable scatter in the data above and below the trend lines. Some of this scatter can be explained by the influence of parameters other than matrix toughness. For example, when two different fabrication procedures were employed to prepare composites using virtually identical starting materials, significant differences in interlaminar fracture energies were obtained. This shows that processing can have an important influence on the fracture resistance, although it may not yet always be possible to explain why.

Another potentially critical factor is fiber-matrix bonding. In Figure 4.34 there is an indication that most of the thermoplastics give a lower interlaminar fracture energy than would be expected based on comparisons with toughened thermosets having similar bulk resin fracture energies. It has been suggested [28] that the

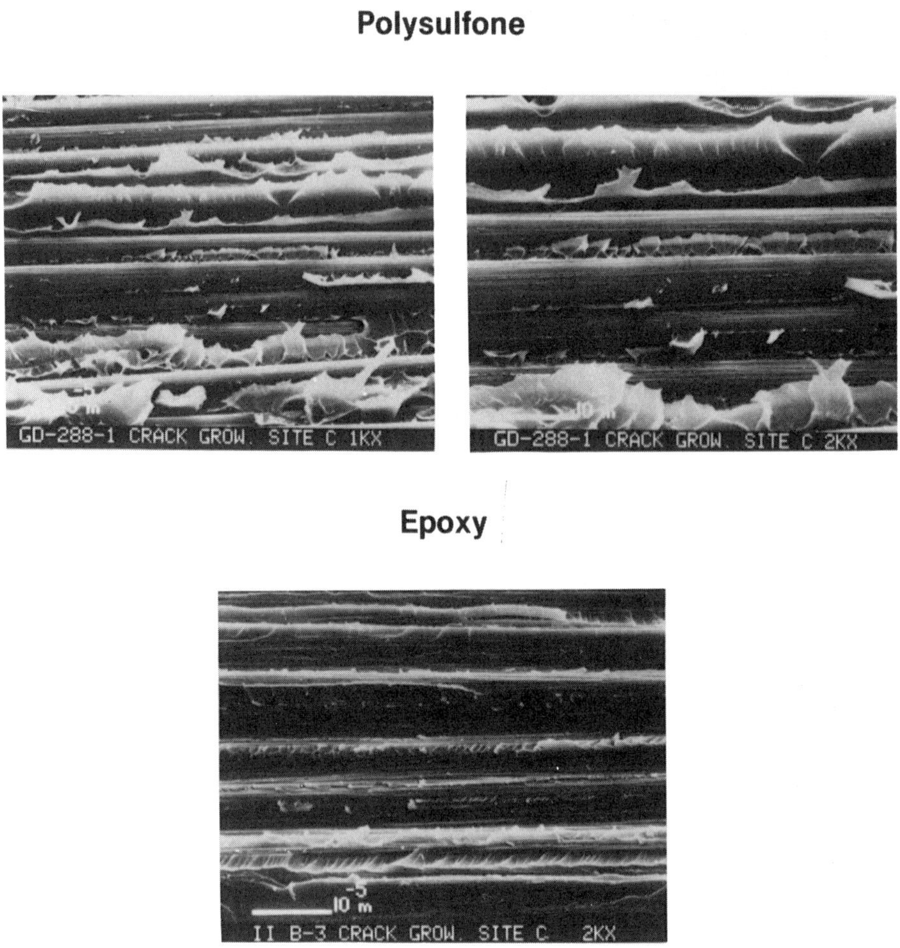


Figure 4.35 Comparison of fracture surfaces for toughened thermoset (epoxy) and thermoplastic (polysulfone).

poor fiber-matrix bonding exhibited by the same thermoplastics is responsible. The hypothesis is that the poor bonding provides an easy path for the crack to advance along, and thus the full potential of the resin toughness is not achieved. SEM pictures of the fracture surfaces in DCB specimens support this hypothesis. For example, Figure 4.35 shows surfaces for a thermoplastic and a toughened epoxy. A clear difference can be seen, since the thermoplastic shows extensive failure at or very near the fiber-matrix interface.

Although results such as those in Figure 4.34 are an important first step, considerable study is required before interlaminar fracture can be understood. Of particular importance are efforts under way in a number of laboratories to study and model the micromechanics of fracture in these systems, to expand the work to more complex loading, and to investigate the effects of rate and temperature on the fracture behavior. Such studies will certainly lead to a major improvement in our understanding of composite failure.

## VII. CONCLUSIONS

The elastomer-modified epoxies represent an exciting class of materials which exhibit many interesting properties. The most important and intriguing aspect of their behavior, however, is the dramatic increase that can be obtained in fracture energy with a minimum sacrifice in other properties. Extensive studies have now elucidated the mechanism of toughening in these materials and empirically developed models to describe the effects of temperature and loading rate on their fracture properties. An approach has also been outlined to establish relationships between the morphology of these two phase systems and their failure behavior.

The application of modified epoxies as adhesives has been studied in mode I and mixed-mode loading. The understanding of mode I behavior has been highly developed and includes many aspects of the rate, temperature, and bond thickness effects. Failure in mixed-mode loading has also been investigated, but much more remains to be done in this area.

The use of toughened materials as matrix resins in composites has also shown considerable promise. Tougher polymers have been found to give composites with improved interlaminar fracture behavior. The relationship between the two properties has been established for mode I loading under simple laboratory conditions. Work to model this behavior and extend these studies to more complex loading and test conditions is under way.

## REFERENCES


1. H. T. Corten, *J. Adhes.*, *3*, 103 (1971).
2. E. J. Ripling, S. Mostovoy, and R. L. Patrick, *ASTM*, *STP*, *360*, 5 (1963).
3. S. Mostovoy and E. J. Ripling, *J. Appl. Polym. Sci.*, *10*, 1351 (1966).
4. S. Mostovoy and E. J. Ripling, *J. Appl. Polym. Sci.*, *13*, 1083 (1969).
5. R. L. Patrick, J. A. Brown, L. E. Vehoeven, E. J. Ripling, and S. Mostovoy, *J. Adhes.*, *1*, 136 (1969).

6. S. Mostovoy and E. J. Ripling, *J. Appl. Polym. Sci.*, *15*, 641 and 661 (1971).

7. E. J. Ripling, S. Mostovoy, and H. T. Corten, *J. Adhes.*, *3*, 107 (1971).

8. S. Mostovoy, E. J. Ripling, and C. F. Bersch, *J. Adhes.*, *3*, 125 (1971).

9. E. J. Ripling, S. Mostovoy, and C. F. Bersch, *J. Adhes.*, *3*, 145 (1971).

10. R. L. Patrick, W. G. Gehman, L. Dunbar, and J. A. Brown, *J. Adhes.*, *3*, 165 (1971).

11. S. Mostovoy and E. J. Ripling, Fracture Characteristics of Adhesive Joints, Final Report, NAVAIR N00019-72-C00250, Materials Research Laboratory, Glenwood, Ill., December 31, 1972.

12. W. D. Bascom, C. O. Timmons, and R. L. Jones, *J. Mater. Sci.*, *10*, 1037 (1975).

13. W. D. Bascom, R. L. Cottington, R. L. Jones, and P. Peyser, *J. Appl. Polym. Sci.*, *19*, 2545 (1975).

14. W. D. Bascom and C. O. Timmons, in "Adhesion Science and Technology," Vol. 9B (L. H. Lee, ed.), Plenum, New York, 1975, p. 501.

15. W. D. Bascom and R. L. Cottington, *J. Adhes.*, *7*, 333 (1976).

16. W. D. Bascom, R. L. Cottington, and C. O. Timmons, *J. Appl. Polym. Sci.*, *Appl. Polym. Symp.*, *32*, 165 (1977).

17. W. D. Bascom, R. Y. Ting, R. J. Moulton, C. K. Riew, and A. R. Siebert, *J. Mater. Sci.*, *16*, 2657 (1981).

18. W. D. Bascom, J. L. Bitner, R. J. Moulton, and A. R. Siebert, *Composites*, *11*, 9 (1980).

19. W. D. Bascom and J. Oroshnik, *J. Mater. Sci.*, *13*, 1411 (1978).

20. W. D. Bascom, D. L. Hunston, and C. O. Timmons, *Org. Coat. Plast. Prepr.*, *39*, 179 (1978).

21. W. D. Bascom, R. J. Moulton, E. J. Rowe, and A. R. Siebert, *Org. Coat. Plast. Prepr.*, *39*, 1964 (1978).

22. D. L. Hunston, J. L. Bitner, J. L. Rushford, J. Oroschnik, and W. S. Rose, *Elast. Plast.*, *12*, 133 (1980).

23. J. L. Bitner, J. L. Rushford, W. S. Rose, D. L. Hunston, and C. K. Riew, *J. Adhes.*, *13*, 3 (1981).

24. W. D. Bascom and D. L. Hunston, in "Adhesion 6" (K. W. Allen, ed.), Applied Sciences, London, 1982, Chap. 14.

25. D. L. Hunston and W. D. Bascom, *Comp. Tech. Rev.*, *5*, 118 (1983).

26. A. J. Kinloch, S. J. Shaw, D. A. Tod, and D. L. Hunston, *Polymer, 24,* 1341 and 1355 (1983).

27. W. D. Bascom, G. W. Bullman, D. L. Hunston, and R. M. Jensen, *Proc. SAMPE Symposium,* Reno, April 1984.

28. D. L. Hunston, *Comp. Tech. Rev., 4* (Winter), 176 (1984).

29. D. L. Hunston and W. D. Bascom, in "Rubber-Modified Thermoset Resins," ACS Symp. Ser. 208 (C. K. Riew and J. K. Gillham, eds.), American Chemical Society, Washington, D.C., 1984, Chap. 7.

30. D. L. Hunston, A. J. Kinloch, S. J. Shaw, and S. S. Wang, in "Adhesive Joints" (K. Mittal, ed., Plenum, New York, 1984.

31. D. L. Hunston and G. W. Bullman, *Int. J. Adhesion Adhesives,* 5(2), 69 (1985).

32. D. L. Hunston, in "Tough Composite Materials," Noyes Press, Park Ridge, N.J., 1985, pp. 1-15.

33. J. C. Bolger, in "Treatise on Adhesion and Adhesives," Vol. 2 (R. L. Patrick, ed.), Marcel Dekker, New York, 1973, p. 1.

34. J. N. Sultan, R. C. Laible, and F. J. McGarry, *J. Appl. Polym. Sci., 6,* 127 (1971).

35. J. N. Sultan and F. J. McGarry, *Polym. Eng. Sci., 13,* 29 (1973).

36. H. Liebowitz, ed., "Fracture, An Advanced Treatise," Vols. I-VII, Academic Press, New York, 1968.

37. A. S. Tetelman and A. J. McEvily, "Fracture of Structural Materials," Wiley, New York, 1967, p. 51.

38. G. R. Irwin, in "Handbuch der Physik," Vol. 6 (S. Flugge, ed., Springer, Berlin, 1956, p. 551.

39. K. Arin and F. Erdogan, *Int. J. Eng. Sci., 9,* 213 (1971).

40. P. D. Hilton and G. D. Gupta, Stress and Fracture Analysis of Adhesive Joints, ASME Paper 73-DE-21, 1973.

41. S. S. Wang, J. F. Mandell, and F. J.McGarry, Fracture of Adhesive Joints, R76-1, MIT, Cambridge, Mass., 1976.

42. T. G. F. Gray, *Bull. Mech. Eng. Educ., 10,* 151 (1971).

43. J. F. Knott, "Fundamentals of Fracture Mechanics," Butterworths, London, 1973.

44. A. J. Kinloch and R. J. Young, "Fracture Behavior of Polymers," Applied Science, London, 1983.

45. G. G. Trantina, *J. Compos. Mater., 6,* 191 (1972).

46. G. G. Trantina, *J. Compos. Mater., 6,* 371 (1972).

47. G. G. Trantina, Combined Mode Crack Extension in Adhesive Joints, T&AM Rep. No. 352, Univ. of Illinois, Urbana, 1971.

48. T. R. Brussat, S. T. Chiu, and S. Mostovoy, Fracture Mechanics for Structural Adhesive Bonds, LR 27614-9, LR 27614-10, LR 27614-11, Air Force Materials Laboratory, Wright-Patterson AFB, Ohio, 1977-1978.

49. S. S. Wang, private communication.

50. A. F. Yee and R. A. Pearson, Toughening Mechanism in Elastomer-Modified Epoxy Resins, NASA CR-3718 (1983) and NASA-3852, 1984.

51. A. C. Meeks, *Polymer, 15,* 675 (1974).

52. A. V. Pocius, W. J. Schultz, and R. E. Adam, *Proc. Adhes. Soc.*, Hilton Head, N.C., February 1986, p. 4.

53. F. N. Kelley, personal communication.

54. J. P. Berry, in "Fracture Processes in Polymeric Solids (B. Rosen, ed.), Interscience, New York, 1964, p. 257.

55. R. L. Patrick, in "Treatise on Adhesion and Adhesives," Vol. 3 (R. L. Patrick, ed.), Marcel Dekker, New York, 1973, p. 163.

56. R. A. Gledhill, A. J. Kinloch, S. Yamini, and R. J. Young, *Polymer, 19,* 574 (1978).

57. D. C. Phillips, J. M. Scott, and M. Jones, *J. Mater. Sci., 13,* 311 (1978).

58. A. J. Kinloch and J. C. Williams, *J. Mater. Sci., 15,* 987 (1980).

59. S. Mostovoy and E. J. Ripling, *Appl. Polym. Symp., 19,* 395 (1972).

60. D. L. Hunston and A. J. Kinloch, unpublished data.

61. E. H. Rowe, A. R. Siebert, and R. S. Drake, *Mod. Plast., 47,* 110 (1970).

62. A. R. Siebert and C. K. Riew, *Org. Coat. Plast. Prepr., 31,* 555 (1971).

63. C. K. Riew, E. H. Rowe, and A. R. Siebert, Rubber Toughened Thermosets, presented at the Symposium on Toughness and Brittleness of Plastics, Division of Organic Coatings and Plastics Chemistry, 168th ACS Meeting, Atlantic City, N.J., September 1974.

64. S. Kunz-Douglas, P. W. R. Beaumont, and M. F. Ashby, *J. Mater. Sci., 15,* 1109 (1980).

65. W. D. Bascom, D. L. Hunston, and J. Rushford, unpublished data.

66. J. A. Sayre, R. A. Assink, and R. R. Lagasse, *Polymer, 22,* 87 (1981).

67. S. C. Kunz, J. A. Sayre, and R. A. Assink, *Polymer, 23,* 1897 (1982).

68. D. L. Hunston and A. J. Kinloch, unpublished data.

69. E. Plati and J. G. Williams, *Polym. Eng. Sci., 15,* 470 (1975).

70. A. J. Kinloch and S. J. Shaw, *J. Adhes., 12,* 59 (1981).

71. D. L. Hunston, unpublished data.

72. S. S. Wang, personal communication.

73. D. L. Hunston and J. L. Rushford, unpublished data.

74. S. Mostovoy and E. J. Ripling, in "Adhesion Science and Technology" (L. H. Lee, ed.), Plenum, New York, 1975, Part B, p. 513.

75. J. Romanko and W. B. Jones, *Proc. International Conference on Adhesion and Adhesives,* pp. 11.1-11.7, Plastics and Rubber Institute, London, 1980; and reports for the Air Force under contract on Fatigue Behavior of Adhesively Bonded Joints, USAF-F33615-76-C-5220.

76. W. S. Johnson and S. Mall, ASTM Spec. Publ. 876, p. 185 (1985).

77. S. Mall and W. S. Johnson, ASTM Spec. Publ. 893, p. 322 (1985).

78. W. D. Bascom and S. Mostovoy, *Org. Coat. Plast. Prepr., 38,* 152 (1978).

79. G. R. Irwin, *Trans. ASME J. Appl. Mech., 24,* 361 (1957).

80. F. Erdogan and G. C. Sin, *Trans. ASME, J. Basic Eng., 85D,* 519 (1963).

81. W. D. Bascom and C. O. Timmons, unpublished results.

82. T. R. Brussat, S. T. Chiu, and S. Mostovoy, Fracture Mechanics of Structural Adhesive Bonds, LR 27614-5, Air Force Materials Laboratory, Wright-Patterson AFB, Ohio, 1976.

83. G. B. McKenna, J. F. Mandell, and F. J. McGarry, *J. Soc. Plast. Ind. Annu. Tech. Conf.*, RPD 1974, Sect. 13-C.

84. J. M. Scott and D. C. Phillips, *J. Mater. Sci., 10,* 551 (1975).

85. C. C. Chiao, R. L. Moore, and T. T. Chiao, *Composites, 8,* 171 (1977).

86. J. M. Whitney and C. E. Browning, *Exp. Mech., 25*(3), 294 (1985).

87. M. K. Tse, M. F. Hibbs, and W. L. Bradley, in *ASTM Symposium on Toughened Composites,* in press.

88. D. L. Hunston, R. J. Moulton, N. J. Johnston, and W. D. Bascom, in *ASTM Symposium on Toughened Composites,* in press.

# Five

# Adhesion of Carbon Fibers to Epoxy Matrices

LAWRENCE T. DRZAL *Michigan State University, East Lansing, Michigan*

Adhesion between graphite fibers and epoxy matrices is a necessary and sometimes controlling factor in achieving optimum performance. Manufacturers' proprietary fiber surface treatments promote adhesion without providing a basic understanding of the fiber surface properties altered through their use. The study described here combined fiber surface chemistry, morphology, interfacial strength measurements, and fracture characterization to elucidate the role of surface treatments. The results of this investigation lead to the conclusion that surface treatments designed to promote adhesion to epoxy matrix materials operate through a two-part mechanism. First, the treatments remove a weak outer fiber layer initially present on the fiber. Second, surface chemical groups are added which increase the interaction with the matrix. Increases in fiber surface area are not an important factor in promoting fiber-matrix adhesion. In some cases the upper limit to fiber-matrix interfacial shear strength is the intrinsic shear strength of the fiber itself.

Reinforcing fibers are also available from various manufacturers with matrix-compatible "finishes" applied to them. Usually these finishes or coatings are 100 to 200 nm thick resin layers applied after surface treatment. It has been hypothesized that their function is to enhance adhesion by either protecting the fiber from handling damage, protecting the fiber surface reactivity, or improving fiber wettability. This study of finished and unfinished graphite fibers concludes that the mechanism by which an epoxy-compatible finish operates is different from what has been hypothesized to date. The finish layer creates a brittle interphase layer between the fiber and matrix which increases the interfacial shear strength but at the expense of changing the failure mode from interfacial to matrix.

## I. INTRODUCTION

The use of composite materials in aerospace applications is increasing because of the significant weight, cost, and performance advantages these materials have over conventional structural materials. As the degree of sophistication of the aircraft designer increases, composite applications will be extended to a realm where current materials will not meet the load and environmental constraints efficiently. This will result in a twofold requirement. First, composite properties which do not depend exclusively on the properties of the reinforcement, e.g., off-axis properties, will become more important in composite structural design. Second, new generations of composite materials will be required where different fibers or matrices are needed to extend the operating environment of existing composites. To answer these needs, a fundamental understanding of the fiber-matrix interphase and its effects on structure-property relationships in a composite is necessary.

Work in the early 1970s expanded the concept of the fiber-matrix interface which exists as a two-dimensional boundary into that of a fiber-matrix interphase that exists in three dimensions [1]. The complexity of this interphase can best be illustrated with the use of a schematic model which allows the many different characteristics of this region to be visualized, as shown in Figure 5.1 [2].

The interphase exists from some point in the fiber where the local properties begin to change from the fiber bulk properties through the actual interface into the matrix, where the local properties again equal the bulk properties. Components of this region can be identified. Many literature citations exist on this concept. There may be morphological variations near the fiber surface which are not present in the bulk of the fiber [3,4]. The surface area of the fiber can be much greater than its geometric value because of pores or cracks present on the surface [5,6]. The atomic and molecular composition of the fiber surface can be quite different from that of the bulk of the fiber [7]. Surface treatments can add surface chemical groups and remove the original surface, giving rise to a chemically and structurally different region [8]. Exposure to air before composite processing can result in adsorption of chemical species, which may alter or eliminate some beneficial surface reactivity. These adsorbed materials may also desorb at the increased temperatures seen in composite fabrication and may be a source of volatiles which disrupt the interface if not removed [9]. Thermodynamic wetting of the fiber surface by the matrix is a necessary condition for fiber-matrix adhesion and is determined by the free energies of the components [10]. Both chemical and physical bonds exist at the interface, and the number and type of each strongly influence the interaction between fiber and matrix. The structure of the matrix in the interphase can be influenced by the proximity of the fiber surface. Changes in reactivity due to adsorption of matrix components can alter the local morphology. Unreacted matrix components and impurities can diffuse to the interphase region, altering the local structure [11].

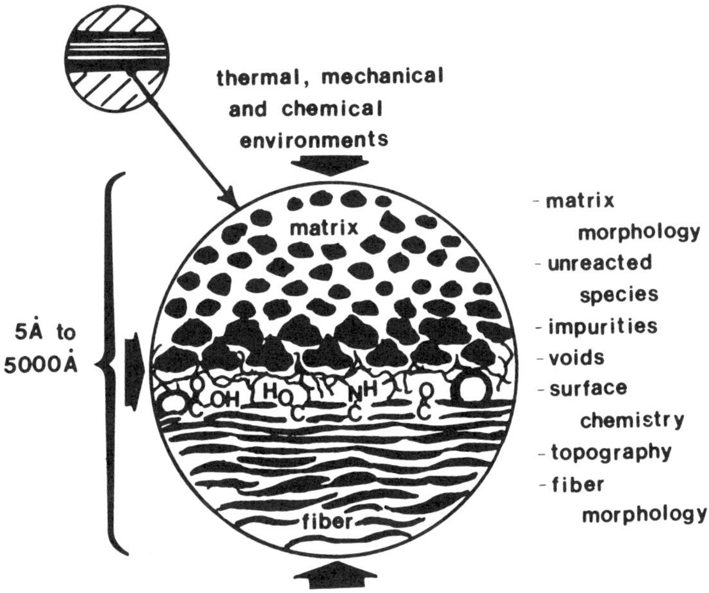


Figure 5.1 Schematic representation of the components of the three-dimensional interphase between fiber and matrix.

These phenomena can vary in magnitude and can occur simultaneously in the interphase region. Depending on the system, the interphase itself can extend from a few to a few thousand nanometers in depth. The structure of this region can have profound effects on the performance of the composite in terms of its mechanical strength and chemical and thermal durability. The exact nature of this region must be understood if accurate predictive models of interphase behavior are to be developed and integrated into a model of composite performance. Because of the complexity of this region, a need for a phenomenological approach has been recognized. Only when the exact nature of this region is understood will the interphase be considered as a material variable to be optimized in composite performance.

Carbon or graphite fibers are usually given surface treatments to enhance their adhesion to polymeric matrices. A variety of types of treatments are available which act by ostensibly different methods to produce better fiber-matrix adhesion (oxidation, reduction, coatings). The first part of this study is directed at determining the mechanism by which fiber surface treatments are effective in increasing adhesion.

Reinforcing fibers are available not only with surface treatments designed to increase the chemical interaction between fiber and matrix but also with surface finishes or sizings applied to enhance compatibility with the host matrix. Surface-treated graphite fibers can be obtained with "matrix-compatible" finishes, for example. Usually these finishes are applied as very thin coatings (e.g., 100 to 200 nm) from solution. Compositionally, they can vary from consisting of the pure resin component without a cross-linking agent and therefore unpolymerized to being a mixture of polymers partially cross-linked. The usual explanation for the use of these finishes is that they

1. Protect the fiber surface from damage
2. Improve the wetting of the fiber by the matrix
3. Protect fiber surface reactivity

The purpose of the second part of the study is to elucidate the mechanism by which surface finishes operate and to assess their effect on fiber-matrix adhesion.

## II. EXPERIMENTAL

### A. Graphite Fibers

Type A graphite fibers were chosen as the representative fibers for this study. These fibers were made from polyacrylonitrile and have mechanical properties representative of fibers being used in composite materials. The A fiber has a tensile modulus of about 35 Msi (245 GPa).

The main structural elements of the fibers are graphitic ribbons which are oriented roughly parallel to the fiber axis. These ribbons are formed of graphitic crystallites which increase in size with increasing graphitization temperature (i.e., 13 graphitic layers and 40 Å wide for the 1500°C type A fiber). The ribbons twist and undulate along the fiber axis as shown schematically in Figure 5.2. This would produce a surface having not only graphitic basal planes but also corners and edges of the crystallites. Detailed discussions of the structure of the fiber are available in the literature [12].

Untreated fibers (AU), fibers surface-treated by the manufacturer for adhesion to epoxy (AS), and surface-treated fibers coated with 100 to 200 nm of DER-330 epoxy resin (AS-1C and AS-4C) were used in this study. A previous study measured surface areas of fibers produced in a similar manner and concluded that an increase in surface area does not accompany the fiber surface treatment used here (see Table 5.1).

### B. Interfacial Shear Strength

A mechanical test capable of testing the fiber-matrix interphase under pure shear loading was required to assess the interactions occurring between fiber and matrix as a function of altered surface properties of the fiber. Composite design usually places the fiber-matrix interphase under shear loading, but conventional composite testing methods evaluate a combination of types of loading as well

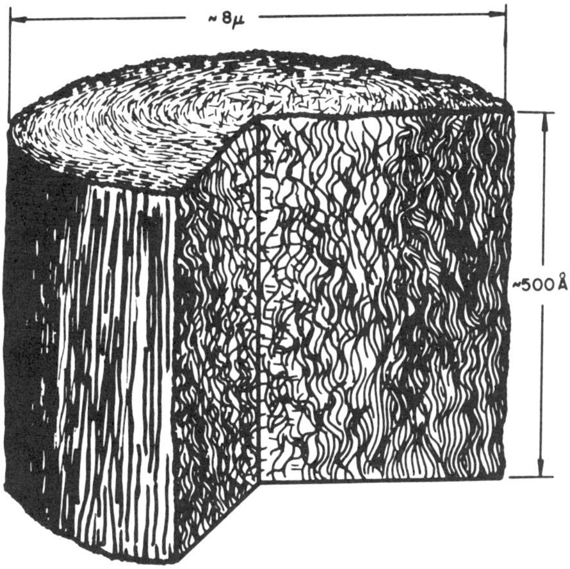


Figure 5.2 Schematic three-dimensional structural model of polyacrylonitrile-based fiber with properties similar to type A.

as the synergistic effects of matrix and fiber on composite properties. This makes it difficult to delineate the effects of the interphase alone.

Single-filament techniques are potentially less complex and lend themselves to application of pure loading modes. The commonly used fiber pullout technique has received wide usage with glass fibers but is difficult to use with the brittle and small graphite fibers of this study. A technique used with metals was adapted for this study and designed to evaluate the interfacial shear strength directly.

Table 5.1 Surface Properties of Various Graphite Fibers Used in This Study

| Fiber | Surface area ($m^2/g$) | Surface oxygen (%) | Surface free polar ($mJ/m^2$) | Energy total ($mJ/m^2$) |
|---|---|---|---|---|
| AU | 0.44 | 9 | 24 | 51 |
| AS | 0.48 | 20 | 30 | 56 |
| AS (750°C/$H_2$) | 0.48 | 3 | 12 | 45 |
| ASC | 0.22 | — | 11 | 40 |

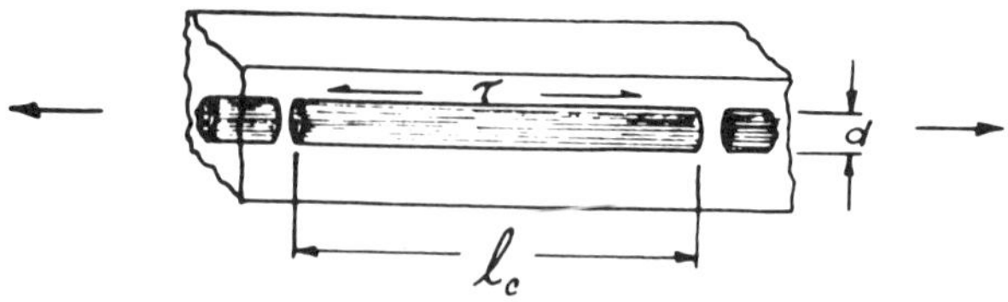


$$\tau = \frac{\sigma_f}{2}\left(\frac{d}{\ell_c}\right)$$

$$\tau = \frac{\sigma_f}{2\beta}\,\Gamma\left(1-\frac{1}{\alpha}\right)$$

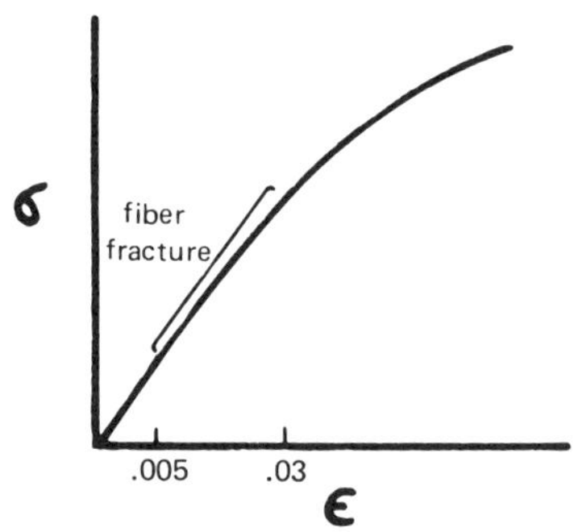


Figure 5.3 Fiber configuration and illustration of fiber fragmentation process during determination of interfacial shear strength.

A single graphite fiber is aligned axially in the cavity of a tensile dogbone coupon in a silicone mold. The desired matrix (in this case an epoxy) is cast around the fiber and the tensile coupon is loaded in tension (Figure 5.3). As first proposed by Kelly [13], the matrix material, which has a high strain to failure, can be loaded in tension. The brittle fiber encapsulated within the coupon is loaded in tension through the transfer of shear forces from the coupon operating within the fiber-matrix interphase. The tensile loads on the fiber build up and exceed the fracture strength of the fiber, causing it to fracture within the tensile coupon. This fiber-fracturing process continues to create shorter and shorter fiber fragments until a length is reached which is no longer sufficient to support enough shear at the interphase and cause further fiber fracture. This limiting length is known as the critical transfer length $l_c$. A simple force balance assuming ideal fiber properties and uniform distribution of loads leads to the relationship

$$\tau = \frac{\sigma_f}{2} \frac{d}{l_c} \tag{1}$$

where d is the fiber diameter, $\sigma_f$ is the fiber fracture strength at the critical length $l_c$, and $\tau$ is the interfacial shear strength.

In practice, a distribution of lengths is achieved because of the variations in fiber properties, e.g., defects, surface heterogeneities, and geometric variations. Therefore a statistical evaluation of the fiber length data must be made. This leads to a modified expression

$$\tau = \frac{\sigma_f}{2\beta} \Gamma\left(1 - \frac{1}{\alpha}\right) \tag{2}$$

where $\Gamma$ is the gamma function and $\alpha$ and $\beta$ are the Weibull shape and scale factors of the two-parameter Weibull distribution [14]. The actual fragment lengths are measured and "best fit" to a Weibull distribution with the aid of a computer.

Evaluation of Equation (2) requires the values of the fiber fracture strength at the critical length. Since this length can often be a fraction of a millimeter, 25-mm gauge length values can be misleading. In this study the fiber fracture strength at the critical lengths was determined directly through the use of a Tecam microtensile testing machine (Tecam Instruments, Princeton, N.J.). This device measures fracture stress directly by an optical-mechanical method. The fracture strength is significantly higher at the critical lengths than at the 25-mm gauge lengths.

The single-filament interfacial shear strength method has been used for glass fiber/polymer systems [15]. Fiber fragment lengths were measured after fracture through elimination of the matrix by combustion. Glass fiber fragments were then counted and measured microscopically. Because of some concern about fragment damage in this process and the desire to obtain more information about the stress operating at the interphase, the sample lengths were measured directly on the stage of a microscope in the specimen as it was loaded. A hand-operated tensile loading device was constructed which kept the sample in place under the microscope and at the same time allowed the load to be increased incrementally. This had the advantage of allowing the fracture process to be observed directly; critical length attainment could be verified before sample failure and fiber fragment lengths could be measured in situ with the aid of a calibrated filar eyepiece. Likewise, by use of transmitted polarized light the stress occurring with the fiber-matrix interphase could be observed and recorded photographically.

Qualitative differences in the stress pattern resulting from interphase changes could be observed. A transmitted-light microscope with polarizing attachment and long-working-distance objectives was used for this purpose. The polarizer and analyzer were set at extinction. Normal chromatic cycles were observed in the epoxy at low strains but disappeared at higher loadings. Fringes or chromatic variations were not observed under the high-strain conditions present at the fiber fragment ends.

## C. Microtoming

The 25-mm gauge section of the 62.5 mm long, 3 mm wide, 1.5 mm thick tensile coupon containing the embedded fiber also contained information about the mode of fracture in the interphase. Conventional optical microscopy could not discriminate events at the resolution level required. Scanning electron microscopy could not penetrate into the specimen. Ultramicrotoming offered the possibility of examing the interphase directly under the high-magnification capabilities of the transmission electron microscope (TEM).

Segments of 1.5 mm were cut from the tensile coupon gauge length section of both unstrained and strained samples. They were trimmed to trapezoidal shape and were mounted in an ultramicrotome (E. I. duPont, Wilmington, Del.) with the fiber axis parallel to the diamond blade. Sections were obtained with a 55° diamond knife with fiber orientation a few degrees off the fiber axis. Sections 60 to 90 μm thick were cut progressively through the sample. These sections were floated onto a water bath and picked up onto 200-mesh Formvar-coated grids (Ladd Research Industries, Burlington, Vt.). They were dried on filter paper and stored in a vacuum desiccator until TEM observation.

The grids were placed directly into a JEOL-100CX TEM for viewing (JEOL, Medford, Mass.). The knife direction in all microtomed photomicrographs is perpendicular to the fiber-matrix interphase. The homogeneous featureless region is the epoxy, and the textured area is the graphite fiber thin section. Each fiber-matrix combination was sectioned prior to straining to verify that interfacial artifacts from the sectioning process are not induced in the sample. All TEM micrographs were taken at 100 kV. Sections included in this chapter are representative of features observed for each fiber-matrix combination.

## D. Matrix

An epoxy matrix was chosen for this study to model structural matrix materials without including the complexity of the commercial chemical formulations [16]. Epon 828 (Shell Chemical Co., Houston, Tex.) cured with the stoichiometric amount of *meta*-phenylenediamine (mPDA) [17] of 14.5 phr was selected as the matrix material. A master batch of both components was obtained. It was stored under dark, dry conditions to prevent deterioration or contamination with time.

A cure cycle and processing cycle were established and followed rigorously throughout this study. Both Epon 828 and mPDA were weighed in separate containers and heated to 65°C to melt the mPDA. The components were mixed, debulked for 2 min at full vacuum, and poured into silicone molds containing the aligned fibers. The molds were processed for 2 hr at 75°C and 2 hr at 125°C, followed by a slow overnight cooldown. Since gelation occurs during the lower-temperature cycle, residual stresses resulting from cure shrinkage were kept to a minimum [18].

## III. PREVIOUS WORK

### A. Surface Spectroscopy

X-ray photoelectron spectroscopy (XPS) or ESCA was conducted to provide atomic and molecular information about the graphite fiber surface selected for this study [10]. The data (Table 5.1) indicate that the surface oxygen content more than doubles and the nitrogen content more than triples with surface treatment. Subsequent elevated temperature treatments were carried out in vacuum and with hydrogen (H) remove the surface oxygen and nitrogen species. The highest temperature treatment with hydrogen, i.e., 750°C, successfully removes almost all of the elements except for a small amount of oxygen.

Other investigators have attempted to deduce the chemical functionality of the surface chemical groups detected on the fiber surface by ESCA. The most probable functionalities present are phenolic, carboxylic, lactone, and carbonyl groups [8,19]. Using peak assignments based on ESCA of other organic materials leads to the conclusion that the surface oxygen species were primarily carboxylic with a small amount of phenolic functionalities [20]. Ehrburger et al. [21] determined the acid strength of groups on the surface of graphite fibers and concluded that strong acidic groups have the largest effect on interfacial shear strength. Nitrogen present on the surface may be chemisorbed as a C-O-N complex [22]. Trace levels of other species are most probably present between the graphitic basal layers and not necessarily on the surface.

The spectra of the ASC (epoxy-coated) fibers are quite different from the spectra of the AS fibers. The carbon 1*s* peak exhibits only two components instead of three (the small peak at 291 eV is a loss peak due to the presence of aromatic carbons). These two components are due to carbon-carbon bonds at 284.6 eV and carbon-xoygen bonds at 286.2 eV. The oxygen 1*s* peak is shifted 0.4 eV higher in binding energy and is much narrower than that found on the AS fiber. This correlates well with the absence of the carboxylic component of the carbon 1*s* peak at 288.7 eV, since the doubly bonded oxygen would be found at a lower binding energy and is consistent with the known structure of the epoxy molecule. Spectra obtained on thick films of the epoxy cast from solvent are quite similar to the spectra obtained here. The very low atomic percent of nitrogen and sodium detected on the ASC fiber surface indicates that the original AS fiber surface is completely covered by the finish layer.

The ASC fiber was placed in a Soxhlet extractor with methyl ethyl ketone (MEK) for 100 hr to remove the epoxy finish that was not chemisorbed to the fiber surface. The maximum temperature seen by the fiber during this extraction and in the subsequent drying operation was 75°C. The extraction with methyl ethyl ketone (MEK ext) produces a fiber surface that exhibits an XPS spectrum with features of both the AS and ASC spectra. There are three carbon 1*s* components, but present in different intensities, from the AS surface. The oxygen 1*s* spectrum is again shifted to higher binding energy but now has a definite broadening of the low-energy side. The full widths at half-maximum (FWHM) for both peaks are between those of the AS and ASC fibers. The surface concentration of oxygen is slightly greater than that obtained for the AS fiber before the application of the finish. This is probably due to retention on the fiber surface of some epoxy whose oxygen content contributes to the total surface oxygen level. The nitrogen intensity is reduced from that found on the AS fiber. This also suggests that these atoms are not detected because of epoxy adsorption on the surface even after extraction. Overall, it can be concluded that some small amount of epoxy remains on the ASC fiber after extraction with methyl ethyl ketone.

### B. Surface Free Energy

A necessary condition for fiber and matrix compatibility is thermodynamic wetting. This quantity can be measured by determination of the contact angle between fiber and matrix. Further insight into the specific interactions at the interface can be obtained by measuring the contact angle on the fiber surface of a variety of liquids having different ratios of the polar to dispersive components of their surface free energies [10].

A change in fiber surface free energy was detected for the surface-treated fibers compared to the untreated fibers (Table 5.1). The total surface free energy increased for the AS fibers over that for their untreated precursors. The measured change is primarily in the polar component of the surface free energy.

The total surface free energy for each fiber is greater than or equal to that for the epoxy used in this study, so fiber wetting by the matrix will occur.

## IV. RESULTS AND DISCUSSION

### A. Fiber Surface Treatment

Table 5.2 contains a summary of the results of the interfacial strength determinations. Column 1 lists the fibers and their specific treatments. Column 2 contains the critical length-to-diameter ratio determined from a computer fit of the fragment data to a two-parameter Weibull function.

Table 5.2 Interfacial Shear Strengths

| Fiber | $l_c/d$ | $\sigma_f$ ksi | $\sigma_f$ GPa | $\tau$ ksi | $\tau$ MPa |
|---|---|---|---|---|---|
| AU | 108 | 581 | 4.00 | 3.50 | 24.1 |
| AS | 41.7 | 681 | 4.69 | 10.7 | 74.0 |
| AS (300°C) | 43.6 | 645 | 4.45 | 10.4 | 71.7 |
| AS (600°C) | 37.2 | 580 | 4.00 | 9.87 | 68.0 |
| AS (750°C/$H_2$) | 46.1 | 630 | 4.35 | 9.54 | 65.8 |

The value of the critical length-to-diameter ratio listed in column 2 represents the experimentally observed parameter. However, note must be made of the fracture events which occur near the critical length. As the fiber fragment lengths approach the critical length, some fragments will be slightly more and some just slightly less than the critical length. The shorter fragments will not longer fracture, but those just slightly greater than the critical length will fracture again, giving fragments approaching one-half the critical length. Therefore fragment lengths are expected to be distributed between the critical length and one-half the critical length. The experimentally determined value would then be the median value in this interval.

Columns 5 and 6 list the interfacial shear strength. These values are sensitive to and reflect the changes detected within the fiber-matrix interphase due to changes in the fiber surface. Comparing the A fibers among themselves highlights the changes occurring as a result of surface treatments. The untreated AU fiber gives an initial value for the interfacial shear strength of 3.50 ksi. The value for the interfacial shear strength on the same fiber after surface treatment (AS) increases by a factor of three. The surface composition and energetic analysis of this fiber show that the oxygen and nitrogen concentration on the AS surface is increased by a factor of two over the AU fiber and that the surface behaves in a more polar manner. The values of interfacial shear strength determined for the other AS fibers, which have been treated in various ways to reduce the surface oxygen content, decrease with decreasing surface oxygen from the value determined for the AS fiber. The interfacial shear strength for the hydrogen-reduced fiber surface (9.54 ksi) is less than that for the AS fiber (10.7 ksi) but greater than that for the untreated AU fiber (3.50 ksi) even though its surface content of oxygen is much less than that of the untreated AU fiber.

The stresses generated within the fiber-matrix interphase are altered with changes in fiber surface treatment. The mode of fracture between fiber and matrix is altered with surface treatment as well. The photoelastic stress pattern observed in each specimen was

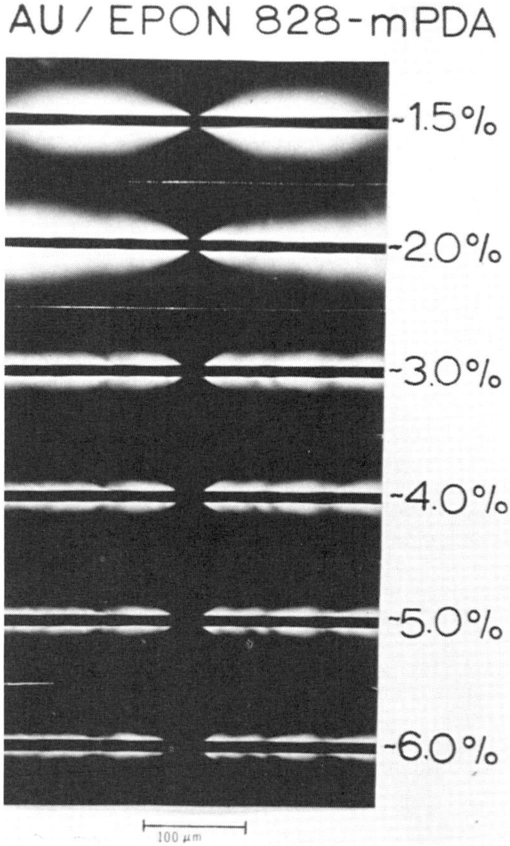


Figure 5.4 Polarized transmitted light micrograph of the AU fiber-Epon 828-mPDA interface at a fiber break and its alteration with increasing strain.

recorded photographically with increasing strain. A collage of these observations at 400× and increasing values of strain was compiled. Figure 5.4 shows the photoelastic data for the AU fiber in the epoxy matrix. The values of strain accompanying the figure are measured from the machine displacement and indicate only the relative changes from one picture to the next. The absolute values of strain would be expected to be smaller than these machine values. At low strains after a fiber break, an intense photoelastic region appears around the ends of the fiber. With increasing strain this region rapidly expands down the fragment away from the break. Dynamic observation of this process suggests that the stresses proceed along the fragment by a stick-slip mechanism. That is, the stresses build up and then appear to release and move ahead an incremental amount repetitively. This results in a stress pattern that has alternating light and dark areas in the micrographs at points where the stress pattern "jumped" ahead. For this particular fiber-matrix combination, each fiber fragment fails interfacially at very low strains. With increasing tensile load, the fiber fragments interact with the matrix only through weak frictional forces. The interfacial shear strength is very low and the fiber exerts very little reinforcing effect on the matrix. Under axial tension, the matrix can extend without being constrained by the presence of the fiber. This results in the gap between fragments extending in length with increasing strain. Subsequent observation of fibers with higher interfacial shear strength shows that this fragment separation is retarded by the presence of a well-bonded fiber.

Ultramicrotomed sections of this specimen demonstrate the mode of interfacial fracture taking place. Figure 5.5 is a TEM photomicrograph of the AU/epoxy interface after straining. The entire interphase has failed and the sectioned specimens show a typical region where separation of the epoxy from the graphite fiber has taken place. Close inspection of these photomicrographs reveals that many small fragments of graphite fiber have been removed to the epoxy side of the fracture path. This suggests that the fracture path was one that involved propagation through the outer layers of the AU fiber as well as interfacial failure between the fiber and epoxy. Many small fragments of graphite fiber can be seen strongly adhering to the epoxy side of the failure path, confirming that failure of a weak fiber outer layer is a major element in interfacial fracture of this fiber-matrix combination.

Figure 5.5 Transmission electron micrograph (TEM) of an ultramicrotomed section cut parallel to the fiber axis of the AU fiber-Epon 828-mPDA interface after straining to 6%.

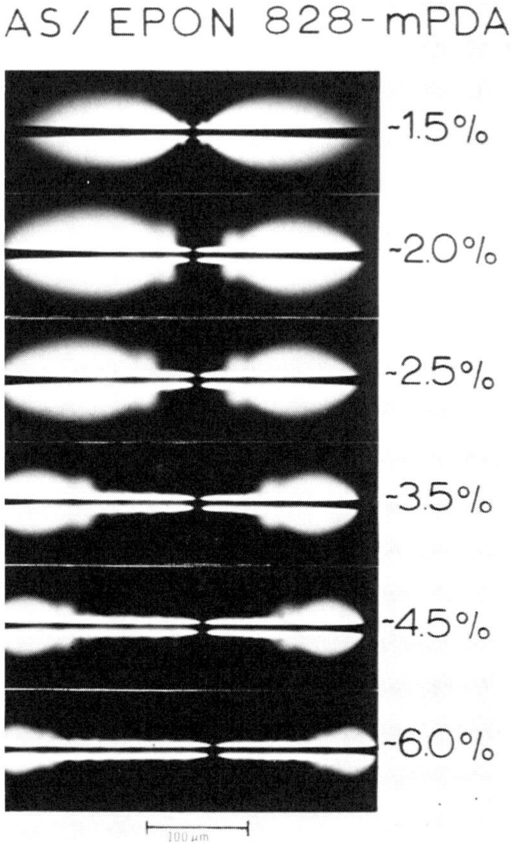


Figure 5.6 Polarized transmitted light micrograph of the AS fiber-Epon 828-mPDA interface at a fiber break and its alteration with increasing strain.

The surface-treated fiber (AS) behaves much differently under polarized light than the untreated AU fiber. Figure 5.6 shows that at the fiber break the stresses immediately build up at the ends of the fiber. With increasing strain, however, a narrow, very intense photoelastic region remains around the fiber while the initial bulbous region moves away from the fiber ends and toward the center of the fragment. The interfacial shear strength measured for this system is more than a factor of three greater than that for the AU fiber, indicating a much greater degree of fiber-matrix interaction. Indeed, the separation at the fiber break remains relatively constant with increasing strain, whereas for the AU fiber the gap between segments widens appreciably.

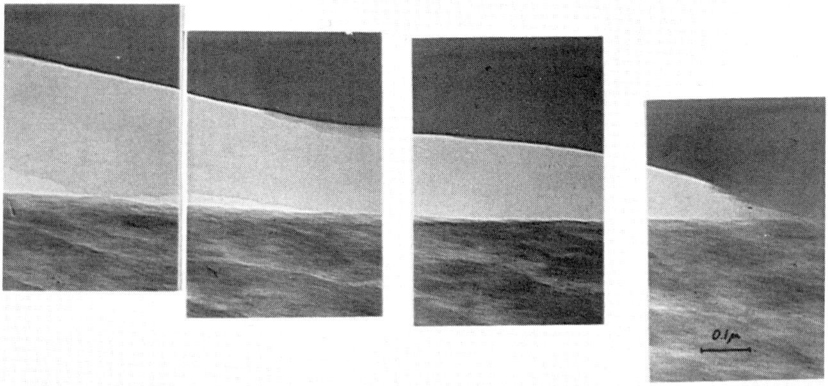


Figure 5.7 Transmission electron micrograph of an ultramicrotomed section of the AS (750°C/H) fiber-Epon 828-mPDA interface after straining to 6%.

The ultramicrotomed sections of the sample show the progagation of an interfacial crack emanating from the fiber fragment ends and progressing along the fragment (Figure 5.7). Comparison of this photomicrograph with the photoelastic stress patterns suggests that the narrow intense stress area that remains around the fiber is a region where an interfacial crack has passed, while the bulbous region at the tip of the photoelastic region appears to be the plastic zone associated with the moving crack tip. The fracture path appears to be strictly interfacial, with no evidence of graphite fiber fragment pullout to the epoxy side as observed with the AU sample.

The photoelastic behavior of the AS fibers with decreasing amounts of surface oxygen groups are identical to that shown in Figure 5.6 for the AS fiber. The critical lengths increased and the value of the interfacial shear strength therefore decreased. The shape of the stress pattern, its movement away from the fiber break, the lack of movement of the fiber fragments, etc. were nearly identical to the observations on the AS fiber.

The ultramicrotomed sections of these samples were also identical to those observed for the AS fiber. Transmission electron photomicrographs of a section of the hydrogen-reduced AS (750°C/H) fiber in the same epoxy show a strictly interfacial fracture path between fiber and matrix, just as in the AS fiber case, with no detectable graphite fiber fragments attached to the epoxy side of the fracture path.

The relationship between the parameters investigated in this study which explain the interaction between fiber surface chemistry and fiber surface treatment and their effect on fiber-matrix interfacial

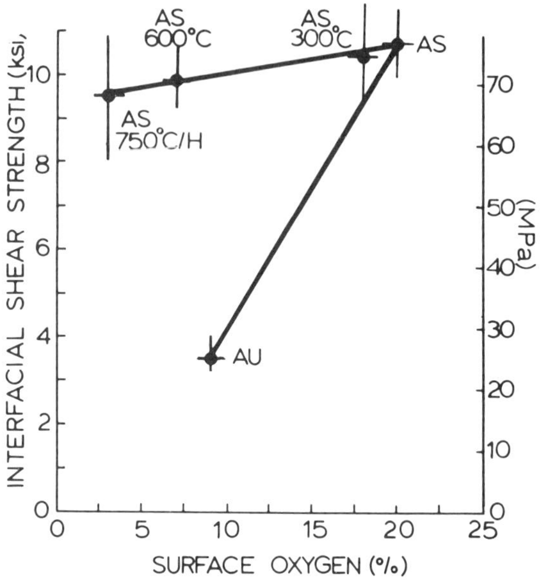


Figure 5.8 Interfacial shear strength plotted as a function of surface oxygen concentration determined by ESCA for the type A fibers.

shear strength is illustrated most effectively by plotting the interfacial shear strength against the fiber surface oxygen content. (Inclusion of atomic species other than oxygen or use of the polar component of the fiber surface free epoxy drawn therefrom.) This plot is displayed in Figure 5.8.

The A-fiber data lead to the conclusion that two separate surface treatment mechanisms are operating. This can be seen by comparing the data starting at the AU fiber data point.

Surface treatment which converts the AU to the AS fiber doubles the oxygen content and causes the interfacial shear strength to rise to a value more than three times that of the AU fiber. However, removal of the surface oxygen groups with the various treatments [i.e., AS (300°C), AS (600°C), and AS (750°C/H)] decreases the interfacial shear strength to a value below that obtained for the AS but still above the value obtained for the AU. Some other change occurred with surface treatment which allows a higher value of interfacial shear strength to be retained even after the surface oxygen groups are removed. Kalnin [23], in a review of graphite fiber surface treatments, proposed that the surface oxidative treatment processes not only add surface chemical groups but also remove some of the outer atomic layers from the fiber surface.

Fitzer et al. [24] and Thomas and Walker [22], in work with phenolic matrix carbon fiber composites, inferred an effect on surface morphology. The work discussed here, which combines the photomicrographs of the fractured interface with the interfacial shear strength and stress patterns, shows that indeed a two-part effect results from fiber surface treatment. There are a mechanical effect due to the removal of the original weak boundary layer and a surface chemical effect due to the addition of surface chemical groups which interact more strongly with the epoxy.

As the fiber goes through its processing cycle while it is being converted from polyacrylonitrile to graphite, the same surface which is in contact with the reactive environments and high temperature becomes the fiber surface. It is reasonable to expect that it has a very disrupted surface layer which might constitute a "weak boundary." Under shear loading in a matrix, interfacial failure at low loads involving separation of the first few fiber layers from the bulk of the fiber is observed because of the inability of this layer to withstand high shear loads on this AU fiber surface. Surface treatment not only adds oxygen groups but also removes this defective surface layer, leaving behind a structurally sound surface able to withstand higher shear loading and having more surface chemical groups capable of interacting with the epoxy. Failure between the epoxy and the AS fiber then becomes interfacial with no observable fiber failure, since the outer weak layer has been removed. The treatments which removed oxygen from the AS fiber surface decreased the interaction with the epoxy but did not affect the ability of the defect-free surface layer to sustain higher shear loads than the AU fiber surface. They therefore gave higher values of interfacial shear strength. For this epoxy matrix and this fiber surface treatment, the surface chemical effect responsible for the increase in interfacial shear strength between the AS fiber and the A fiber without surface chemical groups—AS (750°C/H)—is estimated to be 16%. The effect of removing the defect-laden surface is responsible for the change in interfacial shear strength between the AU fiber and the AS (750°C/H), that is, 84%.

## B. Fiber Finish

Table 5.3 shows data for the interfacial shear strength for two A fibers with and without the surface finish and tested in the Epon 828-mPDA matrix.

Fiber tensile strengths ($\sigma_f$) are given at the critical length for each fiber. These values were interpolated from semilog plots of the fiber tensile strength as a function of gauge length. The Weibull parameters ($\alpha$ and $\beta$) determined from the experimental data are listed, as well as the interfacial shear strength ($\tau$) calculated by use of Equation (2). The coefficient of variation for the value of the interfacial shear strength was between 0.1 and 0.2.

Within either the A-1 fiber or the A-3 fiber data, parallel increases in the interfacial shear strength are observed. The coated untreated fibers have a greater $\tau$ than the uncoated untreated fibers for both fiber types. Likewise, the coated surface-treated fibers have a larger $\tau$ than the uncoated surface-treated fibers for both sets. The ratio of the interfacial shear strength of the coated fiber to that of the same fiber without the finish varies between 1.13 and 1.30 for all the fibers investigated in this study.

Table 5.3 Interfacial Shear Strength ($\tau$) for Graphite Fibers in Epon 828-mPDA

| Fiber | $\sigma_f$ (ksi) | $\alpha$ | $\beta$ | $\tau$ (ksi) |
|---|---|---|---|---|
| AS-1 | 655 | 38.8 | 3.5 | 10.8 |
| AS-1C | 665 | 30.3 | 3.8 | 13.6 |
| AS-4 | 845 | 57.2 | 3.1 | 9.9 |
| AS-4C | 850 | 43.8 | 4.1 | 11.8 |

### C. Failure Mode

The photoelastic stress patterns exhibited by the epoxy near the ends of the fiber fragments created during the interfacial shear strength measurements were recorded as a function of increasing sample strain. Figure 5.9 is a composite of polarized transmitted light micrographs of the AS-1C fibers.

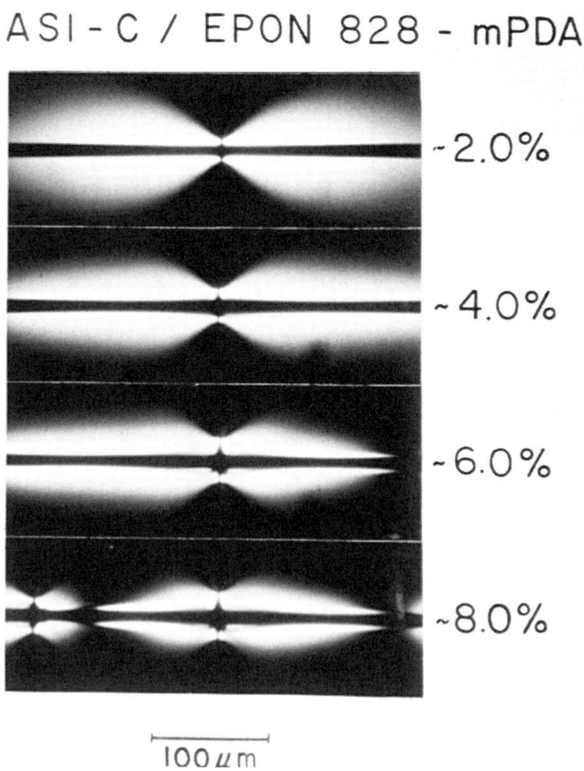


Figure 5.9 Polarized light micrographs (400×) of the AS-1C graphite fiber at increasing strain in the epoxy matrix.

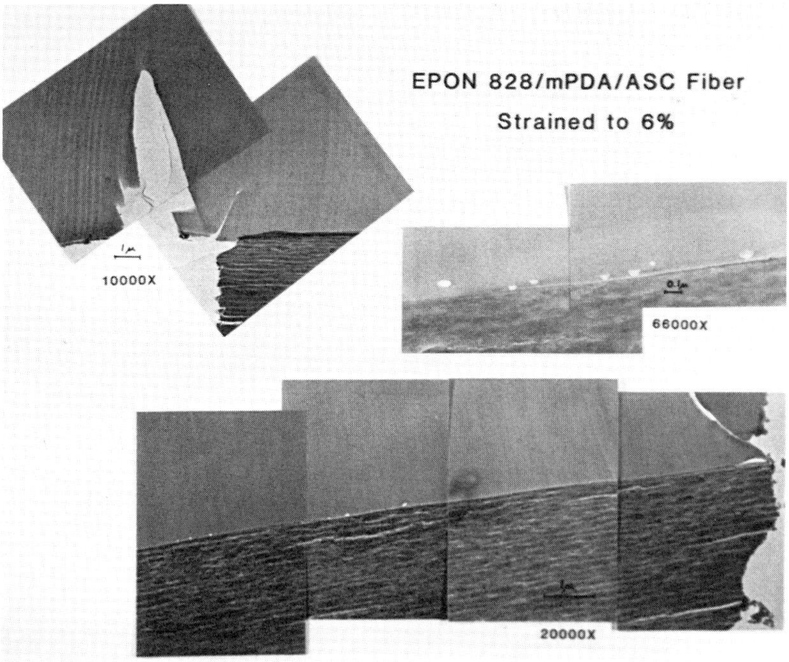


Figure 5.10 Transmission electron micrograph of an ultramicrotomed section of the ASC/epoxy interphase after straining.

The surface-treated and coated fibers (AS-1C and AS-4C) give very intense photoelastic stress patterns. These fibers have the highest interfacial shear strength, and growth of a matrix crack perpendicular to the fiber can be detected at the fiber breaks as seen in the micrograph. Matrix cracking at the fiber break was not observed for any of the other fiber specimens.

Figure 5.10 is a transmission electron micrograph of ultramicrotomed sections of a typical ASC fiber. Sections were made parallel to the fiber axis. Sectioning was completed after straining in order to observe the locus of failure at the fiber breaks. The results of this investigation show that the presence of the 100- to 200-nm epoxy finish increases the interfacial shear strength between fiber and matrix and also changes the failure mode from interfacial to matrix. There are four possible explanations for the effect of the fiber finish on interfacial behavior.

*Damage Protection*

Graphite fibers are brittle materials and are therefore susceptible to strength degradation due to the presence of surface flaws. Handling operations, where fiber-to-fiber contact is present, can introduce surface flaws which would reduce fiber and composite properties. It has been speculated that the fiber finish is beneficial because it prevents fiber-to-fiber contact and hence the introduction of surface flaws.

If the primary function of the fiber finish is to protect the surface from flaws generated by the normal handling procedures, a reduction in tensile properties of the fiber would be observed in the uncoated fibers but not in the coated ones. This is not observed. Within statistical error the fiber strengths of all samples are unchanged by the presence of the coating. This is true at both the 25-mm gauge length and the much shorter critical lengths.

*Wettability*

From a composite viewpoint, complete and thorough infiltration and wetting of the 12,000-filament tows is a necessary condition for good composite properties. Unwetted regions act as stress concentrations which degrade composite performance. Application of a matrix-compatible finish might enhance wetting of the fibers by the matrix and be responsible for the improvement in interfacial shear strength.

Contact angle and surface free energy measurements on the fibers and epoxy matrix used in this study do not indicate that wettability changes would be significant. The fibers used in this work have surface free energies which are greater than the surface free energy of the epoxy and the epoxy/curing agent mixture ($>50$ mJ/m$^2$ for fibers and $<50$ mJ/m$^2$ for epoxy) [10]. The pure epoxy itself has a surface free energy value about the same as that of the epoxy mixture. The epoxy resin is, of course, soluble in the epoxy mixture, and undoubtedly this ensures good wetting and infiltration. Thermodynamically, the presence of the finish alone would not be expected to favor better adhesion.

*Enhanced Surface Reactivity*

Application of a fiber finish may be beneficial to the fiber-matrix adhesion through the creation of a protective environment for the reactive surface groups added to the fiber surface with treatment. Normally, fibers are produced and surface-treated and then wound on a spool for shipment to a composite prepreg or fabrication facility. Environmental exposure might reduce the reactivity of the beneficial surface groups.

Thermal desorption-mass spectroscopic studies of the fibers used in this study showed that the equivalent of up to three monolayers of water, carbon monoxide, and carbon dioxide can be desorbed from these surfaces at temperatures less than 150°C, indicating physical adsorption of atmospheric gases [9,25]. X-ray photoelectron spectroscopy studies of AS fibers manufactured 5 years apart, on the other hand, showed little difference in either quantity or character of the surface species. The XPS analysis of fibers without the finish and fibers that had the finish extracted with MEK showed very little difference. The few percent of the fiber surface that retained the epoxy after extraction, which may therefore be inferred to be chemisorbed, would not be responsible for the 25% increase in interfacial shear strength. Indeed, an 80% reduction in surface oxygen species on the same fiber produced only a few percent decrease in interfacial shear strength. Horie et al.

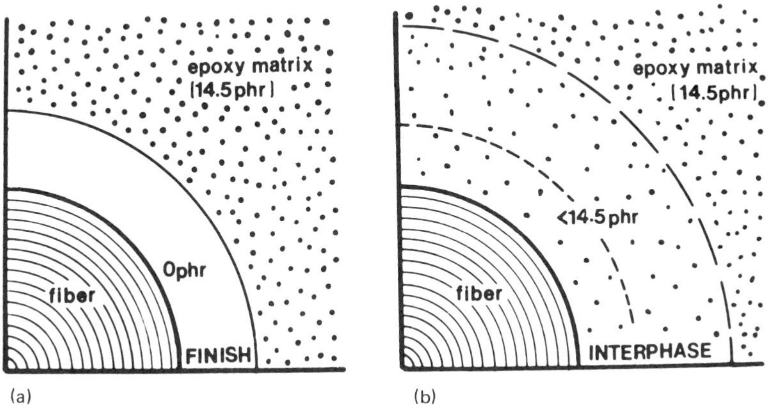


Figure 5.11 Schematic diagram of the interphase region (a) at the initial moment of fabrication and (b) after final cure.

[26] attempted to determine the chemical bonding of the epoxy matrix cured with an amine to graphite fibers. They proposed that hydrogen bonding between the carboxylic surface groups of the fiber and the amine groups of the resin occurs at low temperatures (i.e., 60°C). Since the pure DGEBA epoxy layer prevents amine from coming in contact with the fiber surface, chemical interaction between them is prevented here.

*Creation of an Interphase*

The application of a fiber finish can create an interphase layer around the fiber which could affect the fiber-matrix adhesion. Figure 5.11 is a schematic drawing in quarter view of the fiber and polymer in the interfacial region. Initially the fiber has a 100- to 200-nm finish of pure DGEBA epoxy resin without amine curing agent (Figure 5.11a). This has been applied from solvent and dried at low temperatures, < 100°C. Homopolymerization of the epoxy has not been observed under these conditions [17]. This coated fiber is then placed in a stoichiometric mixture of DGEBA with the proper amount of amine curing agent, in this case 14.5 phr. During the 50 min before gelation the amine is free to diffuse into the amine-free finish at the same time that it is reacting with epoxy functionalities. This would retard free diffusion and create an interphase layer larger than the 100- to 200-nm finish with a gradation in amine concentration decreasing from the bulk of the resin to a very low value near the fiber surface (Figure 5.11b). Whatever amine concentration gradient exists in the interphase, it would have less than the stoichiometric amount necessary for full cure under these conditions. Kardos [27] detected the presence of an interphase in many composite systems, including a glass-Epon 828 composite.

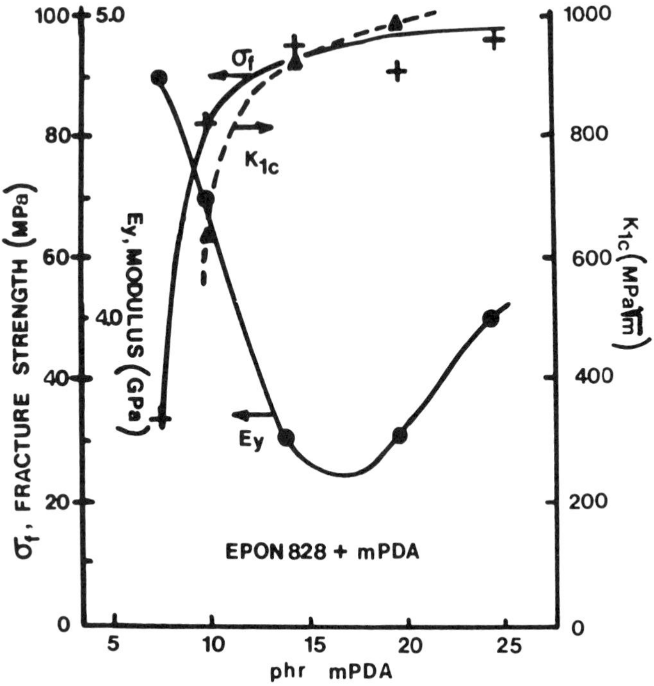


Figure 5.12 Initial tensile modulus ($E_y$), fracture strength ($\sigma_f$), and fracture toughness ($K_{1c}$) of Epon 828-mPDA mixtures as a function of parts per hundred resin (phr) of mPDA.

It is not currently possible to determine the mechanical properties of a polymer region of microscopic dimensions such as that studied here. Therefore, an attempt was made to determine the properties of this region with reduced curing agent by measuring the properties of mixtures of Epon 828 with varying amounts of mPDA cast and cured under conditions identical to those used to fabricate the single-fiber specimens. Figure 5.12 is a plot of the Young's tensile modulus ($E_y$), fracture strength ($\sigma_f$), and fracture toughness ($K_{1c}$) of these epoxy variations. The modulus reported is the initial modulus determined from strain-gauged samples, and the fracture toughness was determined using the compact tension specimen at room temperature and a strain rate of 0.002 in./min. Comparing the epoxy with less than the stoichiometric amount of curing agent to the stoichiometric amount at 14.5 phr indicates that reducing the amine content results in a material with a higher modulus, lower fracture stress, and lower fracture toughness. This indicates that the interphase region would be expected to behave as a more brittle material than the stoichiometric epoxy. Identical results have been reported for similar amine-cured systems [28,29].

The presence of material of higher modulus itself can have an effect on the stress transfer and consequently the interfacial shear strength. Kelly and Tyson [30] compared the effect of an elastic-plastic matrix versus an elastic matrix on the value of critical length for a given fiber. They showed that a material of higher modulus increases the shear stress transfer at the interface and thereby reduces the critical transfer length, giving rise to an increase in the interfacial shear strength calculated by this single-filament method. Others [31,32] derived mathematical relationships for a model single fiber fragment system which show that the interfacial shear strength increases with the shear or tensile modulus of the interphase.

In addition to the increase in interfacial shear strength, the brittle finish layer has a lower fracture toughness. Stress concentrations exist at the fiber breaks [33] which can cause initiation of fracture to occur at the breaks and to grow perpendicular to the fiber axis into the matrix instead of along the fiber-matrix interphase. This can be seen in the micrographs of Figure 5.9 for the ASC fibers. Because fiber-to-fiber distance in an actual composite is on the order of a fiber diameter or less, the presence of a matrix crack perpendicular to the fiber surface could be expected to reduce the fracture toughness properties of the composite when a finish layer is present on the fiber.

## V. CONCLUSIONS

In summary, the following general conclusions can be drawn about the effect of graphite fiber surface treatments on fiber-epoxy matrix shear strength.

1. *Surface treatments.* Graphite fiber oxidative surface treatments tend to increase fiber-matrix shear strength with epoxies through a two-part mechanism. First, an outer weak defect-laden fiber surface layer is removed. This results in a surface which is capable of supporting higher shear loadings. Second, surface oxygen groups are added which can interact with the polar epoxy matrix, contributing to higher fiber-matrix interfacial shear strength.
2. *Surface finishes.* The application of a resin finish to graphite fibers has a twofold effect. First, there is an increase in the interfacial shear strength between fiber and matrix. Second, the mode of failure at the fiber break is altered from growth of an interfacial crack between fiber and matrix to that of a matrix crack perpendicular to the fiber axis.

The mechanism by which this epoxy finish affects both the adhesion and failure mode is through the creation of a brittle interphase around the fiber. The finish layer contains less than the stoichiometric amount of curing agent and creates a layer having a higher modulus along with a lower fracture toughness. This in turn promotes better stress transfer, resulting in a higher interfacial shear strength, but because of the lower fracture toughness the failure mode changes from interfacial to matrix.

## REFERENCES


1. L. H. Sharpe, *J. Adhes.*, *4*, 51 (1972).
2. L. T. Drzal, The Role of the Polymer-Substrate Interphase in Structural Adhesion, AFML-TR-77-129 (1977).
3. R. Bacon and A. F. Silvaggi, *Carbon*, *9*, 321 (1971).
4. R. J. Diefendorf and E. W. Tokarsky, "The Relationship of Structure to Properties in Graphite Fibers," AMFL-TR-133 (1977).
5. V. J. Mimeault and D. W. McKeke, *Nature (London)*, *224*, 739 (1969).
6. V. J. Mimeault, *Am. Chem. Soc. Polym. Prepr.*, *31*, 479 (1971).
7. F. Hopfgarten, *Fibre Sci. Technol.*, *12*, 283 (1979).
8. J. V. Larsen, T. G. Smith, and P. W. Erickson, Carbon Surface Treatments, NOLTR-71-165 (1971).
9. L. T. Drzal, *Carbon*, *15*,
10. G. E. Hammer and L. T. Drzal, *Appl. Surf. Sci.*, *4*, 340 (1980).
11. J. L. Tacich and J. A. Koutsky, in "Chemistry and Properties of Crosslinked Polymers (S. Labana, ed.), Academic Press, New York, 1977, p. 303.
12. R. J. Diefendorf and E. Tokarsky, *Polym. Eng. Sci.*, *15*, 150 (1975).
13. A. Kelly, *Proc. R. Soc. London Ser. A*, *319*, 95 (1970).
14. A. C. Cohen, Jr.,
15. W. A. Fraser, F. H. Ancker, and A. T. DiBenedetto, *Proc. 1975 Conference of the RP/Composites Inst.*, Paper 22A, 1975.
16. D. K. Hadad, J. S. Fritzen, and C. A. May, Exploratory Development of Chemical Quality Assurance and Composition of Epoxy Formulations, AFML-TR-77-217 (1978).
17. H. Lee and K. Neville, "Handbook of Epoxy Resins," McGraw-Hill, New York, 1967.
18. V. B. Gupta, L. T. Drzal, C. Y. C. Lee, and M. J. Rich, *Polym. Eng. Sci.*, *25*, 812 (1985).
19. B. Rand and R. Robinson, *Carbon*, *15*, 257 (1977).

20. D. M. Brewis, J. Comyn, J. Fowler, D. Briggs, and V. A. Gibson, *Fibre Sci. Technol.*, *12*, 41 (1979).
21. P. Ehrburger, J. J. Herque, and J. B. Donnet, *5th London International Conference on Carbon and Graphite*, 1976, p. p. 398.
22. C. R. Thomas and E. J. Walker, *5th International Conference on Carbon and Graphite*, 520 (1978), p. 520.
23. I. L. Kalnin, *Proc. 1st USA-USSR Symposium on Fracture of Composite Materials*, Riga, USSR (G. Sih and V. P. Tamuszs, eds.), Sijthoff and Noordhoff, Amsterdam, 1979, p. 373.
24. E. Fitzer, K.-H. Geigl, W. Huttner, and R. Weiss, *Carbon*, *18*, 389 (1980).
25. L. T. Drzal, J. A. Mecher, and D. Hall, *Carbon*, *17*, 375 (2979).
26. K. Horie, H. Murai, and I. Mita, *Fibre Sci. Technol.*, *9*, 253 (1976).
27. J. L. Kardos, *J. Adhes.*, *5*, 119 (1973).
28. K. Selby and L. E. Miller, *J. Mater. Sci.*, *10*, 12 (1975).
29. S. L. Kim, M. Skibo, J. Manson, R. W. Hertzberg, and J. Janiszewski, *Polym. Eng. Sci.*, *18*, 1093 (1978).
30. A. Kelly and W. R. Tyson, *J. Mech. Phys. Solids*, *13*, 329 (1965).
31. B. D. Agrawal and R. K. Bansal, *Fibre Sci. Technol.*, *12*, 149 (1979).
32. B. W. Rosen, "Fibre Composite Materials," American Society for Metals, Metals Park, Ohio, 1965.
33. J. M. Mahiehi and D. F. Adams, *J. Mater. Sci.*, *21*, 137 (1986).

# Six

# X-Ray Photoelectron Spectroscopy in Understanding Adhesion Phenomena:

## Fundamental Studies and Technological Applications

HERBERT R. ANDERSON, JR. AND KRISHNA G. SACHDEV
*IBM, Hopewell Junction, New York*

## I. INTRODUCTION

The technologically important field of adhesion involving a wide variety of material systems is replete with a number of considerations which dictate the initial bond strength and its durability under the environmental stress conditions. Parameters commonly known to influence initial bond formation include wettability, energetics and morphology of surfaces, mechanical properties of the materials involved in bonding, chemical interactions between contacting surfaces, mechanical stresses across interfaces, and existence of an interphase or a boundary layer which is generally of indeterminate thickness and chemical/mechanical properties. The interfacial regions, which may have unique features differing from the bulk characteristics of respective contacting materials, play a key role in determining the functional attributes such as bond strength, bond durability, and resistance to abrasion, corrosion, and moisture ingress. In view of this, any research and development efforts in the area of adhesion with focus on functional viability of material systems to understand, control, or improve their performance require rapid and reliable analytical methodology to best define the chemistry of surfaces and provide mechanistic insight into the relative importance of the various factors responsible for initial bonding and its durability.

X-ray photoelectron spectroscopy (XPS) or electron spectroscopy for chemical analysis (ESCA) is particularly suited for surface studies designed to provide an understanding of adhesion phenomena. This technique has seen rapid growth since instruments with monochromatized AlK$\alpha$ x-ray sources became commercially available in the early 1970s. An important factor in the popularity of this method is the wealth of experimental and theoretical information generated within the past decade that clearly demonstrates its usefulness for quantitative and nondestructive surface analysis to obtain elemental composition and chemical bonding information in a diverse variety of material systems and its applicability to all elements in the periodic table except hydrogen.

The purpose of this chapter is to provide an overview of the technique, especially in relation to its application in adhesion studies in the area of polymer science and technology, with examples from the literature that highlight its unique advantages in defining the chemical characteristics of surface/interface regions. The material presented here is only a representative segment of the voluminous literature on the applications of XPS in adhesion encompassing varied areas of technology, and it is hoped that this will benefit those involved in the interdisciplinary fields of research and development.

In particular, our focus is on the impact of XPS in studies of polymer-polymer, metal-on-polymer, polymer-on-metal, and polymer-metal oxide contacting surfaces for an understanding of the correlation of surface and interface chemistry with initial adhesion or its durability under environmental stress conditions, and mode(s) of bond failure. Any discussion of bonding mechanism(s) or locus of failure requires precise definition of primary surface characteristics. Therefore selected examples are presented to illustrate the use of XPS to obtain information on the chemical consequences of surface modification through coupling agents, oxygenation/activation by RF plasma treatment, ion bombardment pretreatment, plasma grafting of reactive species such as amination, mild controlled fluorination, and chemical etching, which are some of the commonly practiced methods for improvement of adhesion. Also included are a few examples of the use of XPS in studies of phase segregation in polymers and its effect on adhesion. Before discussion of specific cases in each of these areas, a brief description of the fundamentals of XPS analysis, basic concepts involved in adhesion, and XPS monitoring of surface modification is provided.

## II. INTRODUCTION TO X-RAY PHOTOELECTRON SPECTROSCOPY (XPS)

X-ray photoelectron spectroscopy or electron spectroscopy for chemical analysis is a highly surface sensitive and nondestructive technique for the elucidation of chemical characterics of surface/interface regions in diverse classes of material systems, including metals, metal oxides, organic polymers, and composites. Since the development of this technique by Siegbahn and co-workers [1] during 1955 to 1970, there has been intensive activity in the literature with respect to its theoretical and experimental aspects in studies of surface structure and bonding mechanisms. Several books devoted to the general area of electron spectroscopy describe the basic principles of XPS, including details of instrumentation, the type of information available from the technique, and specific applications in wide-ranging fields of investigation in material science [2-9]. A relatively recent monograph [2] on the practical applications of XPS and Auger electron spectroscopy (AES) covers current progress in the development of these two techniques in addition to the fundamental concepts involved. With the rapid advances and the availability of state-of-the-art instruments designed to provide fast data acquisition, coupled with highly advanced software packages for efficient data processing including deconvolution of overlapping signals, automatic subtraction, and spectral simulation, XPS is now a routine method for surface characterization, especially in relation to polymer adhesion in areas of industry including aerospace, automobiles, and microelectronics, in studies of corrosion phenomena and metal catalyst performance, and in the biomedical applications of polymers. Among the various surface analysis techniques, XPS, AES, and secondary ion mass spectrometry (SIMS) are routinely employed for structure studies, while scanning electron microscopy (SEM) is the most widely used experimental method for study of surface morphology [10]. Several other techniques used to complement these or for specialized applications are ultraviolet-photoemission spectroscopy (UPS), ion scattering spectroscopy (ISS), fast atom bombardment spectroscopy (FABS), Rutherford backscattering spectrometry (RBS), inelastic tunneling spectroscopy (IETS), laser ionization mass spectrometry (LIMS), and vibrational spectroscopies, particularly reflection-absorption IR spectroscopy (RAIR), and attenuated total reflectance (ATR) with the Fourier transform-infrared (FT-IR) method. Several reviews have described various aspects of XPS application, including its scope and limitations vis-à-vis other methods [11-18]. Recently, nuclear magnetic resonance spectroscopy using cross polarization with magic angle spinning (CP-MAS NMR) has been shown to be an extremely useful technique in the study of immobilized organofunctional groups on inorganic surfaces [19].

Analysis with XPS involves excitation of core electrons with a monochromatized beam of soft x-rays, typically $AlK_\alpha$ or $MgK_\alpha$ anode sources providing incident photon energies ($h\nu$) corresponding to 1486.6 and 1253.7 eV, respectively, and determination of electron binding energies (BE) by measuring the kinetic energies (KE) of the emitted electrons. These kinetic energies are given by the following relationship, where BE is the binding energy of the photoelectrons emitted from a given atomic orbital and $\phi$ is the correlation factor for the spectrometer work function:

$$KE = h\nu - BE - \phi$$

The binding energies are characteristic of the atomic orbital of the emitting electrons and are modified by the chemical/electronic environment of a particular element in the overall structural unit, while the number of photoelectrons emitted depends on the number of those atoms in the sample volume. This forms the basis for XPS analysis in elemental identification, quantitative determination of surface elemental composition, and elucidation of details of structural features in the molecular architecture, including information on oxidation states, the nature of chemical bonding, and substituent

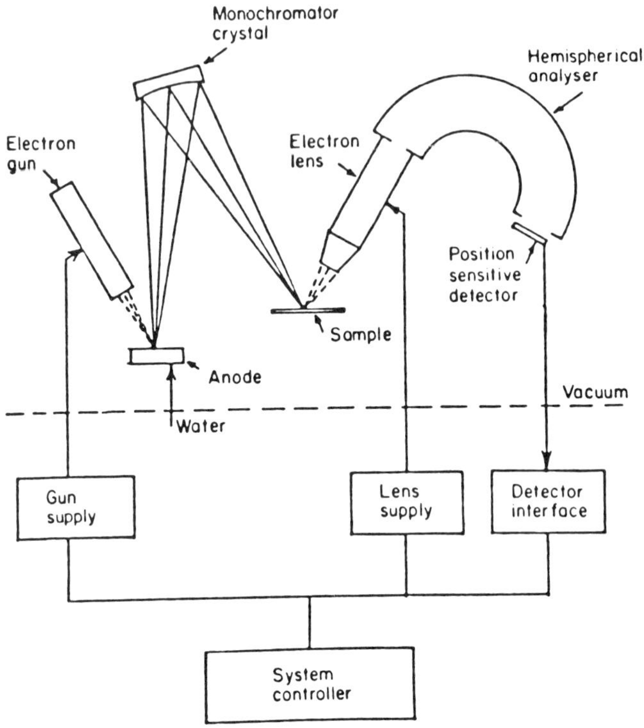


Figure 6.1 Schematic diagram of the 'small spot ESCA.' (Source: Ref. 20a.)

effects in metals, semiconductors, and insulator materials. The XPS technique is known for its excellent depth resolution, and much improved energy resolution has been achieved with monochromatized x-ray sources in modern instruments. Although spatial resolution is still considered to be the major limitation of XPS, innovations in the instrument design continue to result in improved performance in this respect. Advances in instrumentation that provide small-spot analysis capability have been detailed in the recent literature [20]. The X-probe small-spot ESCA instruments, which are designed to provide spatial resolution down to 150 $\mu m$, utilize a spectrally monochromatized and spatially focused Alk$\alpha$ source such that high-intensity x-ray flux levels are produced even at small spot sizes. This feature, in combination with a hemispherical analyzer and multichannel detector, provides rapid data collection and high sensitivity. Schematics of a typical small-spot ESCA are shown in Figure 6.1 [20a]. Specific electronic processes responsible for the various features of an XPS spectrum are illustrated in Figure 6.2 [5].

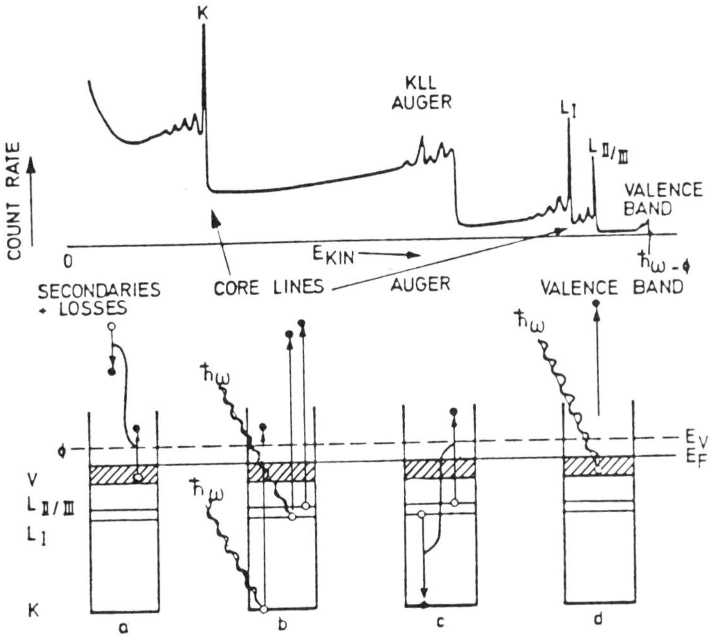


Figure 6.2 Characteristics of an XPS spectrum. (Source: Ref. 5.)

The core level signals which constitute the main features of an ESCA spectrum may be accompanied by π - π* shake-up satellites, which are generally associated with aromatic systems appearing at 6 to 7 eV lower kinetic energy than the main C(1s) photoionization peak. The Auger lines arising from various relaxation processes also serve as useful markers for atomic identification. In studies of organic polymer systems, the most common features of interest are the core level regions of carbon, oxygen, nitrogen, sulfur, silicon, and halogens, while for certain metals, metal oxides, and organometallics, in addition to core level signals, the x-ray-induced Auger lines sometimes offer very useful information.

The practical aspects of XPS analysis involve instrument calibration, energy referencing, data acquisition, peak assignments, deconvolution or curve fitting in case of overlapping signals, and computation of elemental composition from the peak areas with appropriate correction for the relative detection sensitivities of the elements. Factors determining the signal intensity of a given core level for a particular element include the photoionization cross section, the electron mean free path or the escape depth, which is the probability of escape of the photoelectrons without interaction with the solid matrix, and the instrument response function. Empirically derived sets of atomic sensitivity factors relative to F(1s) = 1.00 are available in the literature; they were obtained from measurements with a large number of standard compounds, mostly inorganic fluorides, using the Varian IEE spectrometer with an $AlK_\alpha$ x-ray source [3, 21]. Calculated values of relative sensitivities for various elements are derived from the theoretical photoionization cross sections according to Scofield [22] by including appropriate correction for the kinetic energy dependence of the electron escape depth and the instrument transmission function:

Sensitivity factor = Scofield cross section × correction factor

A tabulation of calculated photoionization cross section values has been made by Scofield [22], and photoelectron binding energies for the various elements have been compiled by King [23]. Apparent sensitivity factors relative to C(1s) = 1.00 for a particular instrument configuration are typically obtained from analysis of standard samples such as polydimethylsiloxane $\left(\text{Si–O}\right)_n$, polytetrafluoroethylene or Teflon $\left(CF_2–CF_2\right)_n$, and Nylon $\left((CH_2)_n\text{NHCO}\right)_n$ through comparison of the measured peak areas with the stoichiometric ratio. Core level binding energy assignments in complex systems are generally derived from both theoretical and experimental studies with model compounds. From spectroscopic studies and theoretical calculations with a variety of organic materials, Clark and co-workers [11,24,25] have provided extensive data on binding energy shifts for C, O, N, S, Si, and halogens. These data show a general correlation with a simple additivity model based on electronegativity and charge density as calculated according to Pauling's concept of partial ionic character [26]. The C(1s) signal of a carbon atom bonded to more electronegative atoms such as O, N, and halogens is shifted to a higher binding energy relative to the C(1s) signal at 285 eV in hydrocarbons. In the absence of directly bonded heteroatoms or secondary shifts due to adjacent electron-withdrawing substituents, the C(1s) binding energy remains quite constant within the ESCA resolution limit (~0.6 eV) regardless of hybridization [11]. Thus any shift from this reference energy serves as an index for changes in oxidation states, atomic charge, nature of substituents, or functional groups in organic and inorganic molecules. In the case of halogen substituents, the magnitude of the shift is determined by the relative electronegativity of F, Cl, Br, and I; since fluorine is the most electronegative atom, it induces the largest primary shift, which is about 3 eV in the C(1s) level relative to C–H, and a secondary shift of less than 1 eV [11]. Based on the available data, the C(1s) binding energy range is of the order of 10 eV, which in most cases allows unambiguous assignments for the commonly encountered substituents. Similarly, for the N(1s) level, the binding energy range is of the order of 5 eV based on studies with a variety of nitrogen compounds, including quaternized nitrogen in which the most positively charged N atom occurs at the highest binding energy [27]. In addition to the experimental data on compounds having different types of substituents, nonempirical calculations provide support for the electronic factors influencing

Table 6.1 Some Binding-Energy Shifts for C(1*s*), N(1*s*), and O(1*s*) ESCA Transitions

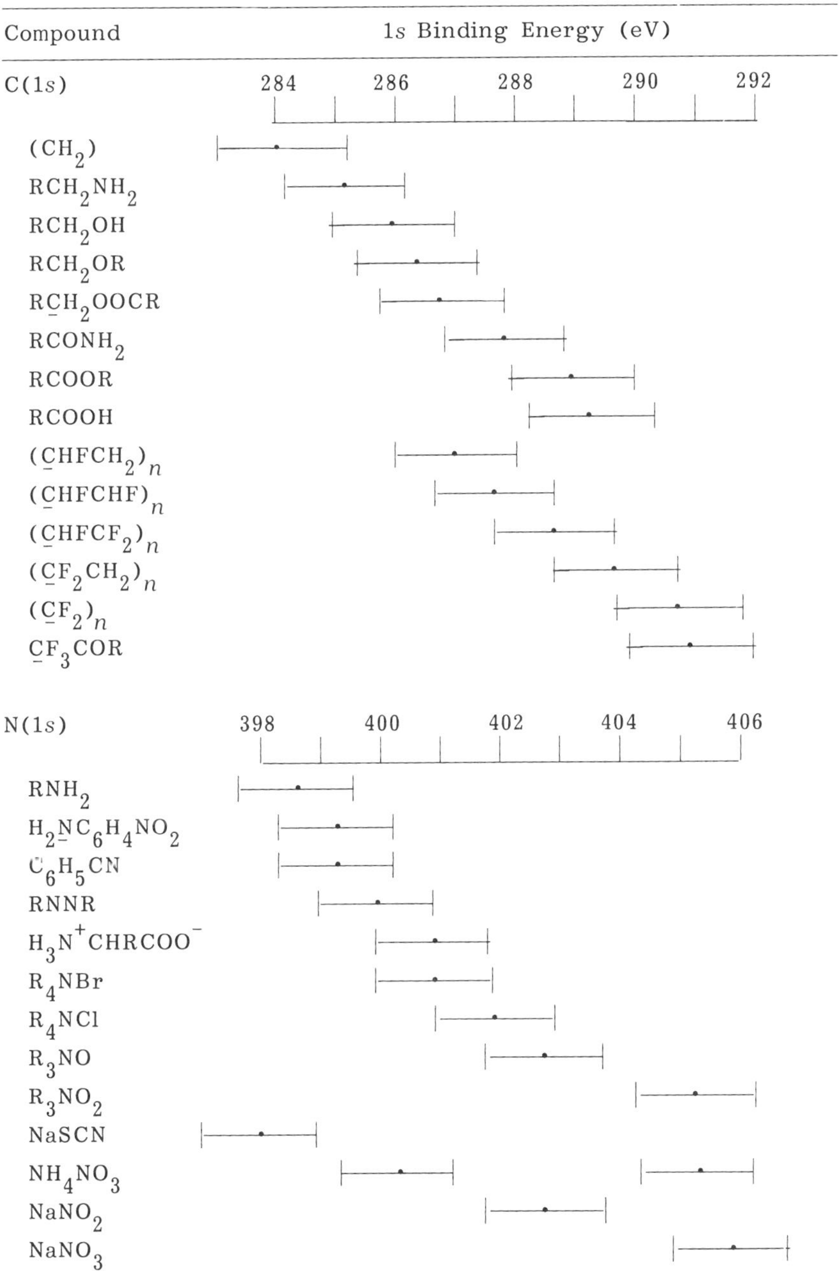

Table 6.1 (continued)

| Compound | 1s Binding Energy (eV) |
|---|---|

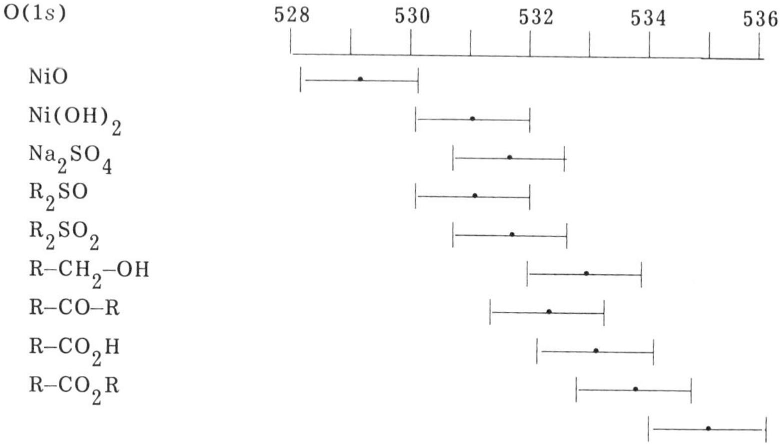


Reproduced from Ref. 68, p. 110.

Table 6.2 Typical Binding Energies for C(1s), O(1s), and N(1s) in Common Bonding States (Reference Energy = 285 eV)[a,b]

| | BE (eV) | | |
|---|---|---|---|
| Functional group | C(1s) | O(1s) | N(1s) |
| C–OH | 286.6 | 533.5 | — |
| C–O–C | 286.6 | 533.5 | — |
| O–C–O | 287.8 | 533.5 | — |
| O<br>‖<br>–C– | 288.3 | 532.6 | — |
| (a) (b)<br>O<br>‖<br>–C–O–C | (a) 289.2<br>(b) 286.6 | (a) 533.4<br>(b) 532.5 | — |
| O<br>‖<br>–C–O⁻ | 289.2 | | |

| Table 6.2 (continued) | C(1s) | O(1s) | N(1s) |
|---|---|---|---|
| $-O-C(=O)-O-$ | 290.8 | | |
| $-(CH_2-CH_2O)-$ | 286.6 | 533.6 | — |
| $-(Si-O-Si-O)-$ | | 533.6 | — |
| $-C-O-NO_2$ | 287.0 | — | 408 |
| $-C-NH_2$ | 285.7 | — | 399.2 |
| $-C-NMe_2$ | 285.3 | — | 399 |
| $-C(=O)-NH_2$ | 287.5 | — | 400.8 |
| $-C\equiv N$ | 285.3 | — | |
| $-C-NO_2$ | 286.8 | — | 407 |
| $-\overset{+}{N}H_3$, $-\overset{+}{N}R_3$ | | | 401.5 |

C(1s) BE shifts due to fluorine substituent(s)[c,d]

| Polymer | C(1s) BE (eV) (b) | C(1s) BE (eV) (a) |
|---|---|---|
| (a) (b) $-(CH_2-CFH)_n-$ | 288 | 285.9 |
| (a) (b) $-(CH_2-CF_2)_n-$ | 290.8 | 286.3 |
| $-(CFH-CFH)_n-$ | 288.4 | 288.4 |
| (a) (b) $-(CF_2-CFH)_n-$ | 289.3 | 291.6 |
| $-(CF_2-CF_2)_n-$ | 292.2 | 292.2 |
| (b) (a) $CF_3-C-CF_3$ | 294.1 | 289 |
| 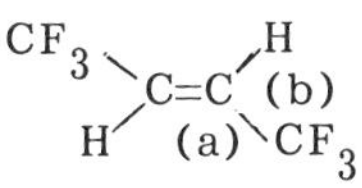 | 294.1 | 287.8 |

Table 6.2 (continued)

[a]BE of Si(2*p*) signals in different bonding states: silicon metal, 99/100 eV, doublet; $SiO_2$, 103.5 eV; silicones (–Si–O–Si–O–), ~102 eV.

[b]BE of S(2*p*) signals in different bonding states: sulfides (R–S–R), 164 eV; sulfones ($R-SO_2-R$), 167.5 eV; sulfonic acid ($RSO_3H$), 169 eV.

[c]Ref. $-(CH_2-CH_2)_n-$, C(1*s*) = 285.00 eV.

[d]These data are based on studies with fluoropolymers.

both the absolute and relative binding energies. Empirical correlation charts on chemical shifts resulting from changes in the electronic environment of the various elements serve as a reference for structure determination for materials under investigation. Some of the chemical shift data available for core level signals of C, O, N, S, and Si as a function of chemical bonding and nature of substituents are collected in Tables 6.1 and 6.2.

## III. DEPTH PROFILING BY ESCA

### A. Argon Ion Sputtering

Analysis by XPS in combination with argon ion sputtering to remove surface layers is a convenient method for probing the depth of nondegradable solid surfaces such as ceramic materials and metals to study chemical composition as a function of increasing distance from the free surface into the subsurface and bulk regions. This method is also useful for selective removal of surface contamination or a weak boundary layer caused by surface adsorption and/or surface segregation of hydrocarbons, fluorocarbons, or silicon-containing species and other components of low molecular weight or lower surface energy. However, the sputter removal technique is not suitable for depth profiling of organic polymers because of ion beam-induced degradation of such materials. Also, application of this method for surface studies of metallic samples with an oxide overlayer is complicated by the possibility that reduction of metal oxide during argon ion sputtering or other artifacts such as preferential sputtering will contribute to the composition profiles.

### B. Angular-Resolved ESCA Technique

In XPS the photoelectron signal observed for a given element represents the average chemical characteristics of the first few tens of angstroms thickness of a surface layer, i.e., the effective sampling depth from which the photoelectron can escape without inelastic scattering. The flux of photoelectrons, I(x), originating at a depth x below the surface and emerging at an angle $\theta$, known as the take-off angle (TOA), relative to the surface is given by

$$I(x) = I_0(x) \exp\left(\frac{-x}{\lambda \sin \theta}\right)$$

where $I_0(x)$ is the flux at depth x and $\lambda$ is the inelastic mean free path (imfp) with 95% of the total signal intensity originating from depth $3\lambda \sin \theta$. Theoretical aspects of photoelectron intensity analysis in the determination of electron mean free path are discussed by Henke [28]. According to angular-resolved analysis as illustrated in Figure 6.3, a particular surface can be profiled at progressively shallower sampling depths by rotating the sample with respect to the analyzer lens with a fixed x-ray source and energy analyzer configuration. At grazing electron exit angles, the XPS signals represent the chemical features of the top 5 to 10 Å as the surface sensitivity is enhanced with decrease in effective sampling depth. At the normal take-off from the surface, the composition profile corresponds to the maximum sampling depth in ESCA studies, which is of the order of 100 Å or so. Details of the fundamentals and applications of angular-resolved XPS have been reviewed in the literature [29]. Recently, application of this method was described in the framework of a flat-layer model used to obtain concentration-depth profiles of the Ag–$Al_2O_3$ and $SiO_2$–Si interfaces and of the surface region of a Cu-Ni alloy [30].

This technique of depth profiling has gained increasing importance in polymer studies, mainly because it does not suffer from the limitations of the sputter-removal method. However, it is well recognized that with the sampling depth being governed by the electron mean free path or the escape depth, accuracy of the data in terms of

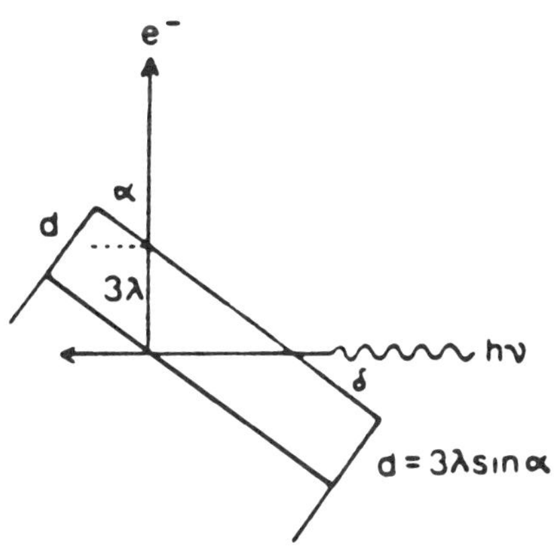


Figure 6.3 Surface sensitivity enhancement by variation of the electron "take-off" angle. (Source: Ref. 2, p. 133.)

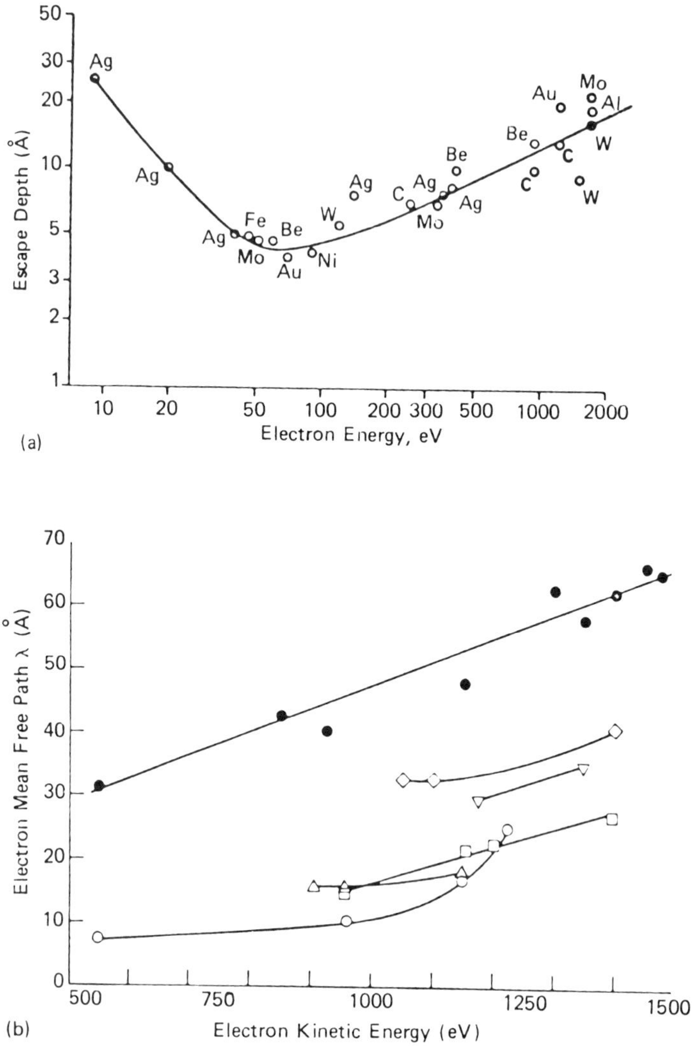


Figure 6.4 (a) Electron mean free path vs. kinetic energy in metals. Source: Ref. 9, p. 108. (b) Electron mean free path as a function of electron kinetic energy through organic materials. Key: ▽, PMMA; □, poly(p-xylylene); ○, plasma polymerized fluorocarbon; •, barium stearate; ◇, cadmium orachidate; and △, carbon.

thickness profiles is limited by the reliability of escape depth measurements of core level photoelectrons. Figure 6.4 shows the well-known dependence of imfp on kinetic energy in the range 100 to 1500 eV for inorganic materials, but whether the same values are directly applicable to organic materials has been a subject of controversy in the literature [31-33] because of the rather large inconsistency in the data from various studies. Most measurements of escape depth are based on the substrate overlayer approach, in which attenuation of the substrate signal is monitored as a function of known overlayer thickness. According to studies reported by Steinhardt et al. [31] for vacuum-deposited carbon films on gold samples, imfp values of 15 Å and 18 Å for electrons of KE 970 and 1169 eV, respectively, have been obtained. However, several other studies reviewed by Powell [32], particularly measurements with Langmuir-Blodgett films, have consistently yielded very large ifmp values. One explanation for the large estimates of imfp with oriented films considers the contribution of the channeling phenomenon for electrons traveling nearly parallel to the long chains of the molecules, especially when the analysis is carried out at a fixed angle normal to the plane of the substrate. A detailed discussion of the various possible causes of the ambiguities in imfp values for organics compared to those for metals and semiconductors has been provided by Clark and Thomas [33], who used the substrate overlayer approach for measurements of imfp from poly(p-xylylene) or Parylene N films. The values reported in this study are 14, 22, 23, and 29 Å for electrons of kinetic energy 969, 1170, 1202, and 1403 eV, respectively, which are not significantly different from those for metals and semiconductors and are in good agreement with the data presented by Steinhardt [31].

## IV. LIMITATIONS OF THE ESCA TECHNIQUE

It is well recognized that chemical characteristics of materials' surfaces are subject to change due to alteration of intrinsic composition profiles caused by any of the following commonly encountered problems.

### A. Surface Contamination

Any adsorption of hydrocarbons and other contaminants such as oxygen-containing species from the ambient prior to or during sample preparation or from the x-ray window when it is in close proximity to the sample may affect the peak shapes and reliability of the data.

### B. Charging Effects

Analysis of insulating materials such as polymer films using a monochromatized x-ray source at a distance from the sample to minimize surface degradation sometimes causes severe charging effects, resulting in signal distortion and broadening accompanied by large binding energy shifts. This problem is generally circumvented by the use of a "flood gun" to provide a low-energy electron flux over the sample surface to effect charge neutralization.

### C. Surface Damage by X-Rays

Among the various organic polymers which are susceptible to x-ray damage, poly(vinyl chloride), poly(vinylidene fluoride), and poly(olefin sulfones) are typical examples, but a large group of polymers are quite stable under the analysis conditions and thus are amenable to surface characterization by the ESCA technique with reasonable accuracy.

## V. FACTORS INFLUENCING ADHESION AND BOND FAILURE

Adhesion is a fundamental requirement for successful fabrication and reliable performance of composite structures comprising polymer coatings in contact with polymers, metals, and semiconductor surfaces. Bond strength between contacting surfaces of various material systems is generally measured by destructive tests and is commonly expressed in terms of the force or work required to mechanically separate one material from another [34]. Through the years, several theories have been postulated to explain the interfacial phenomenon based on wettability, surface energetics, and surface adsorption [35, 36] and general concepts involving acid-base interactions that include hydrogen bonding and electrostatic and covalent bonding [37]. Some of the important factors usually recognized to influence bond strength and interface durability in composite structures as outlined below have been the subject of several reviews [38-40].

### A. Wettability, Surface Energetics

Primary surface characteristics such as wettability, acidity or basicity, surface and interfacial energy, and kinetics of wetting are considered important factors in interface interactions. For example, optimum wetting, which is characterized by a low contact angle in a liquid-solid system, is obtained when the surface tension ($\gamma_{lv}$) of the liquid is less than the surface free energy ($\gamma_{sv}$) of the substrate, although merely satisfying the wetting condition does not necessarily imply good mechanical integrity at interfacial zones. Also, certain surface treatments may sometimes increase adhesion without improving wettability.

In the case of complete wetting, the work of adhesion $W_A = \gamma_{sv} + \gamma_{lv} - \gamma_{sl}$, where $\gamma_{sl}$, the interfacial tension between the solid substrate and the liquid adhesive, is considered to be an important surface energetic criterion in determining adhesion. Thermodynamic considerations related to substrate/adhesive bond durability or interface stability in the presence of liquid environments have been discussed by Kinloch and co-workers [40].

## B. Interpenetrating Network and Mechanical Interlocking

Adhesion considerations usually include the possibility that an important reinforcement mechanism in bonding is interdiffusion and an interpenetrating network, depending on the mutual compatibility of materials. Also, mechanical interlocking across an interface, which depends on surface asperities or roughness is an important contributary factor; it is useful mainly in the adhesion of polymer coatings to porous substrates such as wood and textiles and in the case of electrolytically deposited metals, but it cannot be invoked to explain good bonding of polymers to smooth substrates such as glass.

## C. Acid-Base Interactions and Covalent Bonding

Electrostatic and/or chemical interactions at the interface are considered to be the dominant factors in determining bond strength and bond durability in composite structures. In applying the acid-base concept to polymer adhesion, Fowkes and others [37a] developed a very useful and generalized treatment for interfacial interactions between various polymers and solid surfaces involving hydrogen bonds and both Brønsted and Lewis acid-base pairs. This approach is based on the Drago correlation [41] of enthalpy changes accompanying acid-base complexation in solution as given by

$$-\Delta H_{AB} = E_A E_B + C_A C_B$$

where E and C are empirically derived parameters related to electrostatic and covalent interactions, respectively, between the acid-base pair. In the evaluation of polymer adhesion to acidic or basic surfaces, van der Waals and acid-base interactions including hydrogen bonds are considered to be the dominant factors in determining interface compatibility. The Bolger-Michaels approach to interface interactions between polar compounds and hydrated metal oxide surfaces considers Brønsted acid-base pairs to explain polymer adhesion [37b]. According to the chemical bonding theory, covalent bond formation between contacting surfaces is one of the major factors contributing to the resistance of bonded structures to corrosion and other conditions detrimental to polymer-substrate interface stability in the actual use environment.

## D. Matching of Thermal Coefficient of Expansion (TCE)

Stress-strain and fracture resistance properties of contacting layers determine the extent of stress concentration at an interface caused by TCE mismatch. Since organic polymers typically have 5 to 10 times higher thermal expansion than inorganic materials, this becomes a major factor in bond durability in organic-inorganic systems under conditions of thermal stress. For example, in structures having thermoplastic or thermosetting resin matrices in contact with metals or metal oxides, the TCE mismatch is a major cause of mechanical stress resulting in fracture or delamination under conditions requiring exposure to high-temperature/humidity and chemical environments. In integrated circuits, device fabrication using organic insulators such as polyimides in contact with metals, other insulators, and semiconductors typically involves 400°C cure and multiple thermal excursions; this can result in a highly stressed structure because of thermal mismatch and other factors, depending on material compatibility with the process constraints. The magnitude of stress ($\sigma$) is determined by both the TCE difference ($\Delta\alpha$) and the range of temperature ($\Delta T$) over which the stresses arise and is given by

$$\sigma = f(E,T)\ \Delta\alpha(T)\ \Delta T$$

where E is the modulus of a polymer system.

Another factor generally considered in discussing bond formation and bond failure mechanisms is the possible existence of a weak boundary layer (WBL) at various surface/interface regions. According to this concept as presented by Bikerman [42], a discrete phase-segregated and/or chemically degraded interfacial layer with inferior mechanical properties becomes the primary site of adhesion failure when a composite is subjected to various stress conditions. In terms of the nature of bond failure in composite systems, a "cohesive mode" is defined by separation in a region well within the bulk of either of the contacting materials, whereas in an "adhesive mode" delamination occurs at the interface. In the context of the various factors that influence the nature of failure modes in truly bonded structures, others [36b,43] have also presented arguments which lend little support for failure at the interface in the instance of mechanical separation.

## VI. APPLICATION OF XPS IN ADHESION

Surface and interface characterization by ESCA is a very convenient means of understanding changes in the chemistry of the outer regions of materials' surfaces as a consequence of various treatments; thus it can be employed to develop a correlation of the factors that most influence the adhesion phenomena. This technique has been extensively employed to develop a mechanistic understanding of the interface interactions that correlate with bond strength and bond durability and to identify bond failure loci in a wide variety of material systems. To be able to use this technique for precise evaluation of a mechanism(s) of intrinsic adhesion between contacting surfaces, it is important to establish the prior presence or absence of any extraneous factors that contribute to surface alterations, which otherwise complicate the data interpretation. Among these, surface adsorption of contaminants is a common problem in surface analysis of polymer coatings, metals, metal oxides, etc. Also, commercial polymer resins often carry impurities and additives such as emulsifying agents, antioxidants, low molecular weight plasticizers and/or curing agents, residual polymerization catalysts, or the byproducts thereof. When the commercial polymers are processed in contact with other surfaces, any extraneous materials present can undergo surface segregation and affect the surface and interface properties, which may result in poor wettability of free surfaces and mechanically weak interfaces due to formation of a weak boundary layer. Surface susceptibility to moisture is another factor that can change the surface chemical profiles. For example, polymers having polar groups such as $-NH_2$, $-CONH_2$, $-OH$, and $-\overset{|}{C}=O$ are especially subject to moisture adsorption through hydrogen bonding with these functionalities, which can also result in hydrolytic cleavage in case of moisture-susceptible linkages.

## VII. METHODS FOR SURFACE MODIFICATION TO PROMOTE ADHESION

In order to obtain the necessary surface characteristics in terms of chemical structure and wettability to promote polymer adhesion to polymers, metals, or metal oxides, it is a common practice to prepare modified surfaces by appropriate surface modification schemes. Some of these standard approaches are outlined below.

### A. Adhesion Promoters [44-45]

Surface modification through the application of organofunctional silanes, aluminum chelates, organotitanates, and related organometallics is a well-established method for improving the adhesion of polymeric coatings to siliceous substrates.

### B. Chemical Etching/Oxidation

Wet etching of low surface energy materials such as polyethylene, using chromic acid to cause etching, surface oxidation, roughening, or removal of any weak boundary layer, is a well-known method. The specific nature of the chemical changes resulting from this treatment has been extensively investigated [46].

### C. Surface Activation Through Dry Processing [47-50]

This includes surface modification by $O_2$ plasma treatment, argon ion bombardment, use of an inert gas plasma to reduce surface oxidation and promote higher cross-link efficiency, exposure to an ion beam or electron beam to cause CASING or cross-linking, introduction of reactive functional groups by ammonia and amine RF plasmas, and grafting of other reactive species.

## VIII. ESCA STUDIES OF SURFACE MODIFICATION WITH ORGANOFUNCTIONAL SILANES

Since the early work of Plueddemann and others [44] on silane coupling agents as adhesion promoters for organic coatings on inorganic surfaces, there has been extensive investigation of these materials

Table 6.3 Representative Organofunctional Trialkoxysilanes[a]

| | |
|---|---|
| γ-Aminopropyltriethoxysilane | $NH_2(CH_2)_3Si(OEt)_3$ |
| *N*-Methyl-γ-aminopropyltrimethoxysilane | $MeNH(CH_2)_3Si(OMe)_3$ |
| *N*-(β-Aminoethyl)-γ-aminopropyltrimethoxysilane | $NH_2(CH_2)_2NH(CH_2)_3Si(OMe)_3$ |
| 3-Methacryloxypropyltrimethoxysilane | $CH_2{=}C(CH_3){-}C({=}O){-}O{-}(CH_2)_3Si(OMe)_3$ |
| γ-Glycidoxypropyltrimethoxysilane | $CH_2{-}CHCH_2O(CH_2)_3Si(OMe)_3$ (epoxide ring, –O– bridge) |
| 3-Chloropropyltrimethoxysilane | $Cl(CH_2)_3Si(OMe)_3$ |
| Vinyltrimethoxysilane | $CH_2{=}CHSi(OMe)_3$ |
| 3-Mercaptopropyltrimethoxysilane | $HS(CH_2)_3Si(OMe)_3$ |

[a] Me = $-CH_3$; Et = $-CH_2CH_3$.

for improvement of interface properties. It is well known that organofunctional trialkoxysilanes of the general structure $(RO)_3Si{-}(R){-}X$ (Table 6.3) are effective adhesion promoters for bonding of a variety of polymer systems such as phenolics, epoxies, polyesters, polycarbonates, polysulfones, and polyimides to inorganic surfaces of diverse chemical and physical characteristics including glasses, metals, and metal oxides. Numerous studies have been reported regarding mechanisms that may explain how and why these coupling agents generally lead to adhesion improvement [39,44]. Consideration of the chemical basis for adhesive bonding commonly includes the initial acid-base interactions of silanes with hydroxylated inorganic substrates that result in a modified surface having a chemisorbed and physisorbed silane layer with a preferred orientation and molecular organization and the possibility of a structural gradient in the silane interphase region. In general, covalent bonding of silanes to siliceous surfaces through silanol or alkoxy groups, the formation of an interpenetrating network, and/or interfacial covalent interaction with the polymer matrix through the organofunctional group are considered to be important contributory factors in the reinforcement mechanism for composite structures. In connection with various theories explaining the overall adhesion improvement with organosilanes, several investigations have provided spectral information on the silane-modified surfaces and substrate-silane-polymer interactions and correlations thereof with bond strength and bond failure locus, an aspect greatly benefited by the XPS technique.

In the general class of silane coupling agents, the aminoalkyl-alkoxysilanes are the most commonly employed adhesion promoters. In aqueous solution, these materials undergo rapid hydrolysis and self-condensation, and the kinetics and nature of the products depend on pH, concentration, and temperature [51]. These changes in the solution chemistry coupled with the acidity or basicity of the substrate affect the molecular organization, orientation, and thickness of the silane layer, which influence polymer-silane-substrate interactions and the mechanical properties of the interphase. In various studies of organosilane interactions with inorganic surfaces and their relevance to polymer-substrate adhesion mechanisms, XPS has commonly been employed in conjunction with FT-IR and other surface analysis techniques. In an earlier XPS study of acid-base interactions, Anderson et al. [52] reported that a silicon wafer surface modified with γ-aminopropyltriethoxysilane (γ-APS) showed a doublet at 399 and 400.5 eV corresponding to N(1*s*) signals attributable to the free $-NH_2$ of γ-APS and its protonated form, respectively. These features were rationalized on the basis of proton transfer reactions between the amine functionality and the silanol groups which can be generalized as

$$\equiv Si{-}\overset{\delta -}{O}\cdots\cdots H\cdots : \overset{\delta +}{N}H_2R$$

Similar observations have been reported from a number of other studies dealing with the surface structure of silane films for a mechanistic understanding of possible interactions of silane coupling agents with oxide surfaces and resin matrices. For example, Moses et al. [53], in their studies of silanized surfaces of metal oxide electrodes by XPS in connection with electrochemical applications, showed that the N(1*s*) region is consistent for both $-NH_2$ and its protonated form. Boerio and Williams [54] used the reflection-absorption IR (RAIR) technique in conjunction with XPS analysis to study the molecular structure of γ-APS adsorbed on iron mirrors from aqueous solution and its correlation with durability of iron-epoxy lap joints. For

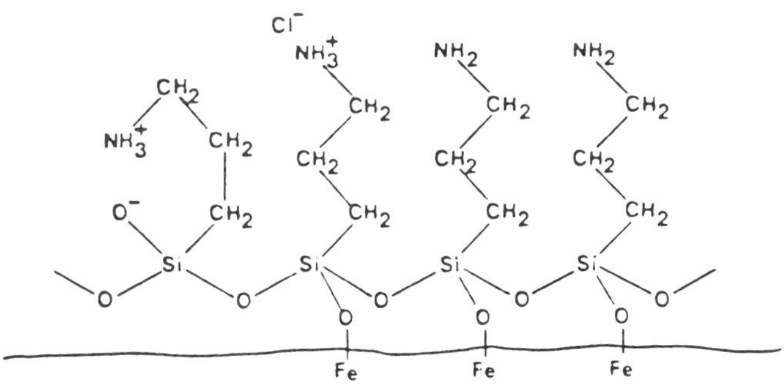


Structure (I) Source: Ref. 54.

films deposited at pH 8, the XPS and IR data were reported to be consistent with structure (I), inferred from the presence of two N(1s) signals at 401.5 and 399.6 eV corresponding to a protonated and free aliphatic nitrogen, while formation of a polysiloxane network was suggested by the presence of an intense IR band at 1135 $cm^{-1}$ due to –Si–O–Si–. These features were reported to correlate with improved iron-epoxy joint strength and durability during an extended water immersion test [54]. Similar analysis of a γ-APS film deposited at pH 12 was reported to show only the N(1s) peak at 399.6 eV, which was correlated with poor adhesion of the lap joint. Using FT-IR and laser Raman spectroscopy, Koenig and Shih [55] studied various interactions of coupling agents with both resin and glass surfaces. In discussing the nature of glass-organosilane interactions, Ishida and Koenig [39b] proposed a generalized model for the interphase structure of the silane layer according to which the bulk is characterized by a solvent-extractable physisorbed oligomeric siloxane network that increases in cross-link density as the substrate surface is approached. Near the glass surface, a chemisorbed layer with a highly cross-linked structure is formed through siloxane linkages and covalent bonding to surface silanols.

## IX. POLYIMIDE ADHESION WITH SILICON SUBSTRATES USING SILANE COUPLING AGENTS AND XPS STUDIES OF LOCUS OF FAILURE

Polyimides are an important class of high-temperature polymers which have been the subject of numerous studies for applications in microelectronics as interlevel dielectrics, passivating layers, and for device isolation. Successful use of polyimides in a technology such as very large scale integration (VLSI) entailing complex fabrication processes for multilayer structures involving contacting surfaces of widely different properties, requires that these materials maintain interface integrity with metals, metal oxides, ceramics, and other polymers at high temperature. Several studies address the question of the adequacy of aminoalkylsilanes as adhesion promoters for polyimide bonding to silicon oxide surfaces and the locus of failure [56-64]. Polyimide (II), which is derived from pyromellitic dianhydride (PMDA) and oxydianiline (ODA) polyamic acid precursor and is representative of the general class, was employed in most of the studies dealing with substrate-silane-polyimide interactions. Polyimides are known for high thermal stability, excellent mechanical properties, low moisture uptake, low dielectric constant, and good bonding to siliceous surfaces primed with coupling agents such as γ-APS. Linde [57] reported on the temperature stability of polyimide bonding to γ-APS-primed silica support based on a study of the thermal degradation of the silica-APS-amic/imide interface by IR and thermogravimetric analysis (TGA). Based on the IR data, it was shown that a surface-bound imide first formed is stable up to 390°C + 10°C under $N_2$, but degradation starts when

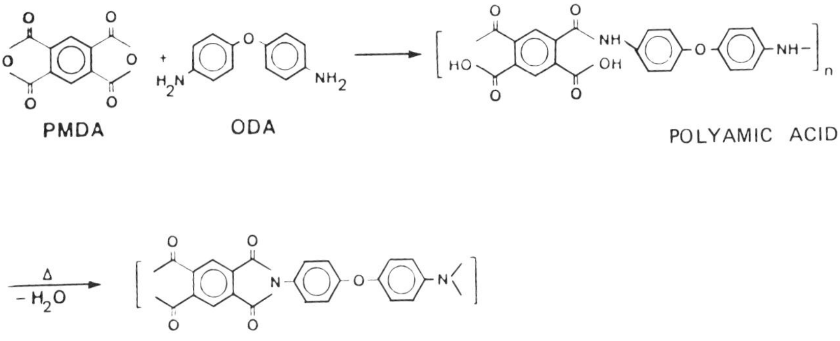


POLYIMIDE II

Structure (II)

it is heated above this temperature in an inert atmosphere or in the presence of oxidizing ambient. The data were rationalized in terms of a chemical bonding mechanism with γ-APS serving as a connecting bridge between the hydroxylated substrate surface and the cured polyimide, as illustrated by structure (III). Saiki and Harada [58] described studies on the peel adhesion of PIQ, a polyimide derived from isoindoloquinazoline dione, using *N*-(β-aminoethyl)-γ-aminopropyldimethoxysilane as an adhesion promoter on $SiO_2$, $Al_2O_3$, and Al surfaces and invoked a similar mechanism involving polyimide-aminosilane bonding. It was reported that the peel strength of the polyimide film on aminosilane-primed oxide degraded extensively when subjected to 350°C, which was attributed to the thermal lability of the interface interactions.

Our recent studies [62] of the thermal stability of bonding of PMDA-ODA polyimide to γ-APS-treated silicon oxide and silicon nitride wafers show that the peel strength remains essentially unchanged even after multiple thermal cycles at 375°C in $N_2$ and forming gas ambient. The locus of failure in mechanical separation of the polyimide film was identified by detailed XPS analysis of both the polymer and the substrate sides exposed in the standard 90° peel test (Figure 6.5). These data show that, regardless of the

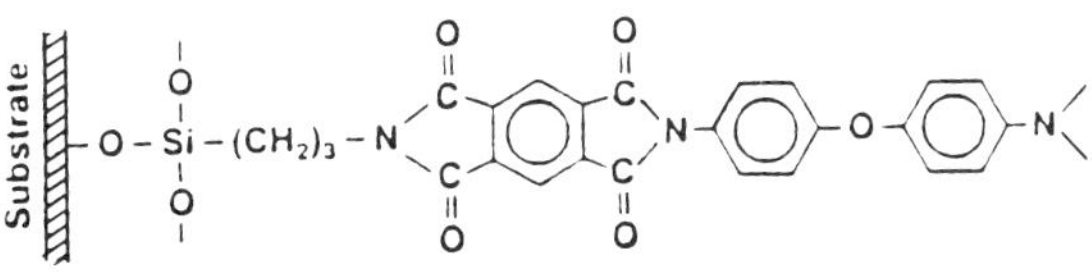


Structure (III)

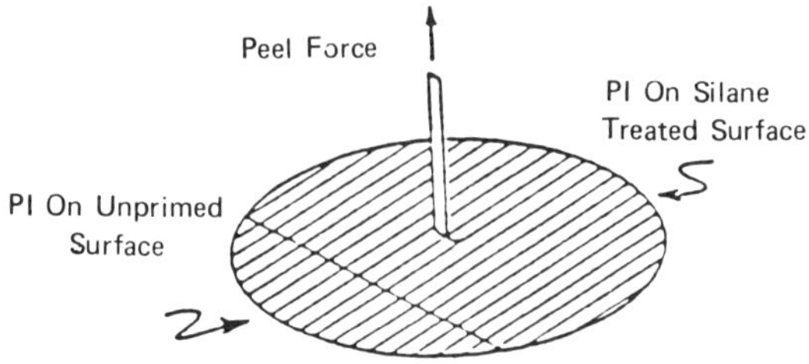


Figure 6.5 90° peel test method for polyimide adhesion.

substrate type, failure occurs by a "cohesive" mode such that the substrate side exposed after peeling retains to a large extent the overall elemental composition and individual spectral features of the C(1s), N(1s), and O(1s) signals typically observed for the PMDA-ODA polyimide reference surface cured to 400°C [62,65,66] (Figure 6.6). In the case of the silicon oxide substrate, deconvoluted spectra of the peeled polyimide side consist of the three C(1s) signals with binding energies corresponding to the reference spectra, with the difference that the nonfunctionalized carbon at 285 eV is significantly increased in intensity along with small though noticeable differences in the relative signal intensity of the higher binding energy C(1s) peaks. Analysis of the substrate side (Figure 6.7) also shows similar spectral features for the various signals, with the lowest binding energy C(1s) peak again of increased intensity compared to the reference spectra. The thickness of the

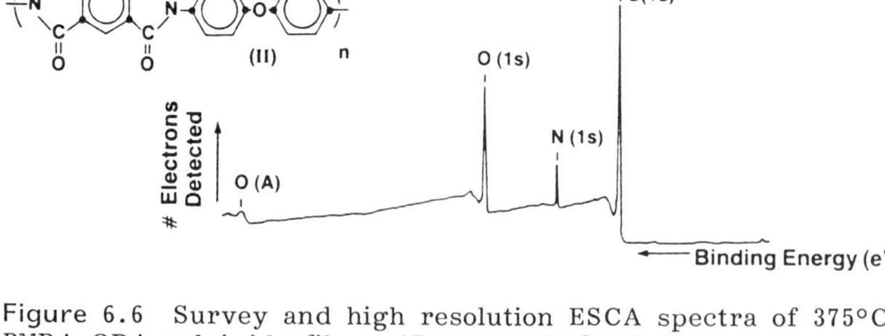


Figure 6.6 Survey and high resolution ESCA spectra of 375°C cured PMDA-ODA polyimide film. (Source: Ref. 62.)

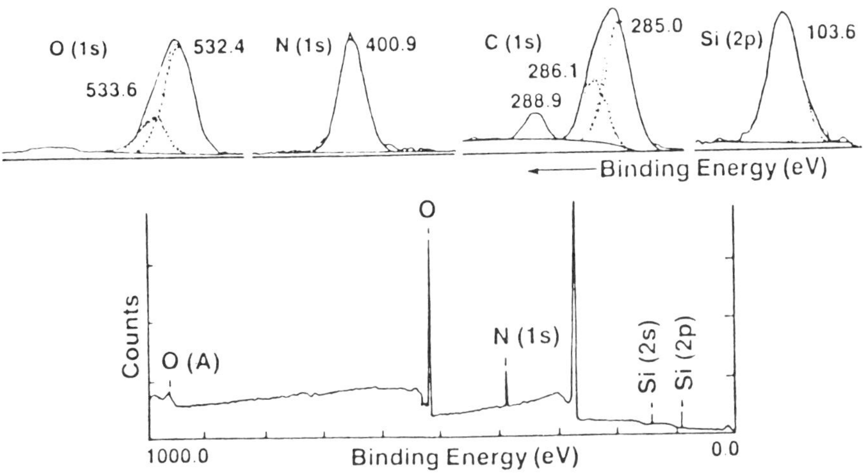


Figure 6.7 ESCA spectra of silicon wafer side after peeling polyimide film. Source: Ref. 62.

organic layer remaining on the substrate side in all cases was estimated to be of the order of 20 to 50 Å based on the Si(2*p*) signal. The observed increase in the nonfunctionalized C(1*s*) peak at 285 eV relative to the signals at 286.1 and 288.9 eV for both the substrate and the polymer side may suggest the presence of a weak boundary layer characterized by alteration or degradation of the polyimide structure at the interfacial zone. These results are supported by other studies of polyimide-substrate adhesion that used XPS to determine failure mode as a function of cure temperature [63,64], which show that the failure mode becomes increasingly adhesive with increase in cure temperature [63].

In most studies of polyimide bonding to siliceous substrates, γ-APS is typically applied from aqueous solution prior to polymer coating. We recently investigated [67] the effect of acidity and basicity of a nonaqueous application solvent, using three different aminoalkylsilanes, γ-APS, *N*-(β-aminoethyl)-γ-aminopropyltrimethoxysilane (AE-APS) and *N*-methyl-3-aminopropyltrimethoxysilane (γ-MAPS), and toluene, pyridine, and acetylacetone as solvents and water as the reference. In all cases, buffered-HF-cleaned silicon wafers were employed as substrates and polyimide coatings were cured to 380°C. Adhesion measurement by the 90° peel test showed that with AE-APS as the adhesion promoter, the highest peel strength was obtained irrespective of the solvent used in its application. However, in the case of γ-APS, the lowest peel value was obtained when it was applied from acetylacetone, an acidic solvent, whereas its *N*-methyl analog (γ-MAPS) provided the lowest value when applied from water. Analysis of the peeled surfaces by XPS again showed that the failure occurred in all cases within the polyimide-silane-substrate interphase, although significant differences were observed in the

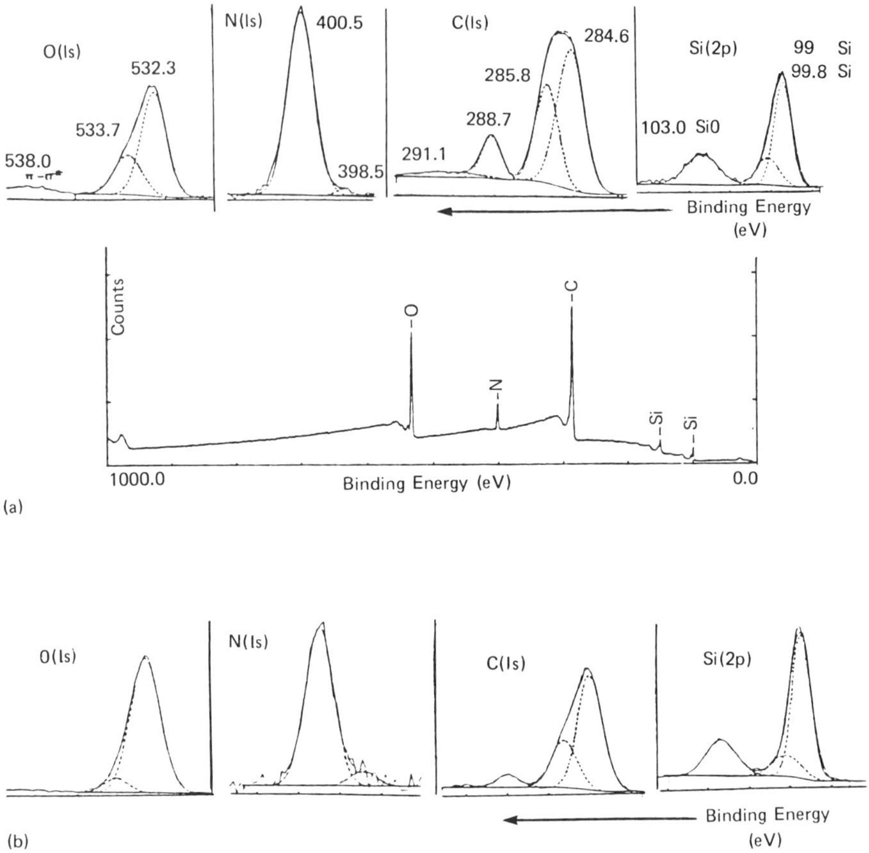


Figure 6.8 (a) ESCA spectra of the wafer side after peeling polyimide film—γ-APS in toluene as adhesion promoter. (b) ESCA spectra of the wafer side after peeling polyimide film—γ-APS in pyridine as adhesion promoter. Source: Ref. 67.

thickness and chemical composition of the film retained on the substrate side. For example, when γ-APS was applied from toluene, the substrate side exposed on peeling showed all the structural characteristics of the polyimide, including the π-π* shake-up satellites, and only 5 to 6% Si due to the underlying substrate; in contrast, 25% Si and evidence for degradation of the polyimide structure were observed when γ-APS was applied from pyridine, suggesting lower thickness of the organic layer on the substrate side in the latter case (Figure 6.8).

## X. XPS MONITORING OF MODIFIED SURFACES USING ION ETCHING AND PLASMA TECHNIQUES

Low surface energy polymers such as polytetrafluoroethylene (PTFE), polypropylene (PP), and polyethylene (PE) generally require surface activation by methods other than silane treatment to obtain the necessary adhesion with various coatings, since coupling agents with such surfaces are known not to be very effective in adhesion improvement. By use of XPS to monitor surface changes accompanying wet etching and exposure to various RF plasma-generated active species [68], it has been possible to correlate the surface chemistry with adhesive properties and provide an understanding of the bonding mechanism(s). The primary effects of these treatments include changes in surface chemical composition that may be caused by removal, oxidation or cross-linking of the outer surface layer; grafting of reactive species; and changes in surface roughness, all of which also affect surface wettability. In the case of polymers having fluorine substituents, such as PTFE, XPS monitoring of the surface changes caused by various treatments is especially convenient because fluorine, being the most electronegative element, induces rather large chemical shifts in the C(1s) signals, which allows the identification of fluorine loss and other changes in surface chemical profiles. For example, an XPS study of PTFE after treatment with sodium-liquid ammonia and also after glow discharge treatment with $NH_3$ and air showed defluorination and incorporation of oxygen- and nitrogen-containing groups at the surface, which resulted in increased wettability and adhesion [69]. Also, XPS studies of polyethylene after corona discharge treatment [70a] showed formation of carbonyl and ether functional groups, as evidenced by the binding energies of the resolved peaks in the C and O regions having C(1s) at 289.1, 287.9, and 286.6 eV corresponding to –COOR, RCOR, and C–O–C, respectively. The O(1s) level showed a doublet of equal intensity at 532.8 and 534.0 eV with an additional feature at 535.2 eV attributable to a peroxy group. XPS studies of surface modification of polyimides by exposure to oxygen and $CF_4$ plasmas have also been reported [65b].

Controlled fluorination of polymers using molecular fluorine in the presence of inert gas diluents has been studied extensively by Lagow and Margrave [71]. Several other studies have used ESCA to monitor the kinetics and chemical nature of the surface species in a variety of polymers fluorinated by this technique [70c,72] and by plasma fluorination [73]. For example, treatment of high-density polyethylene with 10% $F_2$ in $N_2$ for just 0.5 sec resulted in the formation of a monolayer of polyvinyl fluoride according to the XPS analysis, while after 30 sec exposure the surface chemical features were consistent with both the directly bonded and beta fluorine substituent [70c]. In another XPS study of polymer surface fluorination, molecular fluorine diluted with argon was employed for controlled modification of PE, PP, polyethylene oxide, polycarbonate, poly(methyl methacrylate), polysulfones, and plasma-polymerized ethylene and acetylene [72]. According to this method, various surface-bound C–H groups could be selectively transformed to C–F bonds. This was considered to be particularly useful in studying the surface chemical makeup of plasma-deposited hydrocarbon films by conversion to $CF_3$, $CF_2$, and CF groups readily identifiable from the large shifts in C(1*s*) binding energies [11,70c,72]; in the hydrocarbon analogs, $CH_3$, $CH_2$, and CH are nonresolvable. In a separate study, surface fluorination of low-density polyethylene was carried out using the glow discharge technique with a dilute mixture of $F_2$ gas in He [73]. The XPS analysis of the plasma-fluorinated surfaces of the hydrocarbon polymers was reported to be consistent with rapid formation of $-CF_2-$ groups, as indicated by shifts of ~5 to 6 eV from the reference energy (285 eV) within 40 Å depth.

Another well-known technique of surface activation involves plasma grafting of reactive functional groups such as the generation of surface-bound amine groups [74]. This method was described for adhesion improvement in Kevlar-epoxy lamination, which is an important area in the development of high-strength-lightweight composites for metal replacement in the aerospace industry. For example, amination of poly(*p*-phenylene terephthalamide) (PPTA) or Kevlar filaments was carried out by exposure to ammonia and monomethylamine RF glow discharge plasmas [75]. XPS analysis of the aminated surfaces was reported to show a substantial increase in the N(1*s*) signal which was correlated with improved adhesion with epoxy resins. It was shown that PPTA/epoxy laminates reinforced with aminated fabric have higher interlaminar tensile and peel strength, improved mechanical properties, and reduced moisture absorption rates compared to those with the untreated fabric. These differences were attributed to a direct chemical reaction between the surface-bound amino groups and epoxy functionalities resulting in covalent bonds across the interface. Other studies of surface modification of Kevlar fabric involved controlled acid and base hydrolysis of amide linkages followed by anchoring of epoxy groups through reaction of the terminal amine groups with butanediol diglycidylether [76].

## XI. XPS APPLICATIONS IN POLYMER-METAL ADHESION—INTERFACE INTERACTIONS AND CORRELATION OF SURFACE CHEMISTRY WITH ADHESION AND FAILURE MODES

Metal-polymer adhesion is an area of major technological importance in the automobile, aerospace, and microelectronics industries as well as in the solar energy and biomedical fields. Reviews [77,78] on the subject pertain to the application of XPS to develop a mechanistic understanding of the interfacial phenomena in terms of initial bond strength and its degradation as sometimes observed under actual use conditions, which may require repetitive exposure to high temperature and humidity and corrosive chemical environments. The illustrative studies of metal-on-polymer adhesion described here involve thin films which are deposited by evaporation or sputtering, while the studies of polymer-on-metal adhesion involve applications of thick metal coatings obtained by an electroless process followed by electrodeposition. These studies encompass wide-ranging areas of technology.

A striking example of the use of XPS analysis in conjunction with AES and SIMS is in the understanding of polymer-metal interface chemistry in studies by Van Ooij et al. [78] of the rubber-brass bonding mechanism, which is important in the performance reliability of steel-belted radial tires. Although good bonding of natural rubber to Cu-Zn alloys has long been recognized to be due to chemical reaction of the Cu in brass with the elemental S in rubber [79], these studies were the first to provide experimental evidence for the specific nature of interface interactions and establish a correlation between the rubber-brass interface chemistry and the measured adhesion strength. The XPS analysis was used in conjunction with depth profiling by $Ar^+$ bombardment to determine the depth-composition profiles at the fracture locus with samples generated by standard vulcanization and fractured before and after thermal aging. The rubber side of the interface consisted of a $Cu_xS$ layer on top of ZnS and ZnO, while an organic layer was retained on the brass side. The analysis showed S(2*p*) signals corresponding to unbonded S and, as a cross-linking bridge for rubber, $-CS_xC-$ detected at 100 nm depth. Comparison of the signal strength due to C–S groups at the rubber side of the rubber-brass interfacial zone exposed after fracture with that of rubber samples cured in the absence of brass indicated increased cross-link density of the polymer in the former. From the spectroscopic evidence and quantitative bond strength measurements, Van Ooij et al. [78] developed an adhesion model according to which bonding between rubber and brass occurs largely through bond formation between an interfacial film of sulfidic reaction products and cross-linked polymer (Figure 6.9). It was rationalized from these data that metals containing sulfide ions adsorbed at the surface during film growth can have an activating effect on the free sulfur, resulting in an increased cross-link density and thus a higher modulus in the thin polymer interface region. In studies of the effect of steam during processing, XPS analysis was found to be consistent with the presence of $Cu_2S$ and ZnO, indicating an interfacial layer quite different from that formed in the absence of steam. This was also the case with samples fractured after aging in high humidity or steam and showed that adhesion failure is caused by dezincification of unreacted brass leading to a weak boundary layer consisting of $Cu_2S$ and ZnO. It was further shown that addition of a few percent of Ni to brass provided much improved aged adhesion, which was attributed to inhibition of zinc depletion in the ternary alloy containing Ni and consequently suppression of the interfacial film growth during aging.

In previous studies of the adhesion of evaporated metals, typically Ni, Fe, Ti, Al, Au, and Cu to PE and PTFE, Vogel and Schonhorn [80] also discussed the importance of surface cross-linking in adhesive joint strength. They reported that a new Teflon surface generated by evaporation of metal and then its removal had much improved adhesion to standard polymers in comparison to the untreated

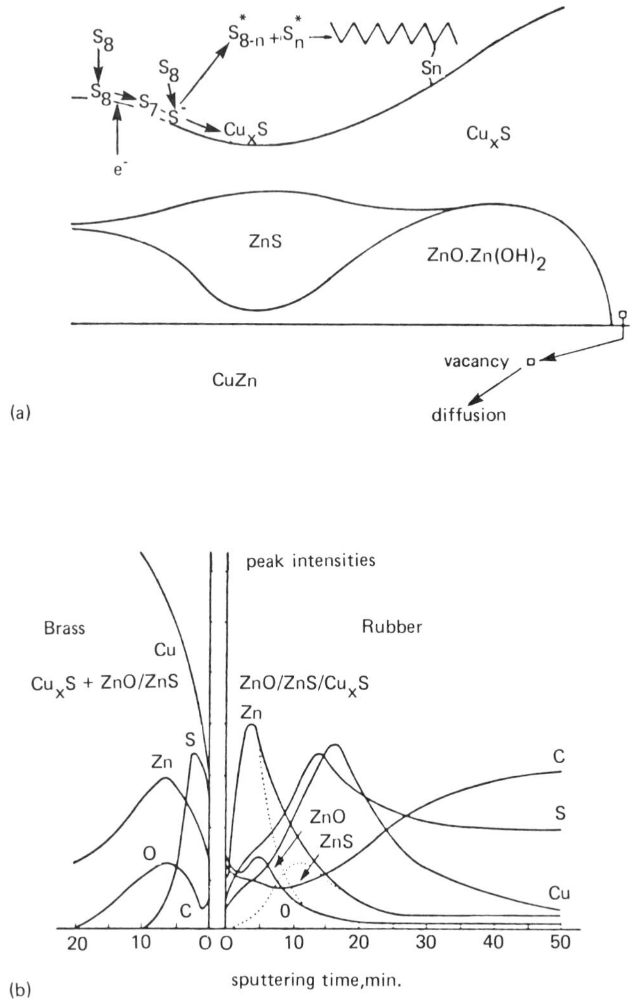


Figure 6.9 (a) Schematic representation of reactions at the rubber-brass interface during vulcanization at 150°C. (b) XPS analysis and argon ion depth profiling of rubber-brass interface; 1 min of sputtering time corresponds to 2 nm material removed. Source: Ref. 78.

Teflon, although no change in wettability was observed. In this work, XPS analysis of the modified surface was consistent with defluorination, formation of organometallic species, and surface cross-linking, and also the peel strength was strongly dependent on the cross-linking depth. A similar correlation between joint strength and cross-link density was obtained for polyethylene by determining the gel fraction in solvent extraction tests, while the XPS data on PE modified by this method showed absence of oxidation or unsaturation. Among the metals, Fe, Ni, and Ti formed the strongest bonds with PE, which correlated with increased cross-link density.

Burkstrand [81-83] reported several studies of metal-polymer adhesion in which XPS analysis was used to provide an understanding of the various chemical interactions at the interfacial region that correlate with improved adhesion of Cu, Cr, and Ni to an $O_2$ plasma-modified polystyrene (PS) surface and to oxygen-containing resins such as polyvinyl alcohol (PVOH). There are numerous examples of oxygen plasma treatment of polymer films causing oxygenation of surfaces with the formation of polar groups, typically C–OH, C–O–C, and –COOR, based on XPS studies of modified surfaces. In the case of metal-polymer adhesion, these surface features are generally reported to accompany improved interfacial bonding attributed to the formation of a metal-oxygen-polymer complex or chelate structure.

The chemical nature of these interactions in various metal-polymer systems can be inferred from the ESCA binding energy shifts and peak shape changes in comparison to the data on bulk samples. For example, it was reported that at 0.5 monolayer coverage of vapor-deposited Cu on oxygen plasma-treated PS, the O(1*s*) signal was shifted to 532.5 eV when the O(1*s*) peak in clean Cu appeared at 529.8 eV under the same conditions. In the case of Cu on a PVOH surface, formation of a Cu–O–C complex at the interface was inferred from the core level binding energies of the Cu($2p_{3/2}$) level and the C(1*s*) and O(1*s*) levels with increasing Cu coverage. Prior to Cu deposition, surface analysis showed only peaks at 285 and 286.8 eV due to $-CH_2-$ and –CHOH–, respectively. At 0.005 monolayer Cu coverage the Cu($2p_{3/2}$) signal appears at 1.2 eV higher binding energy than the bulk Cu value, but with increasing Cu coverage the signal corresponded to the bulk Cu at 932.8 eV, while the O(1*s*) region was characterized by the appearance of a new peak at 530.6 eV relative to the O(1*s*) signal due to CH–OH at 533.0 eV in PVOH. In addition, the C(1*s*) region showed a more rapid decrease in the intensity of the peak at 286.6 eV due to CH–OH relative to $-CH_2-$ at 285 eV. Similar changes supporting metal–O–C complex formation were reported for Ni and Cr deposited on PVOH and on $O_2$ plasma-treated polystyrene surfaces and were accompanied by increased adhesion of metal relative to the unmodified polystyrene. With Ni, the O(1*s*) core line showed a slight broadening in addition to the shift from 533.2 to 532.7 eV, while with Cr it was shifted from the original value of 533.6 to 531.9 eV; both shifts were toward the metal-oxide level. It is known that both Ni and Cr react more strongly with oxygen than Cu and that the dissociation energy of CuO is much lower than that for Ni and Cr.

Related XPS studies of metal-polymer interface interactions, including Cu, Cr, and Ni films evaporated on oxygen-containing polymers such as polyvinyl methyl ether (PVME), polyethylene oxide (PEO), polyvinyl acetate (PVAC), and polymethyl methacrylate (PMMA) provided similar observations in support of the polymer-O-metal complex (Figure 6.10). The XPS results were interpreted in terms of interfacial interaction based on the consideration that large

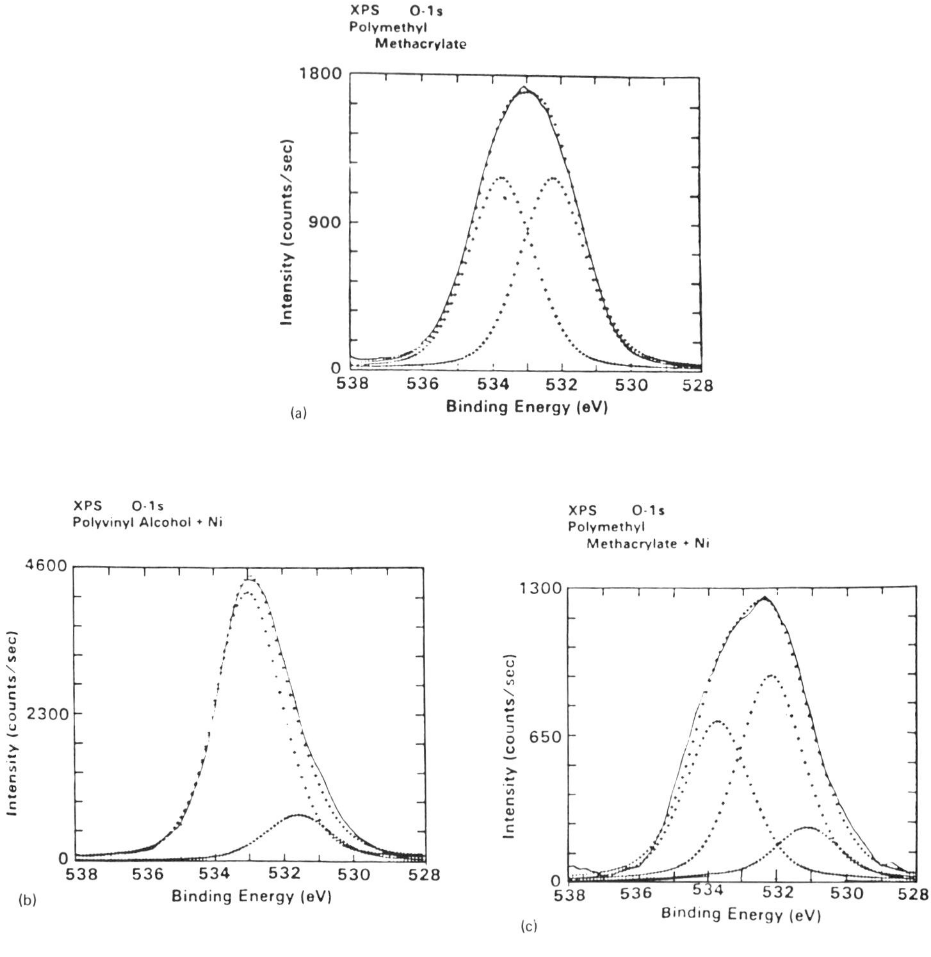


Figure 6.10 (Source: Ref. 82.)

shifts in the core level binding energies of the metal and the complementary polymer ligand would result from strong chemical bonding at the interface. For example, for Cu, Ni, and Cr on PVOH at the lowest metal coverage, the binding energies for the $2p_{3/2}$ electrons of Ni and Cr were reported to be 854.6 and 576.6 eV, corresponding to NiO and $Cr_2O_3$, respectively. For Cu, the lower shift in binding energy was explained by electronic charge transfer to oxygen due to its higher electronegativity compared to Cr [82]. From these studies it was concluded that although no chemical interaction occurs between metals and purely hydrocarbon polymers such

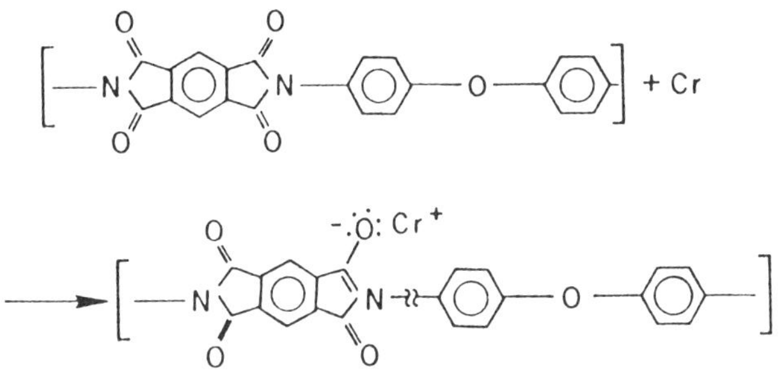


Structure (IV) Source: Ref. 84c.

as polystyrene, interfacial complexes are formed in the case of oxygen-containing polymers because of bond formation at the first monolayer of metal and the polymer surface, as indicated by XPS energy shifts and the lineshape changes in O(1*s*) levels.

Similar studies on the nature of chemical interactions at metal-polymer interfaces have been described for polyimides to provide a mechanistic understanding of the intrinsic adhesion of these materials with metals [84-86]. Chou et al. [84a] employed XPS and transmission electron microscopy (TEM) to examine the interface between Cr and fully cured planar polyimide coatings derived from the PMDA-ODA polyamic acid precursor. With increasing monolayer coverage by a Cr film evaporated at room temperature, the XPS analysis was reported to show a decrease in the polyimide carbonyl carbon (C=O) signal intensity and the appearance of a Cr(2*p*) doublet indicative of metallic Cr against the background of oxidized Cr atoms. These changes were rationalized in terms of interfacial coupling of the polyimide through the C=O functional group with the Cr surface, forming a polymer-O-Cr metal-organic complex (IV) or a partially oxidized monolayer of Cr atoms sandwiched between polyimide and Cr metal. With these data indicative of chemical bond formation at the interface, the Cr/PI adhesion was estimated to increase at least eightfold compared to Cu for which no evidence was obtained for interfacial reaction with PI [84b]. The XPS studies of metal/polyimide interfaces by Ho et al. [85] in the case of Cr, Ni, Al, and Ti were reported to be consistent with the formation of a metal-oxygen-polymer complex involving carbonyl groups of polyimides. In these studies, the Cu-polyimide interface was suggested to be of a more physical nature, with diffusion of Cu into PI leading to the formation of islands and no change in the spectral characteristics of Cu with increasing Cu coverage.

Surface modification of polyethylene by $Ar^{+}$ bombardment to improve adhesion of evaporated Ti has been reported by Bodo and Sundgren [87], who used ESCA to examine the effect of ion bombardment on the polymer chemistry and the nature of bonding at the metal/polymer interface (Fig. 6.11). It was shown that the untreated polyethylene, which had very poor adhesion to Ti, had surface

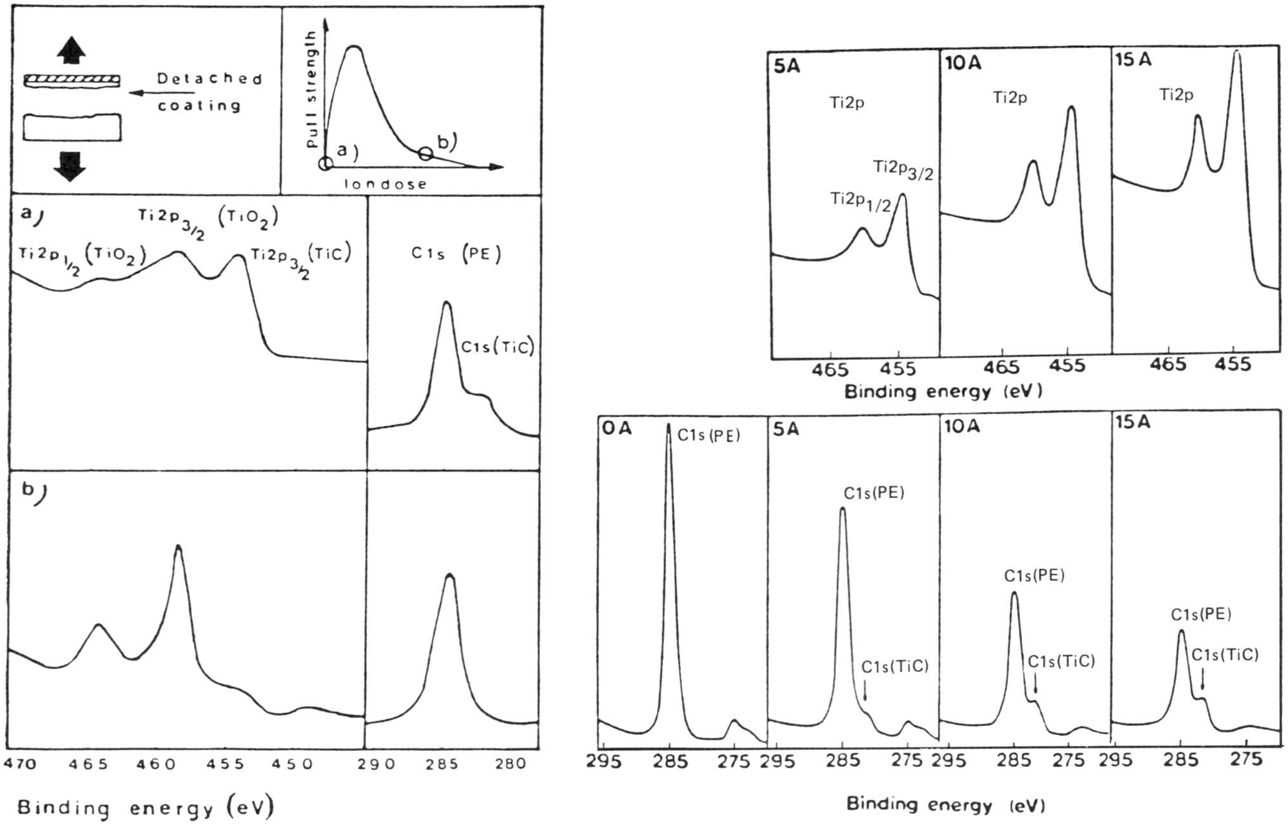


Figure 6.11 Left: Ti2p and C1s spectra of a detached film with low adhesion to (a) unreacted PE, (b) ion-bombarded PE. Right: Ti2p and C1s spectra for increasing thickness of in situ evaporated Ti onto PE. Source: Ref. 87.

contamination from silicon, Na, Cl, and oxygen-containing species. Argon ion bombardment effectively removed these impurities and at the same time led to improved adhesion, which was reported to be dose-dependent. After a dose of $6 \times 10^{14}/cm^2$ the adhesion strength reached its maximum value of 20 MPa, which decreased at higher and at very low doses. In further studies to determine the locus of failure, in situ ESCA examination was performed on the modified polymer surface and Ti/polymer interface. The results were consistent with failure occurring in the bulk polymer in the case of maximum pull strength, whereas interfacial failure was observed in samples having poor adhesion, as was the case with polyethylene exposed to a low $Ar^+$ dose. Analysis of the interface generated by in situ evaporation of Ti by XPS showed that the presence of a carbidelike Ti-C interfacial layer was one of the factors responsible for good adhesion when PE was exposed to the optimum ion dose. Direct bonding at the interface through Ti-C was inferred from the splitting of the C(1s) signal into two peaks with increasing Ti coverage; the lower-energy peak at 281.4 eV was attributed to a carbidic carbon as shown in Figure 6.11. Poor adhesion of Ti on untreated polyethylene was explained in terms of surface impurities interfering with the Ti-C bond density at the interface, while at a higher ion bombardment dose the formation of a $TiO_x$ interfacial layer was considered to be a factor in preventing the polymer-O-Ti type of interaction. Formation of $TiO_x$ on the PE surface exposed to a higher than optimum dose was considered to be due to increased surface reactivity caused by free-radical formation resulting in increased interaction with atmospheric oxygen and moisture.

## XII. XPS STUDIES OF SURFACE CHANGES CAUSED BY PHASE SEGREGATION IN POLYMERIC MATERIALS AND EFFECT ON POLYMER-POLYMER ADHESION

Various plastics and polymers contain emulsifiers, antioxidants, and other stabilizers used in some polymerization processes or additives such as antiblocking agents, which prevent blocking due to formation of a surface layer of low friction coefficient. Such additives frequently migrate to the surface or across interfaces through contact with another surface. This phenomenon may result in the formation of a weak boundary layer between adhering surfaces and influence the failure modes in composite structures. The weak boundary layer (WBL) idea of Bikerman [42] is related to the adhesive behavior of low-energy polymeric systems and to the nature of the fracture locus. In a study by Chan [88] on the mechanism of polymer-polymer adhesion failure, XPS was used to obtain information on the chemistry of fractured surfaces in the case of polyvinylidene fluoride [$-(CH_2-CF_2)_n$; Kynar] laminates using Nylon 11 adhesive. According to the data, the failure locus was characterized by the presence of $(CH_2)_n$ at 285 eV and $(CF_2)_n$ units at 292.2 eV as molecular impurities forming a WBL that was considered to result from migration of these constituents from the bulk to the interface during the bonding process. With samples fractured at liquid $N_2$ temperature, the interface chemistry was consistent with failure occurring at the Kynar-Nylon interface.

In another study of the locus of failure in polysiloxane-metal bonding, XPS analysis of the polymer and metal sides exposed in fractured samples was reported to be consistent with a WBL resulting from surface segregation of additives in commercial polysiloxanes which are commonly employed as encapsulants for semiconductor devices [89]. In these studies, silica-filled polysiloxane resin was molded onto electrolytically prepared Cu, Ag, and Ni surfaces, followed by a pull test to expose both the resin and the metal sides at the fracture locus. The elemental composition of the resin side was very similar to the bulk characteristics of polysiloxane, but on the metal side, in addition to the characteristic signals due to the resin, the presence of Pb was indicated only when the resin had been in contact with Ag or Ni and not to any significant extent with Cu. This was attributed to metal-dependent interfacial segregation of Pb additives in the resin, which are used as polymerization catalysts. The Pb/Si ratio was much higher for Ag and Ni than for Cu, which was directly related to pull strength, the higher level of Pb on the metal side being correlated with lower adhesion strength.

The bonding of polybutadiene to mild steel was studied by XPS in conjunction with ion bombardment for interface analysis and mechanical delamination. The results were reported to be consistent with corrosion-induced cohesive failure, but in the case of cathodically protected samples, separation between two phases was observed [90].

Use of XPS to identify surface segregation of the low surface energy component in solvent-cast films from A-B block copolymers has been described in a number of studies [91-93]. For example, in XPS studies of styrene-alkane copolymers, the relative intensity of $\pi$-$\pi^*$ shake-up satellites due to the styrene component and the core-level signals was used to obtain information on surface structure with respect to the bulk [91]. Similarly, ESCA studies of polyethylene terephthalate (PET) with one side chemically modified for adhesion enhancement showed $\pi$-$\pi^*$ shake-up peaks due to an intact aromatic structure only on the unmodified PET surface. Similar studies of surface segregation in styrene-ethylene oxide block copolymers showed composition profiles consistent with a polystyrene-rich surface indicated by the $\pi$-$\pi^*$ shake-up satellites [92].

In studies of chemical interactions at silicon/noble metal interfaces, XPS was useful in demonstrating a correlation between interfacial bonding and Schottky barrier heights in the case of transition metals and Si (TM = Ni, Pd, Pt) [94]. It was shown that the barrier height correlates with the strength of the TM-Si orbital interaction as well as the charge density around the Si atom.

In biomedical applications of polymers, biocompatibility is crucial to the success of synthetic implants such as intraocular lenses, artificial hearts, and other devices. Blood platelet aggregation is a commonly encountered problem. Various research and development efforts to obtain improved materials in terms of compatibility with biological environments have focused on surface modification to achieve and maintain optimum chemistry, surface energetics, and morphology, which are important factors in determining the suitability of implant devices. In studies of modified surfaces, XPS has been employed in conjunction with other techniques to monitor the extent and permanence of surface modification by plasma oxygenation and grafting of hydrophilic groups and its correlation with blood platelet consumption [95-97]. For example, Ratner [95] described XPS studies of radiation-grafted acrylamide hydrogels onto silicone rubber which demonstrated that these surfaces are subject to changes depending on the degree of hydration. The surface composition of dehydrated samples was consistent with the migration of hydrogels into the substrate, which was inferred from the loss of N(1*s*) signal due to acrylamide and the increase in Si(2*p*) intensity. Also, in the case of polyether urethanes, which are used for extruded tubes, XPS analysis showed variations in surface composition caused by segregation of oligomeric components in these polymers and surface contamination which were correlated with blood platelet consumption.

## XIII. CONCLUSIONS

The ever-increasing need for polymers for specialized applications places high demands on surface-interface properties for reliable performance of products under varied processing and use conditions. With the rapid advances in instrument design and software for XPS and its complementary surface analysis technique, AES, a large body of information has become available on the quantitative aspects of these techniques. The XPS technique is particularly suitable for nondestructive analysis of the chemical structure at polymer surfaces. In this chapter we have attempted to provide an overall perspective on the use of XPS in the study of adhesion phenomena and its important contributions to elucidating the mechanistic aspects of interface interactions and bond failure modes. This technique provides direct spectroscopic evidence for chemical events occurring at the interface regions, which is crucial to the design of approaches to tailor-make surface chemical features for optimal bonding.

Although we have limited our discussion to studies dealing with adhesion of polymers with organic and inorganic surfaces, XPS has played an equally significant role in enhancing our understanding of the interfacial chemistry in other areas, including metal-metal bonding and corrosion phenomena and heterogeneous catalysis.

## REFERENCES


1. K. Siegbahn et al., "ESCA, Atomic, Molecular and Solid State Structure Studied by Means of Electron Spectroscopy," Almqvist & Wiksells, Uppsala, Sweden, 1967; "ESCA Applied to Free Molecules," North-Holland, Amsterdam, 1969.
2. D. Briggs and M. P. Seah, eds., "Practical Surface Analysis by Auger and X-Ray Photoelectron Spectroscopy," Wiley, New York, 1983.
3. C. D. Wagner, W. M. Riggs, L. E. Davis, J. F. Moulder, and J. E. Muilenberg, "Handbook of XPS," Perkin-Elmer, Physical Electronics, Eden Prairie, Minn., 1979.
4. C. Brundle and A. D. Baker, eds., "Electron Spectroscopy—Theory, Techniques, and Applications," Vol. 4, Academic Press, London, 1981.
5. H. Ibach, ed., "Electron Spectroscopy for Surface Analysis," Springer-Verlag, New York, 1977.
6. D. W. Dwight, T. J. Fabish, and H. R. Thomas, eds., "Photon, Electron and Ion Probes of Polymer Structure and Properties," ACS Symp. Ser. 162 (1981).
7. T. A. Carlson, "Photoelectron and Auger Spectroscopy," Plenum, New York, 1975.
8. H. Windawi and F. F.-L. Ho, eds., "Applied Electron Spectroscopy for Chemical Analysis," Wiley, New York, 1982.
9. A. W. Czanderna, ed., "Methods of Surface Analysis," Elsevier, Amsterdam, 1975.
10. R. L. Patrick, in "Treatise on Adhesion and Adhesives," Vol. 3 (R. L. Patrick, ed.), Marcel Dekker, New York, 1973, pp. 163-209.
11. D. T. Clark and W. J. Feist, *J. Macromol. Sci. Rev. Macromol. Chem.*, *C12*(2), 191-286 (1975); D. T. Clark, in "Polymer Surfaces" (D. T. Clark and W. J. Feast, eds.), Wiley-Interscience, New York, 1978, pp. 305-358.
12. W. M. Riggs and M. J. Parker, in "Methods of Surface Analysis" (A. W. Czanderna, ed.), Elsevier, Amsterdam, 1975.
13. C. R. Brundle, *J. Vac. Sci. Technol.*, *11*(1), 212-224 (1974).
14. A. Dilks, in Ref. 4, pp. 277-359; A. Dilks, in Ref. 6, pp. 293-317.
15. M. M. Millard, in "Industrial Applications of Surface Analysis," ACS Symp. Ser. 199 (L. A. Casper and C. J. Powers, eds.), 1982, pp. 143-202.
16. W. E. Schwartz, Jr., *Anal. Chem.*, *45*(9), 789A (1973).
17. W. J. van Ooij, T. H. Visser, and M. E. F. Biemond, *Surf. Interface Anal.*, *6*(5), 197-214 (1984).

18. C. A. Evans, Jr., *Anal. Chem.*, *47*(9), 819A (1975).

19. D. E. Leyden and D. E. Williams, in "Molecular Characterization of Composite Interfaces" (H. Ishida and G. Kumar, eds.), Plenum, New York, 1985, pp. 377-385; also see A. M. Zaper, A. Cholli, and J. L. Koenig, *ibid.*, pp. 299-312.

20. (a) C. D. Wagner and A. Joshi, *Surf. Interface Anal.*, *6* 215 (1984); (b) K. Yates and R. H. West, *Sur. Interface Anal.*, *5*(5), 217-221 (1983).

21. C. D. Wagner, *Anal. Chem.*, *44*(6), 1050-1053 (1972).

22. J. H. Scofield, *J. Electron Spectrosc.*, *8*, 129-137 (1976).

23. R. N. King, Photoelectron Binding Energies, Photoionization Cross-Sections, and Observed Auger Energies for AlK$\alpha$ Radiation at 1487 eV, table compiled by R. N. King, University of Utah, Salt Lake City.

24. D. T. Clark and H. R. Thomas, *J. Polym. Sci. Polym. Chem. Ed.*, *16*, 791-820 (1978).

25. D. T. Clark and A. Harrison, *J. Polym. Sci. Polym. Chem. Ed.*, *19*, 1943 (1981).

26. L. Pauling, "The Nature of the Chemical Bond, Cornell Univ. Press, Ithaca, N.Y., 1960.

27. J. J. Jack and D. M. Hercules, *Anal. Chem.*, *43*(6), 729-736 (1971); D. M. Hercules, *Anal. Chem.*, *42*(1), 20A (1970).

28. B. L. Henke, *Phys. Rev. A*, *6*, 94 (1972).

29. C. S. Fadley, R. J. Baird, W. Siekhaus, T. Novakov, and S. A. L. Bergstrom, *J. Electron. Spectrosc.*, *4*, 93 (1974).

30. M. Pijolat and G. Hollinger, *Surf. Sci.*, *105*, 114-128 (1981).

31. R. G. Steinhardt, J. Hudis, and M. L. Perlman, *Phys. Rev. B*, *3*, 1016-1020 (1972).

32. C. J. Powell, *Surf. Sci.*, *44*, 29 (1964).

33. (a) D. T. Clark and H. R. Thomas, *J. Polym. Sci. Polym. Chem. Ed.*, *15*, 2843-2867 (1977); (b) R. F. Roberts, D. L. Allara, G. A. Pryde, D. N. Buchan, and N. D. Hobbins, *Surf. Interface Anal.*, *2*, 5-9 (1980).

34. J. L. Gardon, in "Treatise on Adhesion and Adhesives," Vol. 1 (R. L. Patrick, ed.), Marcel Dekker, New York, 1967, pp. 269-324; S. J. Croll, in "Adhesion Aspects of Polymer Coatings" (K. L. Mittal, ed.), Plenum, New York, New York, pp. 107-129.

35. F. M. Fowkes, in "Treatise on Adhesion and Adhesives," Vol. 1 (R. L. Patrick, ed.), Marcel Dekker, New York, 1967, pp. 325-449.

36. R. F. Gould, ed., "Contact Angle, Wettability and Adhesion," Adv. Chem. Ser., Vol. 43, American Chemical Society, Washington, D.C., 1964; (a) W. A. Zisman, *ibid.* , pp. 1-51; (b) L. H. Sharpe and H. Schornhorn, *ibid.*, pp. 189-201; (c) J. R. Huntsberger, *ibid.*, pp. 180-188.
37. (a) F. M. Fowkes and M. A. Mostafa, *Ind. Eng. Chem. Prod. Res. Dev.*, *17*, 3 (1978); F. M. Fowkes, in "Adhesion and Adsorption of Polymers" (L.-H. Lee, ed.), Plenum, New York, 1980, pp. 43-52; F. M. Fowkes, in "Physicochemical Aspects of Polymer Surfaces," Vol. 2 (K. L. Mittal, ed.), Plenum, New York, 1983, pp. 583-603; (b) J. C. Bolger, in "Adhesion Aspects of Polymer Coatings" (K. L. Mittal, ed.), Plenum, New York, 1983, pp. 3-18.
38. (a) W. C. Wake, *Polymer, 19,* 291 (1978); (b) K. L. Mittal, *J. Vac. Sci. Technol., 13,* 19 (1976).
39. (a) H. Ishida, in "Adhesion Aspects of Polymer Coatings" (K. L. Mittal, ed.), Plenum, New York, 1983, pp. 45-106; (b) H. Ishida and J. L. Koenig, *J. Polym. Sci. Polym. Phys. Ed.*, *18,* 1931 (1980).
40. R. A. Gledhill and A. J. Kinloch, *J. Adhes.*, *6,* 315 (1974); A. J. Kinloch, in "Durability of Structural Adhesives," Applied Science, London, 1983, pp. 1-39; M. Gettings and A. J. Kinloch, *J. Mater. Sci.*, *12,* 2511 (1977).
41. R. S. Drago, G. C. Vogel, and T. E. Needham, *J. Am. Chem. Soc.*, *93,* 6014 (1971); R. S. Drago, L. B. Barr, and C. S. Chamberlain, *J. Am. Chem. Soc.*, *99,* 3203 (1977).
42. J. J. Bikerman, "The Science of Adhesive Joints," 2nd ed., Academic Press, New York, 1968; 1st ed., 1961.
43. R. L. Patrick, C. M. Doede, and W. A. Vaughan, *J. Phys. Chem.*, *61,* 1036 (1957).
44. E. P. Plueddemann, "Silane Coupling Agents," Plenum, New York, 1982; P. H. Erickson and E. P. Plueddemann, in "Composite Materials," Vol. 6 (L. J. Broutman and R. H. Crock, eds.), Academic Press, New York, 1974.
45. P. F. Cassidy and B. J. Yager, *Rev. Polym. Technol.*, *1,* 1-49 (1972); P. Walker, *J. Coatings Technol.*, *52,* 49 (1982).
46. J. R. Rasmussen, E. R. Stedronsky, and G. M. Whitesides, *J. Am. Chem. Soc.*, *99,* 4736 (1977); D. M. Brewis and D. Briggs, *Polymer, 22,* 716 (1981).
47. J. R. Hollahan and A. T. Bell, eds., "Techniques of Applications of Plasma Chemistry," Wiley, New York, 1974.
48. D. T. Clark and W. J. Feast, eds., "Polymer Surfaces," Wiley, New York, 1978.

49. (a) D. T. Clark, A. Dilks, and D. Shuttleworth, in "Polymer Surfaces" (D. T. Clark and W. J. Feast, eds.), Wiley, New York, 1978, pp. 185-211; (b) H. Schonhorn, *ibid.*, pp. 213-235; D. Briggs and C. R. Ken, *Polymer, 20,* 10 (1979).

50. H. V. Boenig, "Plasma Science and Technology," Cornell Univ. Press, Ithaca, N.Y., 1982.

51. E. R. Pohl, *38th Annual Conf. Reinforced Plastics/Composites,* Society of the Plastics Industry, February 7-11, 1983, Sec. 4-B, pp. 1-4.

52. H. R. Anderson, Jr., F. M. Fowkes, and F. H. Hielscher, *J. Polym. Sci., 14,* 879-895 (1976).

53. P. R. Moses, L. M. Wier, J. C. Lennox, H. O. Finkles, J. R. Lenhard, and R. W. Murray, *Anal. Chem., 50*(4), 576-585 (1978).

54. F. J. Boerio and J. W. Williams, *Appl. Surf. Sci., 7,* 19-31 (1981).

55. J. L. Koenig and P. T. K. Shih, *J. Colloid Sci., 36,* 247 (1971).

56. D. J. Belton and A. Joshi, in "Molecular Characterization of Composite Interfaces" (H. Ishida and G. Kumar, eds.), Plenum, New York, 1985, pp. 187-204.

57. H. Linde, *J. Polym. Sci. Polym. Chem. Ed., 20,* 1031 (1982).

58. A. Saiki and S. Harada, *J. Electrochem. Soc. Solid State Sci. Technol., 129,* 2278 (1982).

59. J. Greenblatt, C. J. Araps, and H. R. Anderson, Jr., in "Polyimides," Vol. 1 (K. L. Mittal, ed.), Plenum, New York, 1984, p. 573.

60. D. Suryanarayana and K. L. Mittal, *J. Appl. Polym. Sci., 29,* 2039 (1984).

61. M. R. Gupta and J. DeLeo, *Proc. Second International Conference on Polyimides,* Ellenvill, N.Y., 1985, p. 738.

62. H. R. Anderson, Jr., M. M. Khojasteh, T. P. McAndrew, and K. G. Sachdev, *Proc. 36th Electronic Component Conference,* Seattle, Wash., May 5-7, 1986, pp. 331-339.

63. R. G. Narechania, J. A. Bruce, and S. A. Fridmann, *J. Electrochem. Soc. Solid State Sci. Technol., 132*(11), 2700-2705 (1985).

64. L. P. Buchwalter and J. Greenblatt, *J. Adhes., 19,* 257-265 (1986).

65. (a) H. J. Leary and D. S. Campbell, *Surf. Interface Anal., 1,* 75 (1979); (b) H. J. Leary and D. S. Campbell, in Ref. 6, pp. 419-433.

66. (a) L. P. Buchwalter and A. I. Baise, in "Polyimides," Vol. 1 (K. L. Mittal, ed.), Plenum, New York, 1984, p. 537; (b) B. D. Silverman, P. N. Sanda, and P. S. Ho, *J. Polym. Sci. Polym. Chem. Ed.*, *23*, 2857-2863 (1985).

67. K. G. Sachdev, S. Cheng, H. R. Anderson, Jr., M. M. Khojestah, and A. V. Patsis, *Proc. 36th Electronic Component Conference*, Seattle, Wash., May 5-7, 1986, pp. 345-354.

68. C. N. Reilley, D. S. Everhart, and F. F.-L. Ho, in Ref. 8, pp. 105-133.

69. D. W. Dwight and W. M. Riggs, *J. Colloid Interface Sci.*, *47*, 650 (1974); G. C. S. Collins, A. C. Lowe, and D. Nicholas, *Euro. Polym. J.*, *9*, 1172 (1973).

70. (a) D. T. Clark and A. J. Dilks, *J. Polym. Sci. Polym. Chem. Ed.*, *16*, 911 (1978); *ibid.*, *15*, 2321 (1977); (b) *ibid.*, *17*, 957 (1979); (c) D. T. Clark, W. J. Feast, W. K. R. Musgrave, and J. Ritchie, *ibid.*, *13*, 857 (1975).

71. R. J. Lagow and J. L. Margrave, in "Progress in Inorganic Chemistry," Vol. 26 (S. J. Lippard, ed.), Wiley, New York, 1975, pp. 161-200.

72. M. Milard, J. Burns, and H. Sachdev, in "Physicochemical Aspects of Polymer Surfaces" (K. L. Mittal, ed.), Plenum, New York, 1983, pp. 773-791.

73. M. Anand, R. E. Cohen, and R. F. Baddour, in Ref. 6, pp. 353-370.

74. J. R. Hollahan, B. B. Stafford, R. D. Falb, and S. T. Payne, *J. Appl. Polym. Sci.*, *13*, 807 (1969).

75. R. E. Allred, E. W. Merrill, and D. Reylance, in Ref. 19, pp. 333-375.

76. T. S. Keller, A. S. Hoffman, B. D. Ratner, and B. J. McElroy, in "Physicochemical Aspects of Polymer Surfaces," Vol. 2 (K. L. Mittal, ed.), Plenum, New York, 1983, pp. 861-879.

77. D. T. Clark, in "Characterization of Metal and Polymer Surfaces," Vol. 2 (L.-H. Lee, ed.), Academic Press, New York, 1977, pp. 5-51; also see H. Yasuda, H. C. Marsh, S. Brandt, and C. N. Reilley, *J. Polym. Sci. Polym. Chem. Ed.*, *15*, 991 (1977).

78. W. J. Van Ooij and A. Kleinhesselink, *Appl. Surf. Sci.*, *4*, 324-339 (1980); W. J. Van Ooij, A. Kleinhesselink, and S. R. Leyenaar, *Surf. Sci.*, *89*, 165-173 (1979).

79. S. Buchan, "Rubber to Steel Bonding," Crosby, Lockwood, London, 1959.

80. S. L. Vogel and H. Schonhorn, *J. Appl. Polym. Sci.*, *23*, 495-501 (1979); H. Schonhorn and R. F. Roberts, *Prepr. Org. Coat. Plast. Chem.*, *36*(2), 223 (1976).

81. J. M. Burkstrand, *J. Vac. Sci. Technol.*, *16*(4) (1979); *ibid.*, *15*, 223 (1978); *ibid.*, *20*(3) (1982).

82. J. M. Burkstrand, *J. Appl. Phys.*, *52*, 4795 (1981).

83. J. M. Burkstrand, in Ref. 6, pp. 340-351.

84. (a) N. J. Chou, D. W. Dong, J. Kim, and A. C. Liu, *J. Electrochem. Soc. Solid State Sci. Technol.*, *131*, 2335 (1984); (b) N. J. Chou and C. H. Tang, *J. Vac. Sci. Technol. A*, *24*, 298 (1984); (c) N. J. Chou and C. H. Tang, *A2*(2), 751-755 (1984).

85. P. S. Ho, P. O. Hahn, J. W. Bartha, G. W. Rubloff, F. K. Legoues, and B. D. Silverman, *J. Vac. Sci. Technol. A*, *3*, 739 (1985).

86. P. O. Hahn, G. W. Rubloff, J. W. Bartha, F. K. Legoues, R. Tromp, and P. S. Ho, *Mater. Res. Soc. Symp. Proc.*, *40*, 247 (1985).

87. P. Bodo and J. E. Sundgren, *J. Vac. Sci. Technol.*, *A*, *2*(4), 1498 (1984); H. Schornhorn and R. H. Hanson, *J. Appl. Polym. Sci.*, *11*, 1461 (1967).

88. Chi-Ming Chan, *J. Adhes.*, *15*, 207-224 (1983).

89. (a) A. Torrisi, G. M. Marletta, O. Puglisi, and S. Pignataro, *Surf. Interface Anal.*, *5*(4), 161-166 (1983); (b) A. J. Kinloch, *J. Mater. Sci.*, *15*, 2141 (1980).

90. J. F. Watts and J. E. Castle, *J. Mater. Sci.*, *18*, 2987-3003 (1983).

91. D. T. Clark and A. Dilks, *J. Polym. Sci. Polym. Chem. Ed.*, *14*, 533 (1976).

92. J. J. O'Malley, H. R. Thomas, and G. M. Lee, *Macromolecules*, *12*, 996 (1979).

93. D. T. Clark, J. Peeling, and J. M. O'Malley, *J. Polym. Sci. Polym. Chem. Ed.*, *14*, 543 (1976).

94. P. J. Grunthaner and F. J. Grunthaner, *Physica B and C*, *5*, Pt. II, 117-118 (March 1983); *Proc. Int. Conf. on the Physics of Semiconductors, 16th*, Montepellier, France, September 6-10, 1982, pp. 831-833.

95. B. D. Ratner, *J. Appl. Polym. Sci.*, *22*, 643 (1978); B. D. Ratner, in Ref. 6, pp. 371-382.

96. R. N. King, G. K. Iwamoto, J. D. Andrade, A. D. Haubold, and H. Shim, *Polym. Prepr.*, *21*, 154 (1980).

97. R. E. Bair, *J. Adhes.*, *20*, 171-186 (1986).

# Seven

# Coupling Agents as Adhesion Promoters in Adhesive Bonding

F. J. BOERIO
*University of Cincinnati, Cincinnati, Ohio*

## I. INTRODUCTION

One of the most important problems associated with adhesive bonding of metals concerns the effects of moisture. In the presence of humid atmospheres or water, even the most carefully prepared adhesive joints gradually weaken and fail near the interface. One approach to this problem, which has been very successful in bonding aluminum and titanium, has been to modify the oxide present on the surface of the metal. Adhesive joints prepared from aluminum adherends anodized in phosphoric acid are extremely resistant to debonding in humid environments [1], probably because of the formation of a thin layer of $AlPO_4$ on the surface of the anodic oxide that inhibits hydration of the oxide [2]. Somewhat similar results have been obtained for titanium adherends anodized in chromic acid [3].

An alternative approach is to pretreat the metal surface with a dilute solution of an organic primer such as an organosilane coupling agent. These compounds were initially developed as pretreatments for glass fibers to reduce the moisture sensitivity of fiber-reinforced composites. However, silane coupling agents are also effective primers for reducing the moisture sensitivity of adhesive bonds to metals.

Most commercial silane coupling agents have the structure

$$R'\text{–Si–}R_3 \qquad (1)$$

where R is a functional group that is easily hydrolyzed and R' is a functional group that is chosen for compatibility with a particular polymer. The coupling agent is usually hydrolyzed in an aqueous solution to form the trisilanol

$$R'\text{–Si–}(OH)_3 \qquad (2)$$

which can adsorb onto an inorganic substrate by hydrogen bonding with hydroxyl groups at the surface of the substrate and then condense to form siloxane polymers. When a polymer is applied to the silanated substrate, it may react through the functional groups R' or it may form an interpenetrating polymer network during condensation of the adsorbed silane.

Several theories have been suggested to explain the mechanisms by which silanes improve the hydrothermal stability of interfaces between organic polymers and inorganic substrates. One of the most widely used, referred to as the "chemical bonding" theory, suggests that silanes form hydrolytically stable bonds between a polymer and a substrate [4]. Other early theories suggested that silanes function by improving the wettability of substrates [5] or forming deformable layers that permit residual stresses at the interface to relax [6]. A recent theory considers silanes as providing a mechanism for the release of stresses at the interface by means of a network of reversible, hydrolyzable bonds [4].

The purpose of this chapter is to describe the characteristics of silane primer films applied to metals and the use of silane primers to enhance the hydrothermal stability of adhesive bonds to metals.

## II. APPLICATIONS OF SILANE PRIMERS IN ADHESIVE BONDING

### A. Iron/Epoxy Adhesive Joints

Gettings et al. [7] prepared iron/epoxy butt joints using a tertiary amine salt curing agent and investigated the use of γ-glycidoxypropyltrimethoxysilane (γ-GPS) primers to improve their hydrothermal stability during exposure to distilled water at 60°C. Samples that were prepared without a primer as a control and tested without exposure to water failed cohesively but near the adhesive/substrate interface. Some adhesive was detected on the substrate failure surface by x-ray photoelectron spectroscopy (XPS) and some oxide was detected on the adhesive failure surface. Joints that were prepared without primer and tested after exposure failed interfacially, with little adhesive remaining on the substrate and little oxide left on the adhesive. Corrosion of the substrate was observed but it was not possible to determine whether the corrosion was associated with the failure or occurred subsequently. Joints that were prepared using the primer and tested after exposure to water also failed cohesively but within the primer. Moreover, little corrosion of the substrate was observed. The improved durability obtained using the γ-GPS primer was attributed to the ability of the primer/oxide interface to resist attack by water.

Boerio and Williams [8] prepared iron/epoxy lap joints using the same tertiary amine salt curing agent as used by Gettings et al. and considered the use of γ-aminopropyltriethoxysilane (γ-APS) as a primer to improve their hydrothermal stability. They found that the effectiveness of the primer was significantly affected by the pH of the aqueous solution from which it was applied. Joints prepared

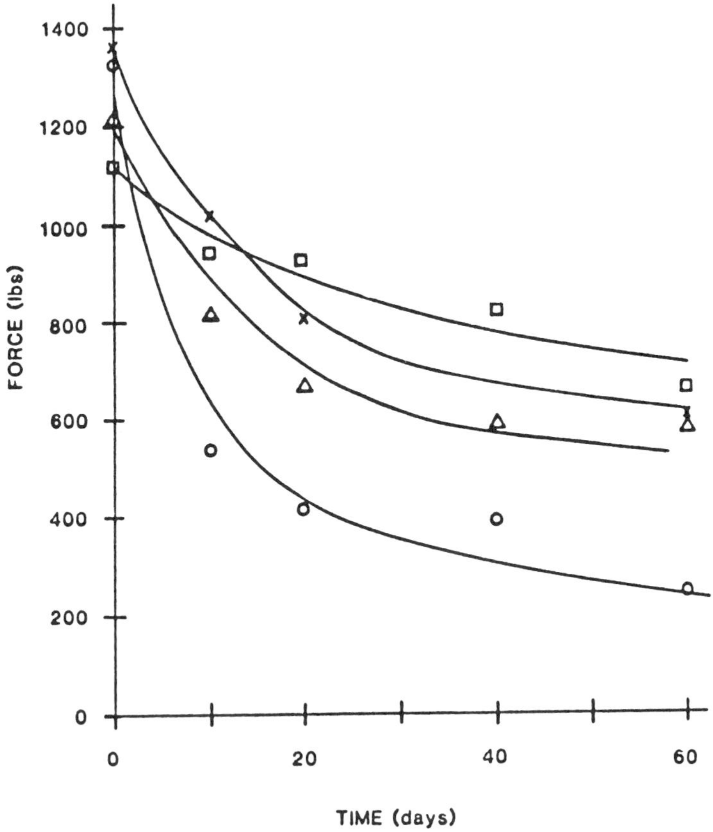


Figure 7.1 Breaking load versus immersion time in water at 60°C for iron/epoxy lap joints. (o) No primer; (△) γ-APS primer at pH 10.4; (x) γ-GPS primer at pH 4.0; (□) γ-APS primer at pH 8.0.

from unprimed substrates retained only about 25% of their initial strength after immersion in water at 60°C for 60 days. Those prepared from substrates pretreated with dilute aqueous solutions of γ-APS at pH 10.4 retained about 50%, but joints prepared from substrates pretreated with dilute aqueous solutions of γ-APS at pH 8.0 were most durable and retained about 75% of their initial strength (see Figure 7.1).

Kinloch [9] considered the factors involved in the environmental failure of simple structural joints and developed a qualitative model suggesting that three states are involved in the failure of such joints. The first is associated with diffusion of water into the joint and accumulation of a critical concentration of water at the interface. The second stage involves the formation of an interfacial crack through, for example, hydrolysis of interfacial bonds or hydration of the oxide and may be accelerated by externally applied stresses or by internal residual stresses. The final stage involves failure of the joint by propagation of the crack along the interface or through the oxide or adhesive.

Gledhill et al. [10] made this model quantitative by assuming that the growth of an interfacial crack was governed by the rate of water diffusion along the interface and applying fracture mechanics concepts to calculate the critical stress required to cause rapid propagation of the crack through the joint. The critical stress required to cause crack propagation in a butt joint is

$$\sigma_c = \frac{K_{1c}}{Q(a\pi)^{1/2}} \tag{3}$$

where $K_{1c}$ is the stress intensity factor, Q is a geometric constant, and a is crack length. Gledhill et al. assumed that $K_{1c}$ for adhesive joints was equal to $K_{1c}$ for crack propagation in the bulk adhesive and that the geometric factor Q was equal to unity.

The interfacial crack length a was assumed equal to the distance into the joint at which a certain critical concentration of water was obtained. Since no relationship between crack length a and critical water concentration was available, an empirical relation was established by using Equation (3) to calculate a from the breaking stress of iron/epoxy butt joints immersed in water at 40°C for 2500 hr. The concentration of water at any distance into the joint was then calculated at the same time and temperature, assuming Fickian diffusion, and the concentration of water at the distance a into the joint was taken as the critical concentration. Once the critical concentration was known, Equation (3) was used to calculate the breaking stress of butt joints after any time in water at any temperature provided that the diffusion coefficient of water in the adhesive at that temperature was known.

According to this model, the loss in strength of an adhesive joint exposed to water can be explained in terms of diffusion of water into the joint and accumulation of a critical concentration of water a certain distance into the joint, causing debonding over that distance. In the case of lap joints, this means that the edges of the joint debond during water immersion and that the size of the bonded region near the center of the joint gradually decreases as the critical concentration is obtained further and further into the joint due to continuing water diffusion into the joint.

Williams and Boerio [11] applied this model to iron/epoxy lap joints by assuming that failure of the joints occurred when a certain critical average shear stress was obtained. The average shear stress in a lap joint is given by $\tau = F/A$, where F is the force applied to the adherends and A is the bonded area. If the joint fails when a critical shear stress ($\tau_c$) is obtained, then the force required to produce failure should be given by $F_c = \tau_c A$. The bonded area (A) can be determined by solving the equation for diffusion of water into a rectangular joint, and the breaking strength ($F_c$) of the joint can be measured experimentally. A plot of $F_c$ versus A should be a straight line passing through the origin and having a slope equal to $\tau_c$.

The equation for water diffusion into a rectangular area $-a < x < a$, $-b < y < b$ is

$$\frac{dc}{dt} = D\left(\frac{d^2C}{dx^2} + \frac{d^2C}{dy^2}\right) \tag{4}$$

where C is the concentration at the point (x,y) inside the rectangular area at time t and D is the diffusion coefficient. The solution to this equation is

$$C(x,y) = C(\infty)(1 - \phi(a,x),\phi(b,y)) \tag{5}$$

where $C(\infty)$ is the maximum water absorption by the adhesive,

$$\left(\frac{\pi}{4}\right)\phi(a,x) = \sum_{n=0}^{\infty} \frac{(-1)^n}{2n+1} \exp\left[\frac{-(2n+1)^2\pi^2 Dt}{4a^2}\right] \cos\left[\frac{(2n+1)\pi x}{2a}\right] \tag{6}$$

and $\phi(b,y)$ has a similar form. Using Equations (5) and (6), the concentration of water at any point in a lap joint can be calculated as a function of time. Moreover, for any time, a contour connecting all points having a given concentration can be drawn.

Gledhill et al. [10] measured the maximum water absorption and the diffusion coefficient for water in an epoxy cured with a tertiary amine salt as 2.20 g per hundred grams of adhesive and $18.1 \times 10^9$ $cm^2/sec$ at 60°C, respectively, and suggested that the critical concentration for debonding the adhesive from the substrate is about 1.37 g of water per hundred grams of adhesive. A computer program was written to solve Equation (5) and contours showing the critical concentration of water in a lap joint as a function of time were constructed. The results for 10, 40, and 60 days are shown in Figure 7.2. The bonded area is the area inside the contours where the concentration is below the critical concentration and is shaded for the case of 60 days immersion. As shown in Figure 7.3 a straight line was obtained when breaking force (F) was plotted against bonded area (A) for iron/epoxy lap joints that were prepared from adherends pretreated with silane coupling agents and then tested after immersion in water at 60°C for varying times.

## B. Titanium/Epoxy Adhesive Joints

Several investigations of silane primers for titanium have been reported. Schrader and Cardamone [12] used an epoxy resin and an anhydride curing agent to prepare single lap shear joints from Ti-6Al,4V coupons that were cleaned in solvent and grit blasted. Prior to bonding, some of the coupons were treated with 1% aqueous

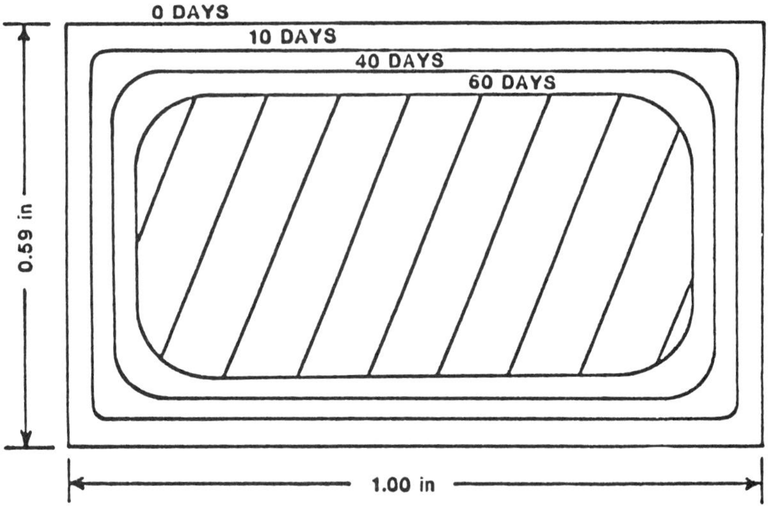


Figure 7.2 Contours showing critical concentration of water in lap joints as a function of time in water at 60°C.

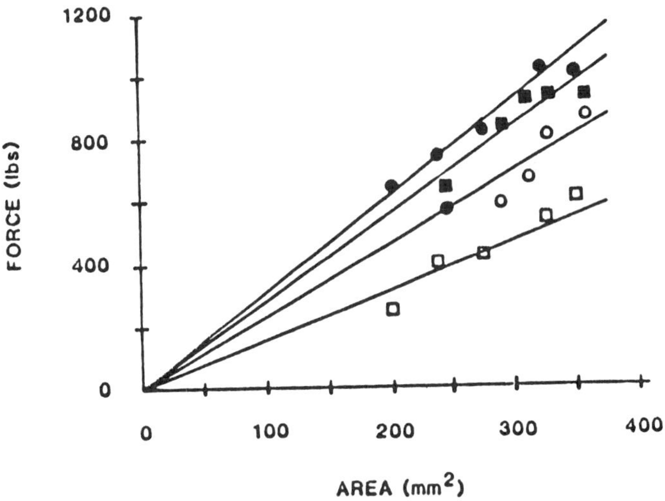


Figure 7.3 Breaking load versus bonded area for iron/epoxy lap joints immersed in water at 60°C. (□) No silane; (o) 1% γ-APS at pH 10.4; (■) 1% γ-APS at pH 8.0; (•) 1% γ-GPS at pH 4.0. The critical concentration was 1.37 g water per 100 g adhesive in (□) and (•) and 1.56 g per 100 g adhesive in (○) and (■).

solutions of γ-APS. All of the coupons were dried in air at 110°C for 20 min and then bonded with an adhesive consisting of an epoxy resin cured with an anhydride curing agent and a tertiary amine accelerator. The initial strength of joints prepared from primed adherends was about 20% higher than the strength of joints prepared from unprimed adherends, and they retained a significantly larger percentage of their initial strength after 24 hr in boiling water. Schrader and Cardamone [12] suggested that metallosiloxane bonds between the coupling agent and the oxidized surface of the substrate were probably responsible for the improvement in bond strength and durability.

Coury et al. [13] obtained similar results using an amine-cured epoxy adhesive and solvent-cleaned titanium adherends. Priming the adherends with amine or epoxy functional silanes before bonding gave initial strengths that were unchanged or somewhat higher (depending on the cure schedule) than those obtained from unprimed adherends. However, the joints prepared from primed adherends retained their strength better after immersion in a boiling pseudo-extracellular fluid for 7 days. The increased strength retention of the joints prepared from primed adherends was attributed to "chemical coupling" between the adherend surface and the adhesive through the coupling agent.

Boerio and Dillingham [14] considered the effects of γ-APS primers on the properties of lap joints prepared from mechanically polished Ti-6Al,4V adherends and an epoxy cured with a tertiary amine salt. Some of the adherends were coated with air-dried films of γ-APS applied from 1% aqueous solutions while others were left unprimed as controls. The pH of the primer solutions was varied from 10.4, which is the natural pH of 1% aqueous solutions of γ-APS, to 8.0 and to 5.5 by the addition of HCl. The initial strength of joints prepared from adherends primed at pH 10.4 or 8.0 was about 50% greater than that of joints prepared from unprimed adherends, while that of joints prepared from adherends primed at pH 5.5 was about 20% higher. Joints prepared from unprimed adherends lost about 30% of their initial strength after immersion in distilled water at 60°C for 60 days, but those prepared from adherends primed at either pH 10.4 or pH 8.0 showed no measurable loss. The strength of joints prepared from adherends primed at pH 5.5 decreased about 20%.

In a separate series of experiments, Ondrus and Boerio [15] used XPS to analyze the fracture surfaces of similar titanium/epoxy adhesive joints. When joints prepared from adherends primed with γ-APS applied at pH 10.4 were tested without immersion in water, fracture occurred cohesively within a thin layer of adhesive near the interface. Bands characteristic of the oxide surface were also observed, indicating that the thickness of the organic material left on the adherend side of the fracture surface was less than the escape depth of the photoelectrons (about 50 Å). Since the Si(2p) spectra indicated the presence of a small amount of primer in the region of fracture, the locus of failure may have been an interphase region consisting of cross-linked adhesive interpenetrated with the polysiloxane primer film.

Some of the joints were immersed in water at 60°C for 7 days and then tested to determine their residual strength. The XPS analysis showed that much less organic material was left on the adherend failure surfaces of joints that were tested after immersion in water than on those of samples tested without immersion in water, suggesting that hydrothermal aging caused the locus of failure to shift to a region closer to the primer/oxide interface.

When γ-APS was used as a primer for steel/epoxy adhesive joints, reducing the pH of the primer solution from 10.4 to about 8.0 before applying the primer resulted in a significant improvement in the strength retention of the joints during hydrothermal aging [8]. Reducing the pH of the primer solution from 10.4 to 8.0 did not have a large effect on the strength retention of titanium/epoxy lap joints [14], but did affect the locus of failure. Priming the adherends with γ-APS at pH 8.0 resulted in a change in the region where failure occurred after hydrothermal aging to a phase very close to the oxide surface which contained very little adhesive but was rich in primer [15].

## C. Aluminum/Epoxy Adhesive Bonds

Several groups have described the use of tapered double cantilever beam (TDCB) specimens to determine the effect of silane primers on the durability of adhesive bonds to aluminum during exposure to humid atmospheres or to water. Patrick [16] investigated the properties of aluminum/epoxy TDCB specimens prepared using an epoxy adhesive cured with an amine. Failure was near the interface when a static load was applied to the beams while the bond line was exposed to water at room temperature. Priming with γ-APS inhibited such interfacial failure as long as the concentration of the solution from which the silane was deposited was close to 0.01%. At higher or lower concentrations, failure near the interface was still obtained.

Kinloch et al. [17] prepared aluminum/epoxy TDCB specimens using an epoxy cured with a tertiary amine salt and determined the time to failure when static loads were applied during exposure to 56% relative humidity or to water at room temperature. Beams exposed to water failed near the interface, but those exposed to 56% relative humidity failed cohesively at times to failure that were greater by about two orders of magnitude for all strain energy release rates that were considered. The characteristics of TDCB specimens prepared from adherends primed with thin films of γ-GPS and immersed in water were similar to those of specimens prepared from unprimed adherends and exposed to 56% humidity. That is, failure was cohesive and the time to failure was increased by about two orders of magnitude.

Boerio and Ho [18] determined the effects of γ-APS primers on the hydrothermal stability of aluminum/epoxy TDCB specimens prepared from polished adherends using the same tertiary amine salt curing agent as used by Kinloch et al. [18]. When TDCB specimens were tested in air by applying increasing loads, the precrack always extended along the center of the bond line at a load of about 75 or

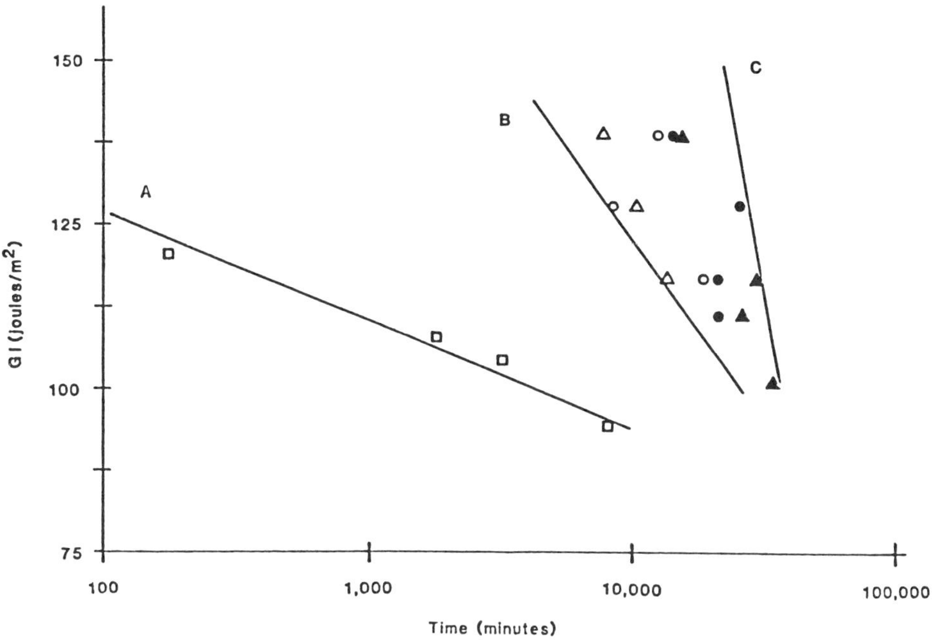


Figure 7.4 Time to failure versus applied fracture energy for aluminum/epoxy TDCB specimens immersed in water at 60°C. (□) No primer; (○) γ-APS primer at pH 10.4; (△) γ-APS primer at pH 8.0; (▲) γ-APS primer at pH 10.4 dried at 100°C for 6 min; (•) γ-APS primer at pH 10.4 dried at 100°C for 10 min. (Reprinted with permission from Ref. 18.)

80 lb, corresponding to $G_{Ic}$ between 1.58 and $1.80 \times 10^5$ mJ/m$^2$, whether a primer was used or not.

When the time to failure was determined as a function of applied strain energy release rate ($G_I$) for TDCB specimens that were statically loaded in water at 60°C, the results shown in Figure 7.4 were obtained. Beams prepared from primed substrates always failed at much longer times than those prepared from unprimed substrates. Results for unprimed beams always fell on a straight line, but there was considerable variation in those for beams prepared from primed adherends. The time to failure for the primed beams always fell within the region defined by the lines labeled B and C in Figure 7.4. The pH of the aqueous solutions from which the primers were applied had little effect on the time to failure, but drying the primer films at 100°C before adhesive bonding had an effect. When the drying time was about 10 min, the longest times to failure, near the line labeled C in Figure 7.4, were obtained. When shorter or longer drying times were used, the time to failure was closer to line B.

The use of γ-APS primers did not affect the mechanism of failure. The precrack was always near the center of the bond line, but during lengthy immersion in water a new crack developed near the interface above or below the tip of the precrack. Failure of the joints always involved the propagation of this secondary crack near the interface.

The results obtained by Boerio and Ho [18] were somewhat different from those reported by Kinloch et al. [17]. As described above, Kinloch et al. found that γ-GPS primers always increased the time to failure for aluminum/epoxy TDCB specimens exposed to water at 23°C by about two orders of magnitude regardless of the strain energy release rate. Boerio and Ho [18] found that γ-APS primers increased the time to failure for TDCB specimens immersed in water at 60°C by an amount that was additive. That is, the slopes of lines B and C in Figure 7.4 were greater than the slope of line A. The difference between the results obtained by Kinloch et al. and those obtained by Boerio and Ho was related to the difference in the temperature of the water in which the TDCB specimens were tested rather than to the difference in silanes.

The durability of adhesive bonds to aluminum in the presence of high humidity or water is related to the stability of the aluminum oxide [19]. Moisture transforms the oxide to boehmite, a hydroxide which adheres poorly to the aluminum substrate. Once boehmite forms, bond strength is greatly reduced.

The formation of boehmite on aluminum during immersion in distilled water is preceded by an induction period that varies with temperature [20]. At temperatures above about 50°C, the activation energy for the process responsible for the induction time is 18.7 kcal/mole. The induction time is about 10 min at 50°C and about 1 min at 80°C. At temperatures below 50°C, the activation energy is much greater, indicating a change in the mechanism by which water interacts with the oxidized surface of aluminum. The induction time is several hours at 40°C and is much greater at lower temperatures.

In the work of Boerio and Ho [18], the time to failure for aluminum/epoxy TDCB specimens immersed in water at 60°C was always long compared to the induction time for the formation of boehmite. It was found that the time to failure was increased by an additive amount through the use of γ-APS primers and that the failure was very near the interface and associated with corrosion of the substrate. The effect of the silane primer was to increase the induction time.

Kinloch et al. [17] found that the time to failure for aluminum/epoxy TDCB specimens immersed in water at 23°C was increased by a factor of about $10^2$ through the use of γ-GPS primers and that the failure shifted from near the interface to near the center of the bond. The times to failure they reported were mostly less than the induction time for formation of boehmite at 23°C. Failure in those experiments may have been within a thin layer of adhesive near the interface and the effect of the silane primer may have been to alter $G_{Ic}$ in that layer.

Films of γ-APS primer that were polymerized by heating at elevated temperatures for a few minutes probably inhibited hydration of the oxide more effectively than films that were less well polymerized. Similar results have been reported for γ-methacryloxypropyltrimethoxysilane (γ-MPS) [21]. However, excessive drying may have resulted in films that were brittle and had poor mechanical properties.

Theidman et al. [22] used wedge tests to determine the effect of silane primers on the durability of bonds between aerospace structural film adhesives (FM-73 and FM-300, American Cyanamid; AF-130 and AF-163, 3M) and clad 2024-T3 aluminum. Most adherends were prepared by solvent wiping with acetone, vapor degreasing with 1,1,1-trichoroethane, and grit blasting with zircon sand. They were then immersed in 1% silane solutions for 20 min, drained, and allowed to dry for 1 hr at 110°C. As a control, some adherends were solvent wiped, degreased, and then (FPL) etched in a mixture of chromic and sulfuric acids at 65°C for 10 min, rinsed in cold tap water for 10 min, and dried at 65°C for 1 hr. Wedge specimens were then prepared in a heated platen press using the recommended cure cycle for the particular film adhesive.

Primer solutions were prepared by hydrolyzing neutral silanes in 0.1% aqueous solutions of acetic acid and then adding enough methanol to provide 95% methanolic solutions. Aminosilanes were used either as solutions in 95% methanol at pH 10.3 or as aqueous solutions at pH 10.3 or 8.0.

Durability tests were carried out by exposing the wedge specimens to 96% relative humidity at 50°C and measuring crack extension as a function of time. With FM-73, the lowest crack extensions were obtained with hydrolyzed γ-GPS primers applied from 95% methanol solutions. Slightly higher crack growth rates were obtained using FPL-etched substrates. Primary aminosilane and acrylic silane primers applied from methanolic solutions gave intermediate extension rates, while diaminosilanes applied from methanolic solutions gave the highest rates.

The crack growth rates obtained using primary aminosilane primers were very interesting. When these primers were applied from methanolic solutions at pH 10.3, the crack growth rates with FM-300 and AF-163 were high, those with FM-73 were intermediate, and those with AF-130 were low.

When primary aminosilane primers were applied from aqueous solutions acidified to pH 8.0 using hydrochloric acid, very low extension rates were obtained with FM-300. Much higher rates were obtained if the solutions were acidified to pH 8.0 using acetic acid or phosphoric acid or if the primers were applied at the natural pH of 10.3.

As discussed above, Boerio and Williams [8] found that primary aminosilane primers (γ-APS) were most effective on iron when applied from solutions that were acidified with HCl, but Boerio and Ho [18] found that pH was not an important factor when γ-APS primers were applied to 2024-T3 aluminum. However, the materials, test methods, and test conditions used by Boerio and Ho were much different from those used by Theidman et al. A careful investigation

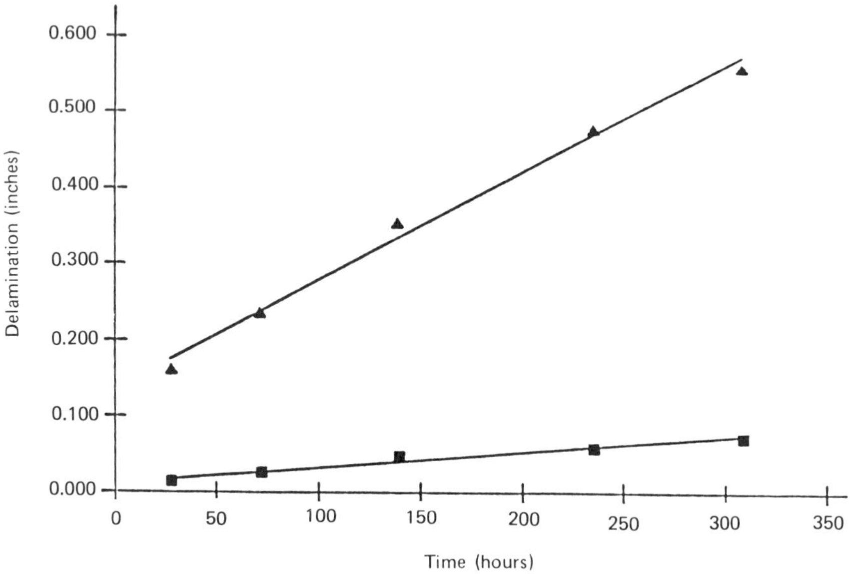


Figure 7.5 Delamination versus time for neoprene/steel strip-blister specimens in 1.0 N NaOH with the steel at a potential of -1.05 volts versus SCE. (▲) No silane in primer; (■) 11.7% aminosilane in primer.

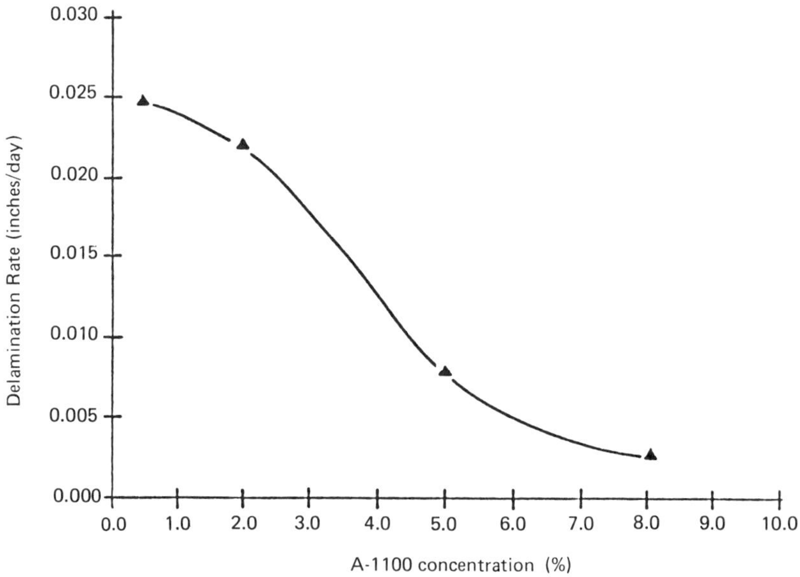


Figure 7.6 Delamination rate versus concentration of silane in the primer.

of the effect of pH on γ-APS primers on aluminum substrates should provide important insight into the mechanisms by which silanes function.

### D. Rubber-to-Metal Bonding

Thornton and Thomas [23] investigated the effect of silanes on cathodic delamination of neoprene rubber from steel in NaOH solutions. "Strip-blister" specimens were prepared by applying a metal primer and rubber adhesive to steel coupons that had been degreased, shot blasted, and degreased again, and then curing neoprene rubber against the coupons by compression molding at 155°C for 45 min. A piece of fluorocarbon tape was placed between the rubber and the adhesive before molding to provide a starter crack. After molding, a fluorocarbon rod was inserted between the rubber and the adhesive to apply a load to the starter crack. The specimens were then immersed in 1.0 N NaOH while the steel substrate was held at a potential -1.05 volts versus SCE and the length of the starter crack was determined as a function of time.

Delamination was usually rapid, with the crack length increasing at a rate of about 0.036 inch per day (see Figure 7.5). However, when 11.7% γ-APS was added to the primer before the primer was applied to the substrate, the rate of delamination was reduced to only about 0.005 inch per day. As shown in Figure 7.6, a minimum was observed in the rate of delamination when the primer contained about 8% γ-APS.

### E. Adhesion of Polyethylene to Sapphire

Silanes are usually selected on the basis of their reactivity with the adhesive. However, silanes may also be useful primers even when there is no obvious mechanism for chemical reaction with the adhesive. Sung et al. [24] investigated the effect of γ-APS primers on the peel strength of adhesive joints prepared by laminating polyethylene onto sapphire (a-$Al_2O_3$) in a hot press at 149°C. They found that the peel strength of the joints depended on the concentration of the primer solutions and on the extent to which the primer films were dried before laminating.

When primers were applied to the sapphire plates from solutions at different concentrations and then dried under the same conditions (1 hr at 25°C under vacuum) before laminating, a maximum in the peel strength was observed at a concentration of about 2%. However, when the concentration of the primer solutions was held constant and the drying time at 110°C under vacuum was increased, the peel strength decreased.

Scanning electron microscopy showed that failure was mostly cohesive within the polyethylene when the primers were dried at 25°C under vacuum for 1 hr before laminating. However, when the films were vigorously dried before laminating, failure was near the silane/polyethylene interface. It was suggested that excessive drying of the primer films before laminating the polyethylene decreased the interdiffusion between the silane and the polyethylene, thereby reducing the joint strength.

## III. CHARACTERISTICS OF SILANE FILMS ON METAL SURFACES

The structure of primer films formed by silanes adsorbed onto metal substrates has been investigated frequently. Boerio and co-workers used reflection-absorption infrared (RAIR) spectroscopy to determine the structure of films formed by γ-APS adsorbed from dilute aqueous solutions onto iron [25], titanium [14,26], aluminum [26], and copper [27]. Films formed by adsorption of γ-APS onto iron from aqueous solutions at pH 10.4 were characterized by strong bands near 1575, 1490, and 1105 $cm^{-1}$ and by weaker bands near 1630, 1330, 1040, and 925 $cm^{-1}$ (see Figure 7.7, spectrum A). The bands near 1105 and 1040 $cm^{-1}$ were attributed to the formation of low molecular weight siloxane polymers on the iron surface. Those near 1630, 1575, 1490, and 1330 $cm^{-1}$ were related to amine bicarbonates formed from atmospheric carbon dioxide dissolved in the solutions. The band near 925 $cm^{-1}$ was attributed to residual silanol groups.

The as-formed films were stable in a dry environment, but when they were exposed to a humid atmosphere or dried in an oven at 60°C for 20 min, the band near 1105 $cm^{-1}$ increased in frequency to about 1150 $cm^{-1}$, the shoulder near 1040 $cm^{-1}$ was resolved, and the band near 925 $cm^{-1}$ decreased in intensity, indicating that additional polymerization to form a high molecular weight polysiloxane had occurred (see Figure 7.7, spectrum B). The bands related to bicarbonates also decreased in intensity.

The structure of films formed by adsorption of γ-APS onto iron at pH 8.0 was considerably different. The spectra of the as-formed films were characterized by strong bands near 1150 and 1040 $cm^{-1}$, indicating that the films were composed of high molecular weight polysiloxanes. Weak bands related to protonated amino groups and to residual silanol groups were observed near 1600 and 1500 $cm^{-1}$ and near 900 $cm^{-1}$, respectively (see Figure 7.8, spectrum A). When these films were dried for 20 min at 60°C, the band near 900 $cm^{-1}$ decreased in intensity because of additional polymerization but few other changes were noted in the spectra (see Figure 7.8, spectrum B).

Boerio and Dillingham [14] found that the structure of films formed by the adsorption of γ-APS onto titanium from aqueous solutions was similar to that of films formed on iron. Low molecular weight siloxane oligomers in which the amino groups were complexed with absorbed carbon dioxide to form bicarbonate salts were formed at pH 10.4. High molecular weight siloxane polymers in which the amino groups were protonated were formed at pH 8.0.

Boerio and Ondrus [28] determined the effect of pH on the reactivity of epoxy resins with γ-APS primer films deposited on iron. Primer films were applied to polished iron substrates from aqueous solutions at pH 10.4 or 8.0, dried for 10 min at room temperature,

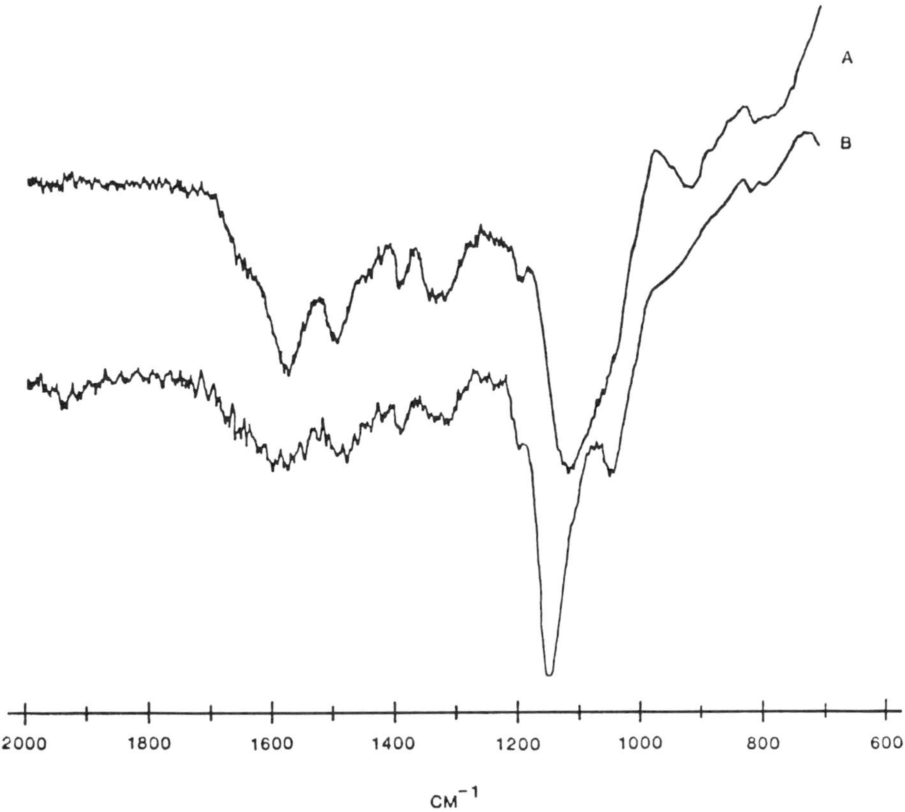


Figure 7.7 Infrared spectra of thin films formed by γ-APS adsorbed onto iron mirrors from 1% aqueous solutions at pH 10.4 (A) before and (B) after heat treating at 100°C for 20 min. (Reprinted with permission from Ref. 28.)

and immersed in containers of epoxy and heated to 100°C. After 1 hr the substrates were removed, rinsed several times with methyl ethyl ketone (MEK), dried, and examined by infrared spectroscopy. A band was observed near 1510 cm$^{-1}$ in spectra obtained from coupons having primer films deposited at pH 10.4, indicating retained epoxy (see Figure 7.9). However, as shown in Figure 7.10, no evidence for retained epoxy was observed in spectra from coupons that were primed at pH 8.0.

Dillingham [29] carried out similar investigations to determine the effect of drying on the reactivity of primer films deposited on titanium from aqueous solutions at pH 10.4. He found that as-formed films reacted readily with epoxies but films that were dried at 100°C for 20 min showed little reactivity.

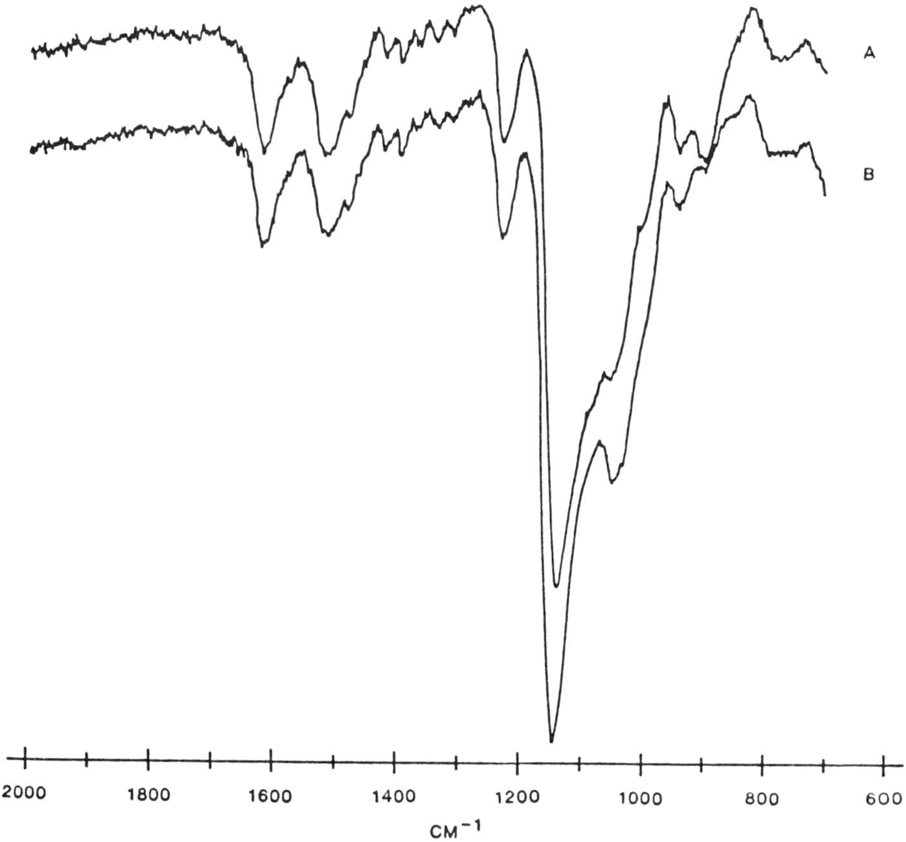


Figure 7.8 Infrared spectra of thin films formed by γ-APS adsorbed onto iron mirrors from 1% aqueous solutions at pH 8.0 (A) before and (B) after heat treating at 100°C for 20 min. (Reprinted with permission from Ref. 28.)

The structure of films formed by adsorption of γ-APS onto 2024 aluminum was sometimes complicated by etching of the substrate [26]. When γ-APS was adsorbed onto 2024 aluminum from aqueous solutions at pH 10.4 for approximately 1 min, little etching was observed and films similar to those formed on iron at the same pH were obtained. However, when the adsorption time was approximately 10 min, etching occurred and low molecular weight polymers were formed on the surface. When γ-APS was adsorbed onto 2024 aluminum at pH 8.0, there was little etching and films similar to those formed on iron at the same pH were formed, regardless of the adsorption time.

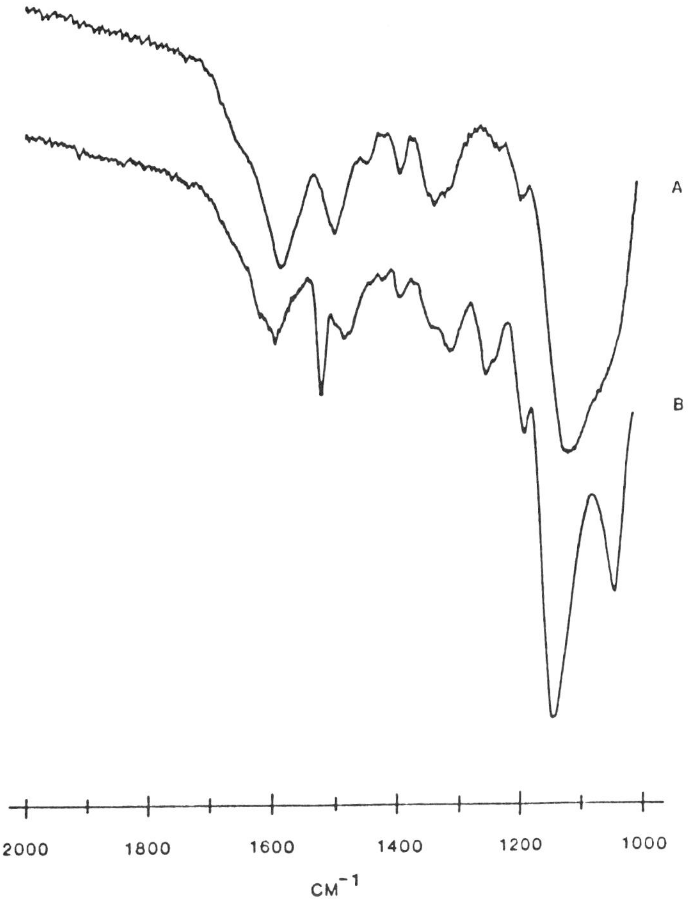


Figure 7.9 Infrared spectra of reaction between air-dried γ-APS film and Epon 828: (A) γ-APS film as applied at pH 10.4 on iron; (B) same film after applying a thin film of Epon 828 (spin coating), heating to 100°C for 1 hr, and rinsing thoroughly in MEK. (Reprinted with permission from Ref. 28.)

Films formed by γ-APS adsorbed onto copper were somewhat similar to those formed on iron [27]. Low molecular weight polysiloxanes that were stable in dry environments but polymerized during exposure to humid atmospheres were formed at pH 10.4. Some etching of the substrate occurred and copper ions accumulated in the primer films. The NH stretching modes, which are usually observed near 3300 and 3400 $cm^{-1}$ for free primary amino groups, were observed near 3150 and 3250 $cm^{-1}$, indicating that the amino groups were coordinated to the copper ions in the films.

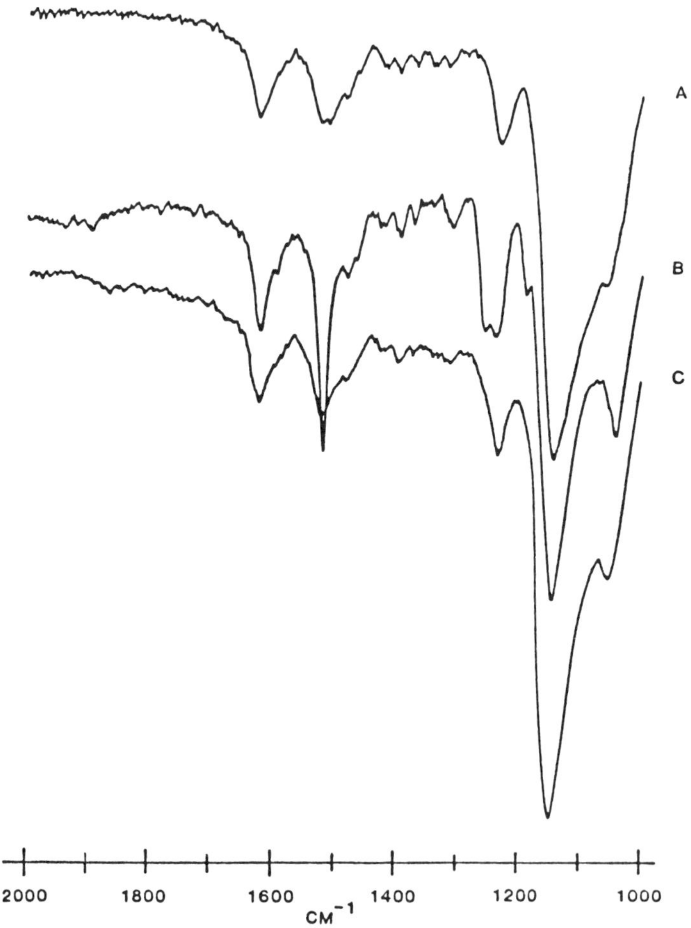


Figure 7.10 Infrared spectra of reaction between air-dried γ-APS film and Epon 828: (A) γ-APS film as applied at pH 8.0 on iron; (B) same film after applying a thin film of Epon 828 (spin coating) and heating to 100°C for 1 hr; (C) same film rinsed thoroughly in MEK. (Reprinted with permission from Ref. 28.)

Several groups have used XPS to investigate the structure of silane primer films applied to metal substrates. Bailey and Castle [30] used XPS to investigate the adsorption of γ-APS and vinyltriethoxysilane (VTES) onto polished iron from methanol solutions. They found that γ-APS was adsorbed with the nitrogen atoms adjacent to the substrate but that the N(1s) binding energy for adsorbed γ-APS was nearly the same as for free primary amines. More VTES was adsorbed than γ-APS.

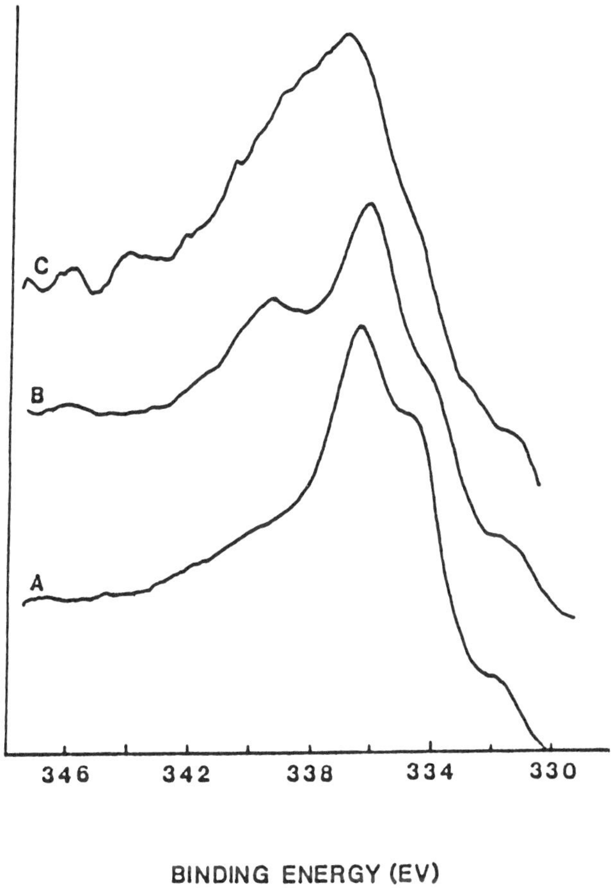


Figure 7.11 Copper $L_3M_{45}M_{45}$ Auger spectra for copper mirrors: (A) "bare" copper mirror; (B) copper mirror with freshly deposited γ-APS film; (C) copper mirror with γ-APS film aged in laboratory atmosphere. (Reprinted with permission from Ref. 27.)

Boerio et al. [27] used XPS to investigate the adsorption of γ-APS onto copper. They found that copper was etched during immersion in aqueous solutions of γ-APS at pH 10.4 and that the as-formed γ-APS films contained Cu(I) ions which were oxidized to Cu(II) during exposure to humid atmospheres. They also found that the N(1s) binding energy was similar to that of free primary amines but noted a significant change in the structure of the x-ray-induced Auger line for copper. As shown in Figure 7.11, the x-ray-induced Auger lines are near 334 and 336 eV for Cu(I) and

Cu(II), respectively, but when γ-APS was adsorbed onto copper, a new component was observed near 339 eV, indicating coordination of the copper ions to the amino groups.

Gettings and Kinloch [31] used XPS and secondary ion mass spectroscopy (SIMS) to investigate the structure of films formed by γ-GPS and γ-APS brushed onto mild steel from aqueous solutions. Results obtained with XPS indicated that γ-GPS formed uniform polysiloxane films over the steel surface but that γ-APS formed incomplete films. When γ-GPS was applied to the steel, the Fe(2p) binding energy shifted upward by about 4.8 eV from its position for the pure metal because of complex bonding between iron, oxygen, and silicon. A smaller shift attributed to oxidation of iron was observed when γ-APS was applied. Gettings and Kinloch found evidence for $FeSiO^+$ radicals in SIMS spectra from steel surfaces treated with γ-GPS but not from those treated with γ-APS, and suggested that γ-GPS formed primary bonds with the oxide but γ-APS did not. The γ-GPS was more effective as a primer for improving the wet strength of iron/epoxy adhesive joints than γ-APS, and it was concluded that the performance of the primer was related to the formation of such bonds.

Boerio et al. [32] used XPS to determine the structure of films formed by γ-APS adsorbed onto 2024 aluminum, which contains approximately 4% copper. They confirmed that the oxide was etched during immersion of 2024 aluminum in aqueous solution of γ-APS at pH 10.4 but not at pH 8.0 and showed that γ-APS films formed at pH 10.4 contained about 4.6% copper, which was coordinated to nitrogen atoms in amino groups, while those formed at pH 8.0 contained only traces of copper.

Sung and Sung [33] investigated the adsorption of γ-APS onto sapphire, a model for the oxidized surface of aluminum, using XPS. They observed that the N(1s) spectra consisted of components near 399.0 and 401.3 eV, which were attributed to free and protonated amino groups, respectively. The component due to free amino groups disappeared after washing with water but that due to protonated amino groups did not, and it was concluded that the protonated amino groups were associated with ionic bonds between the silane and the substrate

Recently, Ondrus and Boerio [34] used RAIR to reexamine the structure of thin films formed by the adsorption of γ-APS onto polished titanium, iron, aluminum, and copper from aqueous solutions at pH 10.4. They found that the RAIR spectra of the as-formed films were similar for all four substrates for adsorption times of approximately 1 min and were characterized by the bands near 1630, 1570, 1470, and 1330 $cm^{-1}$ that are related to amine bicarbonates. However, the SiOSi stretching mode was near 1130 $cm^{-1}$ for films formed on iron and titanium but near 1080 $cm^{-1}$ for those formed on 2024 aluminum and copper, implying that there was less polymerization on 2024 aluminum and copper.

Significant differences were observed when films formed on the various substrates were dried by heating at 110°C for 1 hr. When the films formed on iron and titanium were dried, the bands near

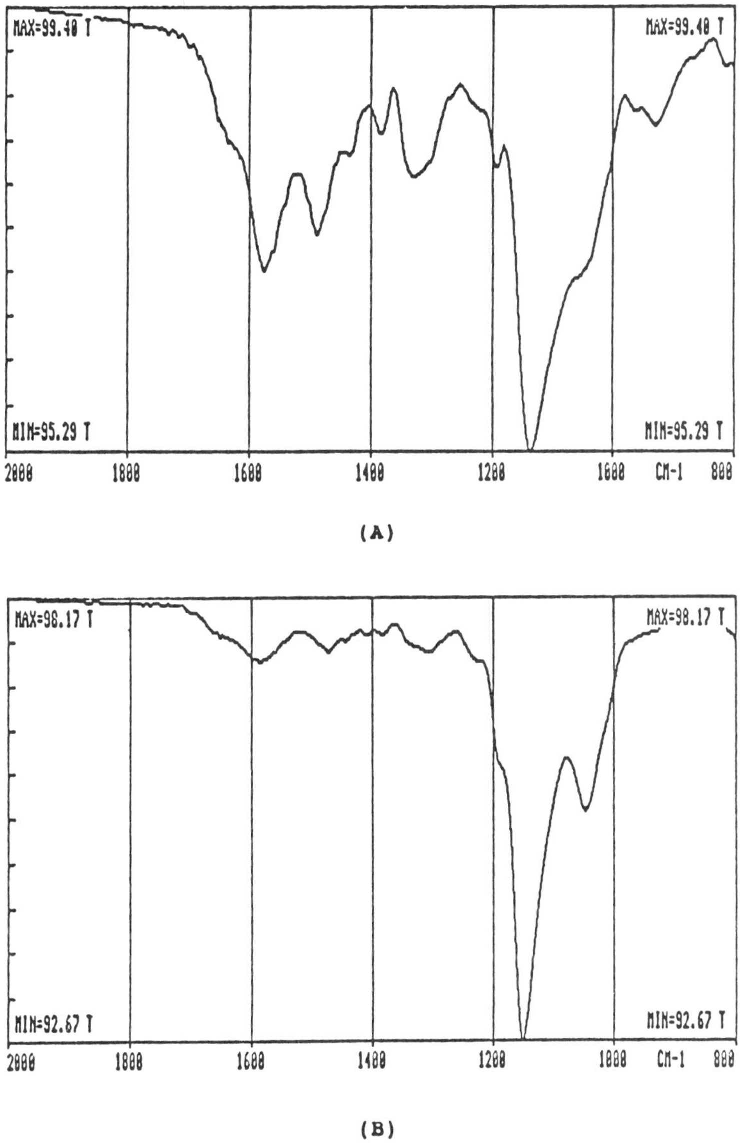


Figure 7.12 Reflection-absorption infrared spectra obtained from polished steel dipped into a 2% aqueous solution of γ-APS for 1 min and then blown dry with $N_2$: (A) after curing for 30 min at room temperature; (B) after postcuring for 1 hr at 110°C. (Reprinted with permission from Ref. 34.)

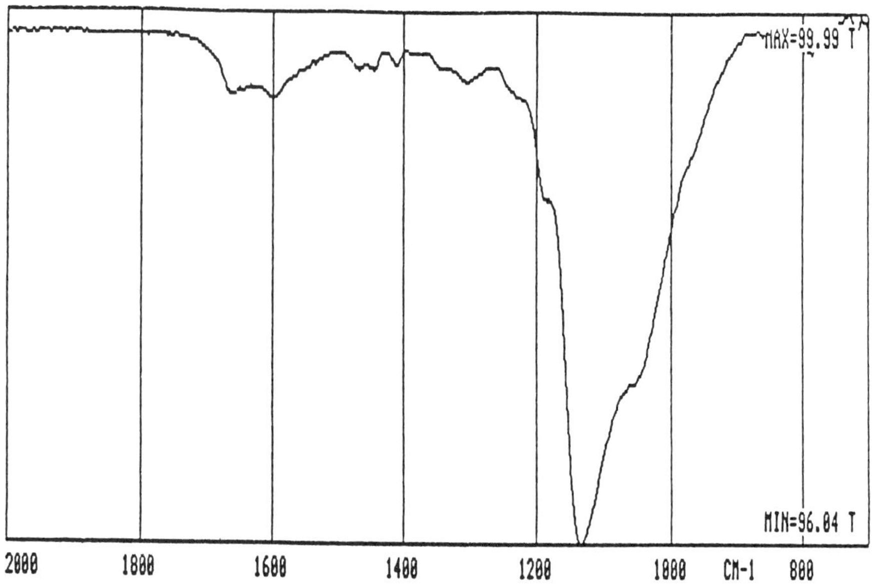


Figure 7.13 Reflection-absorption infrared spectrum from a polished 2024 aluminum mirror dipped into a 2% aqueous solution of γ-APS for 1 min, blown dry with nitrogen, cured at room temperature for 30 min, and postcured at 110°C for 1 hr. (Reprinted with permission from Ref. 34.)

1130 $cm^{-1}$ increased in frequency to about 1150 $cm^{-1}$, showing that additional polymerization took place. At the same time, the bands near 1630, 1570, 1470, and 1330 $cm^{-1}$ decreased in intensity but did not completely disappear, implying that the bicarbonates were dissociated and that carbon dioxide and water were desorbed but then reabsorbed after the films were cooled to room temperature (see Figure 7.12).

When films formed on 2024 aluminum and copper were dried at 110°C, the bands near 1630, 1570, 1470, and 1330 $cm^{-1}$ disappeared and a new band formed near 1660 $cm^{-1}$ (see Figure 7.13) that was attributed to the formation of imines by oxidation of the amino groups according to the reaction

$$2RCH_2NH_2 + O_2 \rightarrow 2RCHNH + 2H_2O \qquad (7)$$

Moreover, films formed on 2024 aluminum and copper showed little tendency to reabsorb carbon dioxide after being dried at 110°C for 1 hr.

The differences between the results for iron and titanium and those for copper and 2024 aluminum were attributed to catalysis of

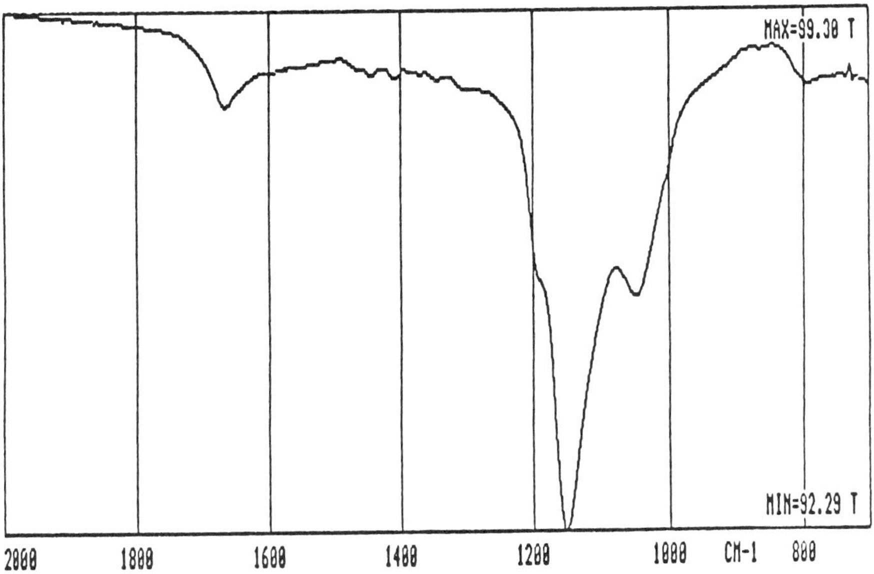


Figure 7.14 Reflection-absorption infrared spectrum from a polished steel mirror dipped into a 2% aqueous solution of γ-APS for 1 min, blown dry with nitrogen, cured at room temperature for 30 min, and postcured at 150°C for 1 hr. (Reprinted with permission from Ref. 34.)

the oxidation reaction described above by copper. When the experiment was repeated using an 1100 series aluminum, which contains no copper, the results were similar to those obtained for iron and titanium. That is, the band near 1660 $cm^{-1}$ was not observed and there was a tendency for the films to reabsorb carbon dioxide after being dried at 110°C for 1 hr.

When similar drying experiments were carried out at a temperature of 150°C, somewhat different results were obtained. As shown in Figure 7.14, films formed on iron were oxidized to form the imine and showed no tendency to reabsorb carbon dioxide after drying. Similar results were obtained for titanium and 2024 aluminum. However, films formed on 1100 series aluminum still showed little oxidation to the imine and did reabsorb carbon dioxide. It was suggested that some aluminum oxides, such as those on 1100 series aluminum, may inhibit free-radical reactions such as oxidation or that iron and titanium may catalyze the oxidation of amines to imines as efficiently as copper at 150°C [34].

Ondrus and Boerio [34] also investigated the effect of temperature on the reaction between γ-APS primers and epoxy resins. In these experiments, the primer films were applied to polished metal substrates from aqueous solutions at pH 10.4 and dried for 30 min at

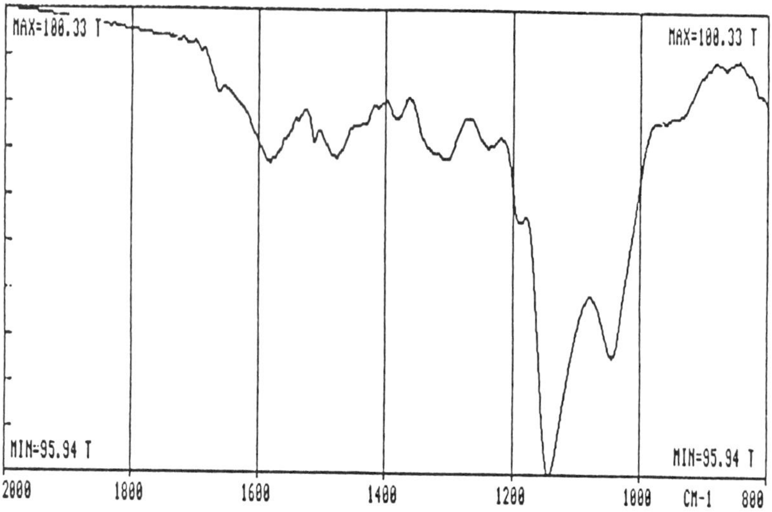


Figure 7.15 Infrared spectrum obtained from a polished steel mirror that was immersed in an aqueous solution of γ-APS at pH 10.4 for 1 min, dried at room temperature for 30 min, and then reacted with epoxy resin at 75°C for 1 hr and rinsed with MEK. (Reprinted with permission from Ref. 34.)

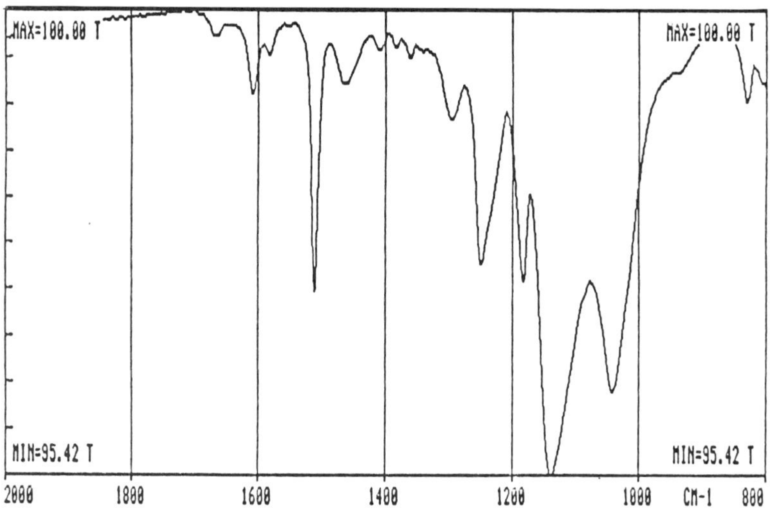


Figure 7.16 Infrared spectrum obtained from a polished steel mirror that was immersed in an aqueous solution of γ-APS at pH 10.4 for 1 min, dried at room temperature for 30 min, and then reacted with epoxy resin at 150°C for 1 hr and rinsed with MEK. (Reprinted with permission from Ref. 34.)

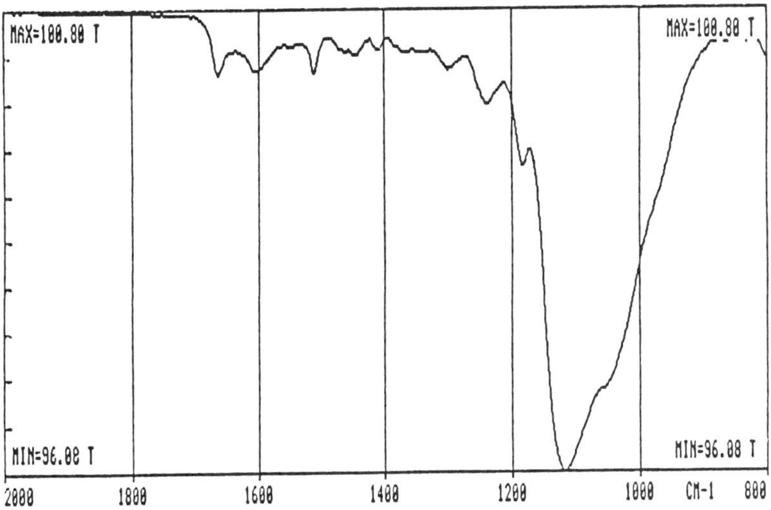


Figure 7.17 Infrared spectrum obtained from a polished 2024 aluminum mirror that was immersed in an aqueous solution of γ-APS at pH 10.4 for 1 min, dried at room temperature for 30 min, and then reacted with epoxy resin at 150°C for 1 hr and rinsed with MEK. (Reprinted with permission from Ref. 34.)

room temperature. The substrates were then immersed in containers of epoxy which were preheated to 75 or 150°C. After 1 hr the substrates were removed, rinsed several times with MEK, dried, and examined with infrared spectroscopy.

When any of the primed substrates were heated in the epoxy at 75°C, the bicarbonates did not dissociate and there was very little reaction between the silanes and the epoxy resin. Only a very weak band characteristic of epoxy resins was observed near 1510 $cm^{-1}$ in the infrared spectra obtained from the substrates after rinsing (see Figure 7.15).

When primed iron or titanium substrates were heated in the epoxy at 150°C, the bicarbonates dissociated but there was very little oxidation of amines to imines. A moderately intense band was observed near 1510 $cm^{-1}$ in infrared spectra obtained from the substrates after rinsing, indicating that there was a significant amount of epoxy retained on the substrates (see Figure 7.16). However, when primed 2024 aluminum substrates were heated in the epoxy at 150°C, oxidation of the amines to imines was observed and, once again, there was very little retained epoxy after rinsing (see Figure 7.17).

## IV. DISCUSSION

Silane coupling agents are extremely useful in primers to enhance the hydrothermal stability of adhesive bonds to metals, but the chemical bonding model does not adequately explain the mechanisms. According to that theory, as little as a monolayer of the coupling agent can bind an adhesive to a metallic substrate by forming covalent bonds that are resistant to hydrolysis between the adhesive and the substrate. However, silanes are effective primers even when there is no obvious reaction possible with the adhesive [24] and it has been argued that metallosiloxane bonds are not particularly stable with respect to hydrolysis [4, p. 20]. Moreover, the simple chemical bonding model can not explain why the pH at which the silane is applied is a significant factor for some substrates but not for others or why extensive drying of the primer films is deleterious.

The extent to which the primer films are dried before the adhesive is applied is important for several reasons. Proper interaction between the primer and the adhesive requires that the primer have a relatively open structure so that the adhesive and primer can interdiffuse and form an interpenetrating polymer network during curing of the adhesive. If the primer film is dried extensively before the adhesive is applied, it will be too cross-linked to be penetrated by the adhesive, but if the primer is not dried at all, it may diffuse away from the surface and into the adhesive.

In the specific case of γ-APS primers applied at pH 10.4, some drying of the films results in an increase in the dry strength of adhesive joints because of an increase in the extent of polymerization of the silane and an increase in the number of amino groups available for reaction with the adhesive due to dissociation of the amine bicarbonates. However, excessive drying of γ-APS films can result in oxidation of the amino groups to imines and reduce the number of amino groups available for reaction, especially when the substrate contains a catalyst for oxidation such as the copper in 2024 aluminum [34].

The pH at which silane primer films are applied is also important. When γ-APS is applied from aqueous solutions that have been acidified with hydrochloric acid, the degree of polymerization of the as-formed films is much greater than that of films applied at the natural pH and the amino groups are protonated to form amine hydrochlorides. As a result, formation of interpenetrating networks by the adhesive and primer is inhibited, resulting in a relatively sharp boundary between the adhesive and the primer and in reduced dry strength of adhesive joints.

The pH at which silane primers are applied may also affect the structure of the primer/oxide interface. Hydroxyl groups present in the uppermost layers of the surface oxides of metals are able to act as Brønsted acids or bases. In the presence of water at high pH values, there will be a net loss of protons from the surface and the surface will have a negative charge. In water at low pH values, there will be a net transfer of protons to the oxide surface and the surface will have a positive charge. At some intermediate value of the pH, referred to as the isoelectric point of the surface (IEPS), there will be no net charge on the surface. The IEPS is, therefore, a measure of the relative acidity or basicity of the surface.

In many silane coupling agents, the only ionizable functional groups are the silanol groups, which have a $pK_a$ of about 3.0 [4, p. 91]. If the IEPS is greater than 3.0, there should be strong forces tending to orient the molecules with the silanol groups adjacent to the surface and the organofunctional groups away from the surface when these silanes are adsorbed onto the oxidized surface of a metal from aqueous solutions at pH values between 3.0 and the IEPS of the oxide.

However, in some silanes, such as $\gamma$-APS, the organofunctional groups are also ionizable and other orientations of the adsorbed molecules may be possible. For example, in $\gamma$-APS the amino groups have a $pK_a$ of approximately 10.0 and the molecules may be adsorbed "upside down" with the amino groups adjacent to the substrate if the IEPS is less than 10.0 and the pH of the silane solution is between the IEPS and 10.0.

In the presence of moisture, the adhesion of a polymer to a silanated substrate may also depend on the IEPS and the $pK_a$ of the functional group adsorbed onto the substrate. Bolger [35] suggested that an ionizable functional group will be stably adsorbed to an oxide surface in the presence of water if the pH of the water is between the $pK_a$ of the functional group and the IEPS of the oxide surface.

The stability of iron/epoxy and titanium/epoxy lap joints prepared from adherends pretreated with $\gamma$-APS may be related to the IEPS of the substrates and to the $pK_a$ values of the amino and silanol groups. The IEPS of polished iron and titanium substrates has been estimated using indicator dyes and is approximately 7.5 and 12.0, respectively [36]. Therefore, during exposure to water at a pH near 6.0, the stability of iron/epoxy joints prepared from adherends pretreated at pH 8.0 should be somewhat greater than that of joints prepared from adherends pretreated at pH 10.4. However, the stabilities of titanium/epoxy joints prepared from adherends pretreated with $\gamma$-APS solutions at pH 8.0 or 10.4 should be similar. That was, in fact, observed.

## REFERENCES


1. J. S. Ahearn, G. D. Davis, T. S. Sun, and J. D. Venables, in *Proc. Symp. Adhesion Aspects Polymer Coatings*, K. L. Mittal, ed., Plenum, New York, 1983, p. 288.
2. G. D. Davis, T. S. Sun, J. S. Ahearn, and J. D. Venables, *J. Mater. Sci.*, 17, 1807 (1982).
3. B. M. Ditchek, K. R. Breen, T. S. Sun, and J. D. Venables, in *Proc. 12th SAMPE Tech. Conf.*, M. Smith, ed., SAMPE, Azusa, Calif., 1980, p. 882.

4. E. P. Pleuddemann, *Silane Coupling Agents*, Plenum, New York, 1982.

5. P. W. Erickson and E. P. Plueddemann, in *Interfaces in Polymer Matrix Composites*, E. P. Plueddemann, ed., Academic Press, New York, 1974, Chap. 1.

6. R. C. Hooper, *Proc. 11th Annu. Tech. Conf.*, SPI Reinforced Plastics/Composites Inst. Sect. 8B, 1956.

7. M. Gettings, F. S. Baker, and A. J. Kinloch, *J. Appl. Polym. Sci.*, 21, 2375 (1977).

8. F. J. Boerio and J. W. Williams, *Appl. Surf. Sci.*, 7, 19 (1981).

9. A. J. Kinloch, *J. Adhes.*, 10, 193 (1979).

10. R. A. Gledhill, A. J. Kinloch, and S. J. Shaw, *J. Adhes.*, 11, 3 (1980).

11. J. W. Williams and F. J. Boerio, *Proc. 37th Annu. Tech. Conf.*, SPI Reinforced Plastics/Composites Inst., Sect. 2D, 1982.

12. M. E. Schrader and J. A. Cardamone, *J. Adhes.*, 9, 305 (1978).

13. A. J. Coury, A. H. Jezne, and P. T. Callahan, in *Proc. Symp. Adhesion Aspects Polymeric Coatings*, K. Mittal, ed., Plenum, New York, 1983, p. 163.

14. F. J. Boerio and R. G. Dillingham, in *Adhesive Joints: Formation, Characteristics, and Testing*, K. L. Mittal, ed., Plenum, New York, 1984, p. 541.

15. D. J. Ondrus and F. J. Boerio, *J. Adhes.*, 22, 1 (1987)

16. R. L. Patrick, in *Treatise on Adhesion and Adhesives*, R. L. Patrick, ed., Marcel Dekker, New York, 1973, Chap. 4.

17. A. J. Kinloch, W. A. Dukes, and R. A. Gledhill, in *Adhesion Science and Technology*, L. H. Lee, ed., Plenum, New York, 1975, p. 597.

18. F. J. Boerio and D. J. Ondrus, in *Surface and Colloid Science in Computer Technology*, K. L. Mittal, ed., Plenum, New York, 1987, p. 155.

19. J. D. Venables, D. K. McNamara, J. M. Chen, B. M. Ditchek, T. I. Morgenthaler, T. S. Sun, and R. L. Hopping, *Proc. 12th National SAMPE Symp.*, SAMPE, Azusa, Calif., 1980, p. 909.

20. R. S. Alwitt, in *Oxides and Oxide Films*, Vol. 4, J. W. Diggle and A. K. Vijh, eds., Marcel Dekker, New York, 1976, Chap. 3.

21. F. J. Boerio, R. G. Dillingham, and R. C. Bozian, *Proc. 39th Annu. Tech. Conf.*, SPI Reinforced Plastics/Composites Inst., Sect. 1A, 1984.

22. W. Theidman, F. C. Tolan, P. J. Pearce, and C. E. M. Morris, *J. Adhes.*, 22, 197 (1987).

23. J. S. Thornton and R. W. Thomas, to be published.
24. N. H. Sung, A. Kaul, I. Chin, and C. S. P. Sung, *Polym. Eng. Sci.*, 22, 637 (1982).
25. F. J. Boerio, L. H. Schoenlein, and J. E. Greivenkamp, *J. Appl. Polym. Sci.*, 22, 203 (1978).
26. F. J. Boerio, C. A. Gosselin, R. G. Dillingham, and H. W. Liu, *J. Adhes.*, 13, 159 (1981).
27. F. J. Boerio, J. W. Williams, and J. M. Burkstrand, *J. Colloid Interface Sci.*, 91, 485 (1983).
28. F. J. Boerio and D. J. Ondrus, in *Surface and Colloid Science in Computer Technology*, K. L. Mittal, ed., Plenum, New York, 1987, p. 155.
29. R. G. Dillingham, M.S. Thesis, University of Cincinnati (1983).
30. R. Bailey and J. E. Castle, *J. Mater. Sci.*, 12, 2049 (1977).
31. M. Gettings and A. J. Kinloch, *J. Mater. Sci.*, 12, 2511 (1977).
32. F. J. Boerio, C. A. Gosselin, R. G. Dillingham, and J. M. Burkstrand, *Proc. Natl. SAMPE Tech. Conf.*, 15, 212 (1983).
33. N. H. Sung and C. S. P. Sung, *Proc. 35th Annu. Tech. Conf.*, SPI Reinforced Plastics/Composites Inst., Sect. 23B, 1980.
34. D. J. Ondrus and F. J. Boerio, *J. Colloid Interface Sci.*, in 124, 349 (1988).
35. J. C. Bolger, presented at the 159th meeting of the Electrochemical Society, Minneapolis, Minn., 1981.
36. R. G. Dillingham, D. J. Ondrus, and F. J. Boerio, *J. Adhes.*, 21, 95 (1987).

# Index